Biology of *Trypanosoma cruzi*

Edited by Wanderley de Souza

Published in London, United Kingdom

IntechOpen

Supporting open minds since 2005

Biology of Trypanosoma cruzi
http://dx.doi.org/10.5772/intechopen.80373
Edited by Wanderley de Souza

Contributors
Kenechukwu C. Onyekwelu, José Luis Vega, Juan C. Sáez, Camila Gutierrez, Camilo Juyumaya, Luis Rodríguez, Juan Güiza, Iván Barría, Emile Barrias, Lissa C. Reignault, Wanderley De Souza, Carlos Robello, Luisa Berná, Sebastián Pita, María Laura Chiribao, Adriana Parodi-Talice, Fernando Alvarez-Valin, Rubem F. S. Menna-Barreto, Marcos André Vannier-Santos, Giselle V. Brunoro, Solange L. DeCastro, Maria de Nazaré C. Soeiro, Jorge González, Juan San Francisco, Bessy Gutiérrez, Marcia Cristina Paes, Marcelo Sousa Silva, Cláudia Jassica Gonçalves Moreno, Johny Wysllas de Freitas Oliveira, Joice Castelo Branco, Laura Araújo, Aline Maria Queiroz, Sílvia Tavares Donato, Nilton José da Silva Júnior, Emilly Thays da Silva Rodrigues, Phileno Pinge-Filho, Maria Isabel Lovo-Martins, Marli Cardoso Martins-Pinge, Alondra Cruz Reyes, José Luis Rosales Encina, Laila Gutiérrez-Kobeh, Arturo A. Wilkins-Rodríguez, Jocelyn Ginette Pérez Lazo, Andrés G. Lescano, Pedro Mayor

First published in London, United Kingdom, 2019 by IntechOpen
IntechOpen is the global imprint of INTECHOPEN LIMITED, registered in England and Wales, registration number: 11086078, 7th floor, 10 Lower Thames Street, London, EC3R 6AF, United Kingdom
Printed in Croatia

British Library Cataloguing-in-Publication Data
A catalogue record for this book is available from the British Library

Additional hard and PDF copies can be obtained from orders@intechopen.com

Biology of Trypanosoma cruzi
Edited by Wanderley de Souza
p. cm.
Print ISBN 978-1-83968-203-2
Online ISBN 978-1-83968-204-9
eBook (PDF) ISBN 978-1-83968-205-6

For EU product safety concerns:
IN TECH d.o.o., Prolaz Marije Krucifikse Kozulić 3, 51000 Rijeka, Croatia, info@intechopen.com or visit our website at intechopen.com.

Meet the editor

Wanderley de Souza was born in Bahia State, Brazil. He graduated in Medicine in 1974 and obtained his MSc and PhD at the Rio de Janeiro Federal University in 1976 and 1978, respectively. He is Professor of Cell Biology and Parasitology. He occupied several administrative positions as Rector of Rio de Janeiro North State University, Secretary for Science and Technology of Rio de Janeiro State, and Vice-Minister of Science and Technology of Brazil. He is a member of several scientific societies as well as academies of science and medicine. He is also a member of the editorial board of several international journals in the area of parasitology. He has published more than 600 scientific papers and 100 articles in many prestigious Brazilian newspapers.

Contents

Preface

Chagas disease is an important life-long infection in humans that can be divided into distinct clinical stages: the acute phase, where patient symptoms can vary from asymptomatic to severe; the indeterminate form, which is usually asymptomatic; and the chronic phase, where cardiomyopathy and/or digestive megasyndromes appear. It is caused by the intracellular parasitic protozoan *Trypanosoma cruzi*, the major focus of this book. There are two major reasons for significant interest in the study of this protozoan and the disease it causes. First, in Latin America, at least 8 million people are infected with *T. cruzi* and 13,000 die each year. In addition, migration patterns are driving the globalization of the disease and there are around 300,000 and 120,000 people infected in the USA and Europe, respectively. Second, *T. cruzi* is an interesting biological model for studying processes such as: (1) cell differentiation, where a non-infective stage transforms into an infective one; (2) cell invasion, where the infective stages are able to penetrate into a mammalian host cell, where they multiply several times and thus amplify the infection; and (3) evasion from the immune system, using several mechanisms.

Taking into consideration the importance of and interest in *T. cruzi* and Chagas disease, this book is organized in four sections. Section 1 deals with the basic aspects of the protozoan biology with an overview of the lifecycle of the parasite in invertebrate and vertebrate hosts, as well as the various developmental states of *T. cruzi* maintained in the laboratory. Available information on the reservoir host of *T. cruzi* in the Amazon rainforest is also discussed. Section 2 deals with basic biochemical data on the parasite, including proteomic and genomic data presently available for various *T. cruzi* developmental stages. Section 3 deals with the interaction of the parasite with the host, covering topics such as entry of the parasite into host cells and mechanisms involved in evasion of the immune response. Section 4 comprises data on experimental chemotherapy.

Wanderley de Souza MD, PhD
Professor of Cell Biology and Parasitology
at the Instituto de Biofísica Carlos Chagas Filho,
Universidade Federal do Rio de Janeiro,
Brazil

Section 1

Basic Biology

Chapter 1

Introductory Chapter: Biology of *Trypanosoma cruzi*

Wanderley de Souza

1. Introduction

Trypanosoma cruzi, an important zoonotic protozoan that causes Chagas disease, is the focus of this book. There are two major reasons for the significant interest in the study of this protozoan and the disease it causes. First, Chagas disease is an important life-long infection in humans that can be divided into distinct clinical stages: the acute phase, where patient symptoms can vary from asymptomatic to severe; the indeterminate form, which is usually asymptomatic; and the chronic phase, where cardiomyopathy and/or digestive mega syndromes appear. In Latin America, at least 8 million people are infected with *T. cruzi* and 13,000 die each year. In addition, migration patterns are driving the globalization of the disease and there are around 300,000 and 120,000 people infected in the USA and Europe, respectively. Second, *T. cruzi* is an interesting biological model for studying processes such as: (a) cell differentiation, where a non-infective stage transforms into an infective one; (b) cell invasion, where the infective stages are able to penetrate into a mammalian host cell, where they multiply several times and thus amplify the infection; and (c) evasion from the immune system, using several mechanisms.

To better understand the information presented in various chapters of this book, let us review some basic information about *T. cruzi* infection [1]. **Figure 1** shows a general view of the life cycle of *T. cruzi* in both vertebrate and invertebrate hosts [2]. The three basic developmental stages (trypomastigote, amastigote, and epimastigote) are schematically shown in **Figure 2a–c**, based on images obtained using transmission electron microscopy [2]. The various structures and organelles found in the protozoan are indicated. **Figure 3** shows a scheme, where the various phases of the interaction of the trypomastigote form of *T. cruzi* with a host cell are indicated [1, 2]. The process starts with adhesion of the infective stage to the host cell surface followed by parasite internalization with formation of a parasitophorous vacuole (PV), lysis of the PV membrane, division of amastigotes in the cytoplasm of the host cell, and transformation of amastigotes into trypomastigotes that are then released into the intercellular space.

IntechOpen

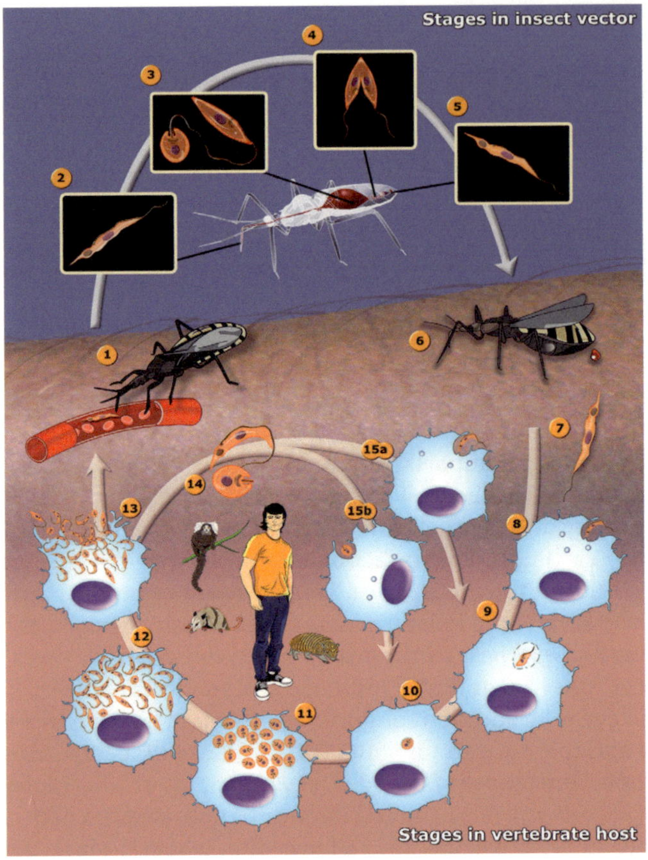

Figure 1.
The life cycle of T. cruzi. 1—The insect vector (female or male) bites a mammalian host and ingests trypomastigotes located in the blood. 2—Metacyclic trypomastigotes. 3—Trypomastigotes transform into epimastigotes and some spheromastigotes. 4—Epimastigotes multiply in the midgut. 5—Epimastigotes transform into metacyclic trypomastigotes in the hindgut. 6—The insect vector passes the metacyclic trypomastigotes in feces near a bite site after feeding on a mammalian host. 7—Metacyclic trypomastigotes form. 8—Metacyclic trypomastigote infects macrophages. 9—Metacyclic trypomastigote transforms into amastigote. 10—Amastigote is released from the parasitophorous vacuole. 11—Amastigotes multiply in the cytoplasm. 12—Amastigotes transform into trypomastigotes. 13—Trypomastigotes burst out of the cell. 14—Amastigotes and trypomastigotes form. 15—(a) Trypomastigotes and (b) amastigotes infect macrophages. In the central portion of the figure, we added the most important animal reservoirs involved in the maintenance of the parasite in the domestic and peridomestic environment (from [2]).

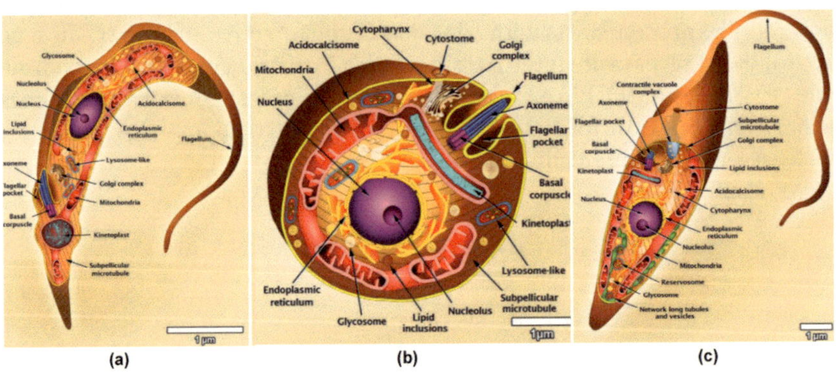

Figure 2.
Schematic representations of T. cruzi (trypomastigote, amastigote, and epimastigote) organelles with 2D models. These images were made based on micrographs of light microscopy as well as scanning and transmission electron microscopy (from [2]).

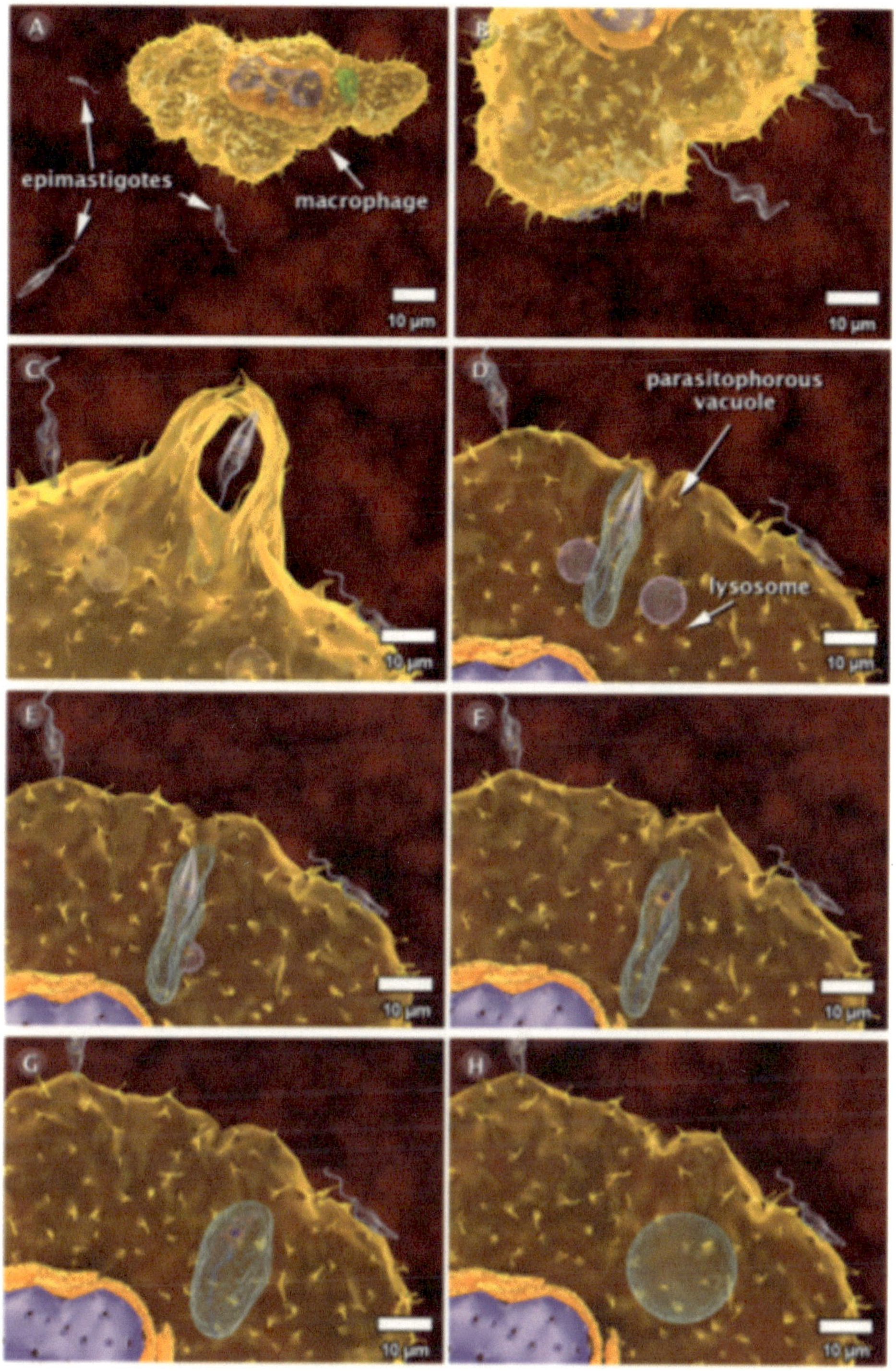

Figure 3.
Schematic representation of the interaction of trypomastigotes with the host cell. The various steps of the interaction are shown with attachment (A, B), internalization with formation of a parasitophorous (PV) (C), lysis of the PV membrane (D), successive division of amastigotes in the cytoplasm (E, F), transformation of amastigotes into trypomastigotes (G) and rupture of the cell releasing a large number of trypomastigotes (H) (from [2]).

Author details

Wanderley de Souza
Instituto de Biofísica Carlos Chagas Filho, Centro de Ciências da Saúde, Universidade Federal do Rio de Janeiro-UFRJ, Rio de Janeiro, RJ, Brazil

*Address all correspondence to: wsouza@biof.ufrj.br

References

[1] Barrias ES, de Carvalho TM, De Souza W. *Trypanosoma cruzi*: Entry into mammalian host cells and parasitophorous vacuole formation. Frontiers in Immunology. 2013;**4**:186. DOI: 10.3389/fimmu.2013.00186

[2] Teixeira DE, Benchimol M, Crepaldi PH, De Souza W. Interactive multimedia to teach the life cycle of *Trypanosoma cruzi*, the causative agent of Chagas disease. PLoS Neglected Tropical Diseases. 2012;**8**:e1749

Chapter 2

Life Cycle of *Trypanosoma cruzi* in the Invertebrate and the Vertebrate Hosts

Kenechukwu C. Onyekwelu

Abstract

Trypanosoma cruzi (*T. cruzi*) is a protozoan parasite that causes Chagas disease, a zoonotic disease that can be transmitted to humans by blood-sucking triatomine bugs. *T. cruzi* is a single-celled eukaryote with a complex life cycle alternating between reduviid bug invertebrate vectors and vertebrate hosts. This article will look at the developmental stages of *T. cruzi* in the invertebrate vector and the vertebrate hosts, the different surface membrane proteins involved in different life cycle stages of *T. cruzi*, roles of different amino acids in the life cycle, carbon and energy sources and gene expression in the life cycle of *T. cruzi*. The author will also look at extracellular vesicles (EV) and its role in the dissemination and survival of *T. cruzi* in mammalian host.

Keywords: *Trypanosoma cruzi*, metacyclogenesis, trypanosomatid, epimastigote, metacyclic trypomastigotes, extracellular vesicles

1. Introduction

The genus *Trypanosoma* has many species of protozoans but only *Trypanosoma cruzi*, *Trypanosoma brucei gambiense* and *Trypanosoma brucei rhodesiense* cause disease in humans. *Trypanosoma cruzi,* a protozoan parasite hemoflagellate is the etiologic agent of Chagas disease [1] while *Trypanosoma brucei gambiense* and *Trypanosoma brucei rhodesiense* causes African trypanosomiasis. Chagas disease also known as American trypanosomiasis affects millions of people throughout the Americas [2]. In 1909, Carlos Chagas first described this disease when he discovered the parasite in the blood of a Brazilian child with lymphadenopathy, anemia and fever [3, 4].

T. cruzi is a member of the family trypanosomatidae in the order Kinetoplastida and belongs to a subspecie called stercoraria (**Figure 1**). The development of stercoraria parasites takes place in the intestinal track of the invertebrate vector and the infection to the vertebrate occur through the feces. *T. cruzi* is carried in the guts of hematophagous triatomine bugs (kissing bugs) and transmission occurs when infected bug feces contaminate the bite site or intact mucous membranes. During feeding, the infected triatomine insect receives a significant amount of blood in its digestive system which forces the elimination of the bulk of accumulated excreta (consisting of feces and urine) which is normally deposited on the skin surface. The released feces contain the metacyclic trypomastigotes, which by

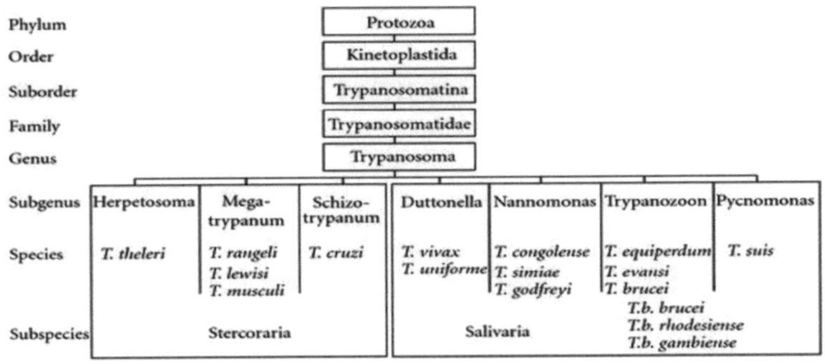

Figure 1.
Classification of trypanosomes. Source: Baral [7].

active movement and release of histolytic enzymes, actively penetrate the skin. Other modes of *T. cruzi* transmission includes through organ transplant, through transfusion and congenitally [5]. The mechanism of transmission of *T. cruzi* contrasts with that of the three subspecies of African trypanosomes that causes human and animal African trypanosomiasis disease, *Trypanosoma brucei gambiense*, *Trypanosoma brucei rhodesiense* and *Trypanosoma brucei brucei*, which are transmitted via the saliva of their vectors (salivarian) and with the mechanism by which the nonpathogenic trypanosome found in the Americas *Trypanosoma rangeli* is transmitted to its mammalian hosts. In addition to colonizing the stomach of their invertebrate vector, salivarian parasites migrate towards the salivary gland of the vector where the infectious form for the vertebrate develops but never pass to the intestinal track. In the process of obtaining a blood meal by the vector, infection of the vertebrate occurs through saliva. Like tsetse fly (the vector of human and animal African trypanosomiasis), the triatomine vector ingests circulating trypomastigotes when it takes a blood meal from an infected mammalian host. *T. cruzi* infects vertebrate and invertebrate hosts during defined stages in its life cycle [6].

2. The life cycle of *Trypanosoma cruzi*

The life cycle of *Trypanosoma cruzi* involves two intermediate hosts: the invertebrate vector (triatomine insects) and the vertebrate host (humans) and has three developmental stages namely, trypomastigotes, amastigotes and epimastigotes [8].

The general view of the life cycle of *Trypanosoma cruzi* is as shown in **Figure 2**. The cycle started with the insect sucking of bloodstream trypomastigotes of the infected vertebrates. Most of the ingested trypomastigotes are broken down in the stomach of the insect while the surviving trypomastigotes transform into either spheromastigotes (spherical stage) or into epimastigote stage few days later [9]. Epimastigotes move to the intestine where they divide intensely and attach to the perimicrovillar membranes which are secreted by intestinal cells of posterior midgut [10, 11]. Attachment to perimicrovillar membranes (PMM) in the insect midgut is an essential step for parasite division and is important in the process of metacyclogenesis which involves the transformation of the non-infective epimastigotes into highly infective trypomastigotes known as metacyclic trypomastigotes [12].

Metacyclic parasite forms express a set of surface glycoproteins that interact with mammalian cells [13, 14]. One of the glycoproteins, a metacyclic-stage-specific 82-kDa glycoprotein (gp82), has been implicated in host cell invasion [15]. The gp82 glycoprotein is an adhesion molecule that binds to host cells in a receptor-mediated manner and triggers Ca^{2+} mobilization [16] which is essential for parasite

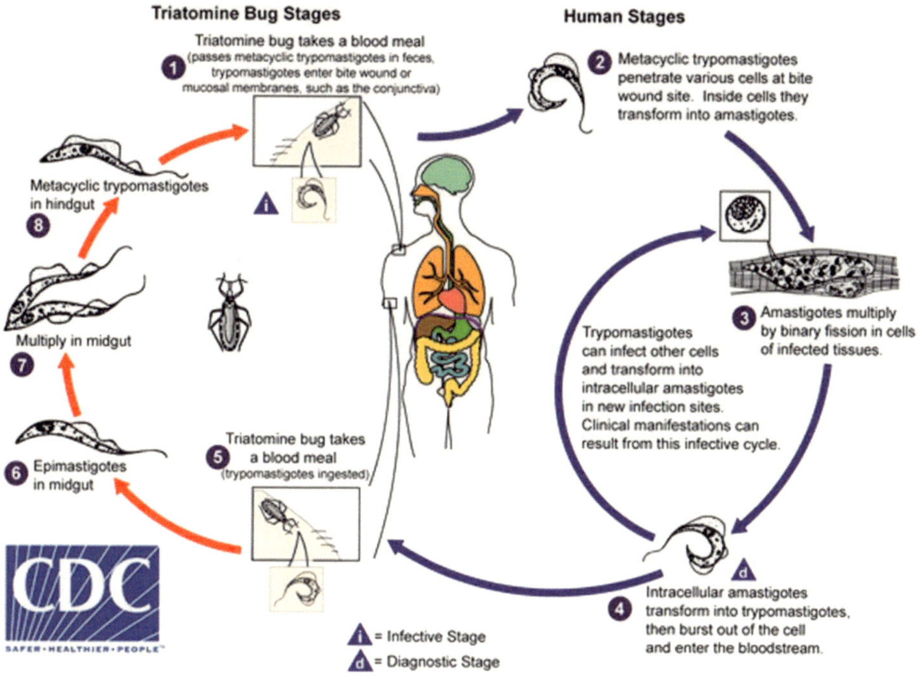

Figure 2.
Life cycle of T. cruzi showing the various forms of the protozoan in the invertebrate (triatomines) and vertebrate (mammals) hosts. Adapted from the Center of Control Diseases homepage.

penetration [17–20]. The gp82 also induces the activation of metacyclic trypomastigote protein tyrosine kinase [21] and an increase in the parasite intracellular Ca^{2+} concentration [22]. Other glycoproteins which are expressed in bloodstream or tissue-culture derived trypomastigotes and which are implicated in mammalian cell adhesion/invasion are gp83 [23], gp85 [24], and Tc-85 [25]. The gp83 has been reported to signal through the mitogen-activated protein kinase pathway to up regulate *T. cruzi* entry in macrophages [26]. Also, surface glycoinositolphospholipids (GIPLs) of the parasite have been shown to be involved in the attachment process [10].

3. Metacyclogenesis

Metacyclogenesis is the fundamental step in the life cycle of *T. cruzi* which involves the differentiation of epimastigotes into metacyclic trypomastigotes and occurs in the midgut of triatomine vector. During metacyclogenesis, bloodstream trypomastigotes differentiate into replicative epimastigotes in the triatomine insect's stomach, which divide in the midgut, migrate to the rectum and adheres to the epithelium through a flagellum prior to differentiation into a non-replicative, infective metacyclic trypomastigote form which are then released into the excreta of the triatomine insect while taking a blood meal on the vertebrate [27, 28]. The factors that trigger metacyclogenesis are still unknown, but might be stimulated by nutritional starvation, cyclic AMP and adenylate cyclase [29]. For instance, while in the midgut of triatomine vector, *T. cruzi* epimastigotes multiply in the nutrient rich environment after obtaining blood meal and as the meal is digested, the parasite density increases, the environment becomes nutrient poor making the epimastigotes become more elongate. On the epimastigotes reaching the insect rectum, it undergoes metacyclogenesis into human infective trypomastigote forms. Metacyclogenesis occurs when epimastigotes from the nutrient poor hindgut attach to the waxy cuticle of the triatomine vector rectum, initiating a dramatic morphological change. Once formed, metacyclics detach from the waxy cuticle and are excreted. Contamination of the triatomine vector bite wound of the mammalian host with these excreta leads to infection, completing the life cycle.

Description of metacyclogenesis can be in two parts, the first leading to the second. Firstly, the trypanosome senses loss of sugars from its environment and responds by elongating its cell body and flagellum and by activating its mitochondrion which leads to the lengthening of the trypanosome flagellar membrane that is rich in sterol and more hydrophobic than the somatic membrane. Secondly, the lengthening of the flagellar permits the trypanosomes to adhere to a hydrophobic surface and it is this interaction that triggers metacyclogenesis. This trigger for metacyclogenesis is cyclic adenosine monophosphate (cAMP) mediated.

Cyclic AMP plays an important role in the control of lower eukaryotes differentiation [30–32]. The relative amounts of cyclic AMP can change according to the surrounding environment, enabling the organisms to adapt quickly to new conditions. The differential balance of cAMP may result in activation of protein kinases [33, 34], transcription of specific genes [35–37] and changes in the cytoskeleton structure [38], which ultimately lead to morphogenetic cell alterations. Cyclic AMP balance could vary as a response to a changing environment leading to differential gene expression and morphological changes allowing the parasite to go through its life cycle. Calmodulin is known to play a direct role in controlling the levels of cAMP in eukaryotic cells [39] and in the case of *T. cruzi*, it has been shown to activate the cAMP phosphodiesterase [40].

4. The developmental stages of *T. cruzi* in vertebrate and invertebrate

The *T. cruzi* developmental stages alternates between infective and noninfective forms. Amastigote and epimastigote are noninfective but replicative stages inside the mammalian host (vertebrate) and in the gut of the insect vector (invertebrate) respectively (**Figure 3**). The bloodstream trypomastigotes found in the blood of the

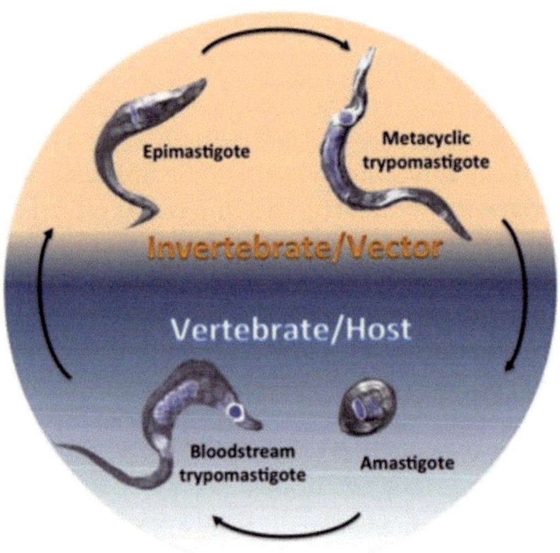

Figure 3.
Developmental stages of T. cruzi in vertebrate and invertebrate. Adapted from: Jimenez [96].

vertebrate host and the metacyclic trypomastigotes found in the rectum of the insect invertebrate vector are considered as the two different trypomastigote infective but nonreplicative developmental stages. The sucking of the blood of vertebrate mammalian host infected with the bloodstream trypomastigotes by the insect started the cycle and inside the stomach of the insect, the ingested trypomastigotes transform into epimastigotes which replicates intensely in the midgut. The epimastigotes transform into metacyclic trypomastigotes in the hindgut of the insect host which are eliminated through feces when the insect vector takes a blood meal from an uninfected host. The excreted metacyclic trypomastigote in the lesioned skin caused by the insect bite leads to *T. cruzi* infection. Once the metacyclic trypomastigote is inside the mammalian host, it invades the host cells at the inoculation site and transform into the replicative amastigote form which transform back into bloodstream trypomastigote form upon completion of a replicative cycle as intracellular amastigotes.

5. Surface membrane proteins

Membrane proteins play an important role in the biology of *T. cruzi*, including the interaction between parasite and host [41–45]. Two thousand seven hundred and eighty four (2784) proteins belonging to 1168 protein groups were identified in the proteomic analyses of different stages in the life cycle of *T. cruzi*. The *T. cruzi* proteins is in relative abundance throughout its life cycle since about 30% of the identified proteins were found at all life cycle stages and at least 248 proteins were only expressed at one stage of the life cycle. The families of surface membrane proteins from *T. cruzi* which are the most abundant and/or relevant during its life cycle are: mucin, trans-sialidase, TcGP63, amastin, TcTASV, mucin-associated surface proteins (MASP) and cruzipain.

5.1 The mucins family

Trypanosoma cruzi is covered by a dense layer of mucin-type molecules which are the major *T. cruzi* surface glycoproteins [46]. These proteins are widely distributed over the cell body, flagellar pocket and flagellum of the different developmental forms [47] and play a key role in the parasite protection as well as in the infectivity and modulation of the host immune response throughout the life cycle of *T. cruzi* [48–51]. The mucins of *trypanosoma cruzi* is divided into two gene families, namely TcMUC and TcSMUG [48, 52] and these proteins are divided into groups based on their central domains: TcMUC is divided into TcMUC I, TcMUC II and TcMUC III [52, 53] . TcMUC I protein is distributed on the amastigote and the bloodstream trypomastigote surface and is the major component in the amastigote form. This protein show internal tandem repeats on their structure with a T_8KP_2 amino acid consensus sequence which are suitable targets for the O-glycosylation pathway in *T. cruzi*, flanked by an N-terminal signal peptide and a C-terminal glycosylphosphatidylinositol anchor signal [48, 55]. TcMUC II protein is also distributed on the amastigote and the bloodstream trypomastigote surface and is found more in membrane lipid rafts of the trypomastigote stage [54]. Like TcMUC I, TcMUC II genes encode proteins that share similar N- and C-terminal but without the T_8KP_2 motifs [56, 57]. The single gene product of the TcMUC III group is termed trypomastigote small surface antigen (TSSA) and has been identified as a mucin-like glycoprotein [58]. TSSA are displayed on the surface of the trypomastigote forms of *Trypanosoma cruzi* and are expressed in vivo as a 20-kDa protein during the mammal-derived stages [59–61].

The second mucin protein family TcSMUG family is divided into two groups: TcSMUG S (small) and TcSMUG L (large) according to their encoded mRNA size

[58, 62] and encodes for very small open reading frame containing a putative signal peptide at the N-terminus and a GPI-anchor signal in the C-terminus. The TcSMUG S group is found in the epimastigote and metacyclic trypomastigote forms and encodes for 35–50 kDa mucins N-glycosylated (Gp35/50 mucins) and they are the major acceptors of sialic acid on the parasite surface by parasite trans-sialidases in *T. cruzi* [63, 64]. In contrast, TcSMUG L group encodes for mucin-type glycoconjugates which are not sialic acid acceptors and only present in the surface of the epimastigote stage [65, 66] and contains one or two additional N-glycosylation signals between the N-terminal region and the threonine-rich region depending on the origin of the encoding allele [65].

5.2 The trans-sialidases (TS) protein family

The trans-sialidases (TS) protein family of *trypanosoma cruzi* are a large superfamily, which includes 1430 gene members, including 693 pseudogenes [67]. Trans-sialidase shares certain characteristics with mucin protein family as they are distributed along the cell body, flagellum, and flagellar pocket of *T. cruzi* [68]. The trans-sialidase superfamily is classified into four groups based on their sequence similarity and functional properties namely: TS 1, TS II, TS III and TS IV [69]. The TS activity involves the transfer of sialic acid from host glycoconjugates to mainly the parasite mucins present in the plasma membrane of trypomastigotes [70, 71]. Trypanosomes are unable to synthesize the monosaccharide sialic acid; therefore they need to scavenge it from the infected host using these TS activities. The sialylation process in *T. cruzi* is crucial for its viability and propagation into the host [72, 73].

The TS I comprises of proteins with trans-sialidase (TS) and/or neuraminidase activities [74]. Neuraminidase activity occurs when nonsuitable acceptor molecules for sialic acid are present and then sialic acid is transferred to water [75]. Neuraminidase activity is involved in the removal of sialic acid from parasites and/or host-cell molecules which is required for parasite internalization [76]. The TS I members incudes: TCNA (neuraminidase), SAPA (shed acute-phase antigen), and TS-epi. SAPA and TCNA proteins are closely related with 84% homology at the amino acid level and have active trans-sialidase and neuraminidase activities and are expressed during bloodstream trypomastigote stage [77]. SAPA and TCNA have two main regions: an N-terminal catalytic region and a C terminal extension, which repeats 12 amino acids (SAPA repeats) in tandem with the consensus sequence. SAPA has only 14 tandem repeats compared to 44 for TCNA and the presence of SAPA repeats increases the half-life of the protein in the blood [78]. Both proteins are attached by glycosylphosphatidylinositol to the parasite plasma membrane and can be found in the serum of deeply infected mammals.

TS-epi, the third member of group TS I is an active trans-sialidase expressed in the insect dwelling epimastigote form at the stationary phase and is different from the trans-sialidase expressed of the blood trypomastigotes. Unlike other members of the group, TS-epi lacks SAPA repeats and is not attached to the membrane by glycosylphosphatidylinositol.

Members of TS group II includes: ASP-1, ASP-2, TSA-1, Tc85, SA85, GP82, and GP90 and they all have been implicated in host-cell attachment and invasion. ASP-1, ASP-2 (both are amastigote surface proteins) and TSA-1 (trypomastigote surface antigen) are targets of *T. cruzi*-specific $CD8^+$ cytotoxic T lymphocytes and they induce strong antibody responses in infected mice and humans. The SA85 glycoproteins are expressed by amastigote and bloodstream trypomastigote forms but only the amastigote form expresses the mannose-binding protein ligand which seems to be involved in the opsonization of the parasite enhancing its infection capability. The Tc85 molecule is an 85 kDa glycoprotein which is found abundantly in bloodstream trypomastigotes and is identified as a ligand capable of binding to different host receptor molecules located on the cell surface of either monocytes, neutrophils, or fibroblasts. The GP82 and GP90 members of TSII are glycoproteins expressed on the surface of the metacyclic trypomastigote form and they are mainly found at the plasma membrane with opposite roles in mammalian cell invasion. GP82 is able to activate a Ca^{2+} signaling pathway in host cells following parasite adhesion, which is required for *T. cruzi* internalization and is also the signaling receptor that mediates protein tyrosine phosphorylation, which is necessary for host-cell invasion. As defined by its reactivity with monoclonal antibodies, GP90 is a metacyclic stage-specific glycoprotein and expressed by metacyclic forms but lacks any enzymatic activity. GP90 is present in the mammalian stages of *T. cruzi* life cycle and has the antiphagocytic effect mediated by the removal of sugar residues necessary for parasite internalization.

TS Group III which is formed by surface proteins present in mammal bloodstream trypomastigotes includes: complement regulatory protein (CRP), surface flagellar protein (FL-160), chronic exoantigen (CEA), and trypomastigote excretory-secretory antigens (TESA) [79]. These surface proteins are recognized by sera from patients infected with Chagas' disease and they are able to inhibit the classical and the alternative pathways of complement activation, which could be a protection from lysis by the host in the trypomastigote form [80, 81]. TESA is distributed on the cell surface membrane of *T. cruzi* [80] whereas CRP, FL160, and CEA are flagellum associated membrane proteins [80, 82, 83].

The TS Group IV is included in the trans-sialidase superfamily because it contains the conserved motif VTVxNVxLYNR, which is shared by all known TS members and is composed of genes encoding trypomastigote surface antigens whose biological function is still unknown [84, 85]. The TsTc13 protein, a member of TS Group IV has been shown to be highly antigenic and is present in the infective metacyclic trypomastigote form [86].

5.3 TcGP63 family

This protein is expressed by trypanosomes and *Leishmania* species and is a family of cell surface-localized, zinc-dependent metalloproteases also known as GP63 proteins, major surface proteases or leishmanolysins. They serve as ligands for host cell attachment and protect the parasite from intraphagolysosomal degradation in *Leishmania* while they function to release variant surface glycoproteins from the cell surface during antigenic variation in the bloodstream form of the African trypanosome (*Trypanosoma brucei*). *Trypanosoma cruzi* possesses GP63-like genes (TcGP63) which are differentially regulated, suggesting its functional importance at multiple stages in the parasite life cycle [87]. The TcGP63 family is made up of two groups of proteins namely: TcGP63-I and TcGP63-II [88]. The *TcGP63-I* has low gene copies of 5–10, whereas *TcGP63-II* has 62 gene copies [87]. The TcGP63-I group is present in all the life-stages/cycles of *T. cruzi* and possess metallopeptidase activity and are bound to the protozoan's membrane by a C-terminal glycosylphosphatidylinositol- (GPI-) anchor signal. In *T. cruzi*, TcGP63-I exist in two isoforms: the glycosylated and the nonglycosylated isoforms. The 61 kDa glycosylated isoform is present in similar levels in both epimastigote and amastigote forms and is irregularly expressed on the surface membranes of the epimastigote while the 55 kDa nonglycosylated isoform is present in the metacyclic trypomastigote and is located intracellularly near the kinetoplast and the flagellar pocket [88]. Members of TcGP63-II are two transcripts of 2.6 and 2.8 kb protein.

5.4 Amastin family

The amastin family which is a group of transmembrane glycoproteins consists of small proteins of about 180 amino acids. The genome of *trypanosoma cruzi* has two groups of amastin family: the β-amastin and the δ-amastins. The β-amastin group has two members namely: β1-amastin and β2-amastin. Genes encoding for the β1-amastin and β2-amastin are localized in the chromosome 32 of *T. cruzi*. The δ-amastins group has δ-amastin and δ-ama40/50 as members and are found on chromosomes 34 and 26, respectively. β1-amastin and δ-amastins are located at the cell surface. In addition to their surface localization, β2 amastin shows a disperse distribution within the cytoplasm [89]. Even though the exact biological function of amastin is still unknown, as a transmembrane proteins, amastins could play a role in proton or ion traffic across the membrane [90]. The β-amastin transcripts are more abundant in epimastigotes than in amastigotes or trypomastigotes while transcript levels of δ-amastins are upregulated in amastigotes from different *T. cruzi* strains and β-amastins may be involved in the parasite adaptation to the insect vector [91].

5.5 *T. cruzi* trypomastigote alanine, serine and valine (TcTASV) family

The TcTASV protein family is conserved among all the *T. cruzi* lineages analyzed so far and has no orthologues in other species, including the closely-related trypanosomatids *T. brucei*, *T. rangeli* and *Leishmania sp* and belong to a medium-size multigene family of ~40 members that was identified from a library of trypomastigote-enriched mRNAs [92]. TcTASV proteins are expressed mainly in the trypomastigote stage and its function is still unknown. In TcTASV proteins, the N- and C-terminal regions respectively possess a signal peptide and a consensus for a glycosylphosphatidylinositol (GPI) anchor addition, and display the highest level of conservation, while the central region presents more variability. The TcTASV protein family is divided into 4 groups: TcTASV-A, TcTASV-B, TcTASV-C and TcTASV-W, with each group defined by the primary amino acid sequence and length of polypeptides. TcTASV protein family can be distinguished by the common amino acid motif *tasv_all* that starts approximately at amino acid 42 ($Vx_1x_2x_3[CES]x_4x_5TDGx_6Lx_7Wx_8x_9x_{10}x_{11}Ex_{12}x_{13}Wx_{14}x$-$_{15}Cx_{16}x_{17}x_{18}P$). Each group has certain amino acid in-between the *tasv_all* motif. The TcTASV-B contains serine and arginine, and TcTASV-W has alanine at X_4 and glutamic acid at X_5 while TcTASV-C and TcTASV-A both have proline and glycine at positions X_4 and X_5.

5.6 Mucin-associated surface proteins (MASPs) family

The MASP family is characterized by having highly conserved N and C-terminal domains and a variable and repetitive central region, with a maximum expression in the human infective stages of the parasite. MASP are expressed simultaneously in bloodstream trypomastigotes as well as in amastigotes and epimastigotes and MASP molecules are the most abundant antigens found on the surface of the infective trypomastigote stage of *T. cruzi*. MASP family plays an important role in the invasion of the mammalian host cell, but could also be crucial for the survival and the establishment of the parasite in the invertebrate host. The overexpression of MASPs in the intracellular parasites prior to the division of the amastigotes located in the plasma membrane suggests that some of the proteins of this extensive family play a major biological role in the survival and multiplication of intracellular amastigotes.

5.7 Cruzipain family

This glycoprotein is synthesized as a zymogen that is activated by cleavage of the N-terminal pro-domain to generate the mature protease and belongs to the mammalian papain superfamily but contains, as other cysteine proteases (CPs) from trypanosomatids, an unusual C-terminal extension. Cruzipain family has many groups which include: native-cruzipain (N-cruzipain), recombinant-cruzipain 1 (R-cruzipain 1) and recombinant cruzipain 2 (R-cruzipain 2). Cruzipains are expressed on all the body surface of epimastigotes and amastigotes forms while in the trypomastigote form, cruzipain is expressed only in the flagellar pocket region. Cruzipain plays a role in the process of *T. cruzi* internalization into mammalian cells. Cruzipain is not only essential for parasite survival but also generates a strong immune response in infected individuals.

6. Extracellular vesicles in the life cycle of *T. cruzi*

Extracellular vesicles (EVs) typically consist of a lipid bilayer membrane containing integral membrane proteins and a luminal cavity that is loaded with a variety of soluble proteins and nucleic acids. *T. cruzi* parasites, like many other cells, release extracellular vesicles (EV) that are involved in cell-cell communication or in the modulation of the host immune responses to promote the establishment of an infection [93]. In *T. cruzi* parasites, two classes of extracellular vesicles have been characterized based on size. These include exovesicles also referred to as ectosomes and exosomes. Ectosomes which bud directly from the plasma membrane have a size of 100–1000 nm while exosomes which are vesicles that are secreted into the extracellular environment following the fusion of multivesicular endosomes with the plasma membrane, typically occurring at the flagellar pocket membrane have a size of 30–100 nm. Analysis of extracellular vesicles released by epimastigotes and metacyclic trypomastigotes in culture demonstrated the presence of two populations of extracellular vesicles containing plasma membrane and intracellular proteins, and also nucleic acids. The *T. cruzi* small membrane proteins (TcSMP), a family of proteins or phosphatases detected on *T. cruzi* EVs has been shown to trigger Ca^{2+} signaling and lysosome mobilization/exocytosis, events that promote formation of parasitophorous vacuoles and parasite invasion. In the early stages of *T. cruzi* infection, parasites promote the release of plasma membrane vesicles from the host cell, which may contribute to parasite survival in the circulatory system, an event thought to help mediate host cell invasion.

Immune cells are one of the main targets of extracellular vesicles. Extracellular vesicles secreted during acute and/or chronic *T. cruzi* infection should play a role in the dissemination and survival of this parasite in the vertebrate mammalian host since the released extracellular vesicles from virus-infected cells, bacteria, fungi or parasites have been demonstrated to play a pivotal role in the modulation of the immune system. Several types of extracellular vesicles are promoters of the innate and acquired immune response and defined as types of pathogen-associated molecular patterns (PAMPs) which could be formed by a wide range of macro-molecules such as lipids, proteins, carbohydrates, or nucleic acids and are recognized by pattern recognition receptors (PPRs) such as toll-like receptors (TLRs) present in leukocytes and various non-immune cells, which will in turn initiate a signaling cascade that leads to the activation of an immune response against the pathogen [94]. In *T. cruzi*, several PAMPs have already been described, for instance, parasite cytidine-phosphate-guanosine (CpG)-DNA released from lysed intracellular parasites stimulates TLR7 and TLR9 activation and production of T helper type 1 (Th1) proinflammatory cytokines or parasite α-Gal-containing glycoconjugates such as mucins or gp85/trans-sialidase recognized by TLR2/6 leading to tumor necrosis factor (TNF-α) production in macrophages and inhibition of IL-12 in dendritic cells. Studies on *T. cruzi* extracellular vesicles have shown that these vesicles could act as an agonist of TLR2 signaling, which leads to the secretion of proinflammatory cytokines (TNF-α and IL-6) and nitric oxide which could be explained by the presence of GPI-anchored molecules like mucins, mucin-associated surface proteins (MASPs), or trans-sialidases (TSs) on the extracellular vesicles surface.

7. Carbon and energy sources in *T. cruzi* life cycle

The life cycle of *T. cruzi* alternates between glucose-rich and glucose-poor environments having to adapt to different sources of energy and carbon. The differentiation of epimastigotes, the non-infective dividing forms found in the digestive tract of the invertebrate host into metacyclic trypomastigotes (metacyclogenesis) occur in an amino-acid-rich and carbohydrate-poor medium. In vertebrate, the trypomastigotes differentiate into the dividing forms called amastigotes occur in a medium poor in free glucose.

Trypanosomatids can use either glucose or amino acids as main carbon and energy source, although one cannot rule out the use of fatty acids as well. Amino acids, especially L-proline and L-glutamine which are abundant in the hemolymph and tissue fluids of the blood sucking vector are the main source of carbon and energy in the insect stages. L-Proline seems to be involved in several mechanisms of resistance to oxidative, nutritional and thermal stress and is important in metacyclogenesis for the differentiation of intracellular epimastigotes to trypomastigotes in *T. cruzi*. Several amino acids such as proline, aspartate and glutamate are actively transported and oxidized by *T. cruzi*. The catabolism of aromatic amino acids appears to be related to the cytosolic $NADH^+$ reoxidation. Also, the presence of at least 60 genes belonging to a single family of amino acid transporter in *T. cruzi* further reinforces the relevance of amino acids in the biology of these organisms. While some trypanosomatids metabolically prefer glucose to amino acid when grown in a medium rich in glucose and amino acid, as seen in proline and glucose metabolism in *T. brucei*, other trypanosomatids like *Crithidia deanei*, a monoxenic and non-pathogenic trypanosomatid living in the insect gut preferentially metabolize amino acids irrespective of the glucose content of the culture medium [95].

Amino acids are crucial nutrients during the *T. cruzi* life cycle; apart from their use as carbon and energy sources, they participate in several biological processes that help the parasite adjust to their changing environment. Arginine metabolism is linked to *T. cruzi* growth and is involved in the management of cell energy under nutritional stress condition. Certain amino acids such as proline, glutamate, and aspartate are essentials in the process of metacyclogenesis. Apart from being involved in the process of metacyclogenesis, proline and glutamate seems to have a broad variety of functions. While proline is involved in fulfilling the energy requirements for host cell invasion, differentiation from the intracellular transient epimastigote-like stage to trypomastigote forms and resistance to oxidative stress, glutamate is directly involved in osmoregulation and cell volume control.

8. Changes in gene expression during the life cycle of *T. cruzi*

As a result of changing environments during the life cycle, *T. cruzi* undergoes rapid and significant changes in gene expression, which are achieved essentially at the post-transcriptional level through modulation of messenger RNA (mRNA) stability and translational control mechanisms. In order to adapt to the different environment they find within one or the other host species, *T. cruzi* also undergoes drastic morphological and biochemical changes and are also able to differentiate from proliferative to nonproliferative cells within the same host. All these changes are orchestrated by the differential expression of stage-specific genes.

Cellular differentiation is controlled at multiple levels including, for most eukaryotic cells, initiation of gene transcription. The discriminatory mechanisms for the initiation of transcription at individual loci is largely absent in trypanosomatids and most protein-coding genes lack promoters and are transcribed as long polycistronic units that are processed into individual mRNAs. Consequently, trypanosomes rely on post-transcriptional processes such as translational efficiency, mRNA stability and post-translational modification to coordinate developmental transitions and other adaptive responses encountered throughout their complex life cycles.

In eukaryotes, protein-coding genes are transcribed into monocistronic pre-mRNA transcripts containing coding sequences (exons) and non-coding sequences (introns) that are processed into mature mRNAs through *cis*-splicing reactions. RNA polymerase II is the enzyme responsible for the transcription of protein-coding genes while RNA polymerase I transcribes ribosomal RNA. In trypanosomatids, however, transcription is polycistronic, there are no introns and, therefore, no *cis*-splicing reactions. Processing of pre-mRNA into single gene units is effected by *trans*-splicing reactions, a process that has been found to operate only in trypanosomatids and other organisms like *Euglena*, nematode and trematode worms.

Granules of mRNA such as processing bodies (P bodies) and stress granules (SGs) are involved in post-transcriptional regulation of gene expression. P bodies are constitutively present in the cell and can grow in size and number when cells are perturbed while SGs only arise under cellular stress. P bodies contain mRNA and proteins involved in translational repression, mRNA decapping, $5' \rightarrow 3'$ mRNA decay, nonsense-mediated decay (NMD) and the miRNA (microRNA) pathway. P bodies were initially thought to be the place where mRNA was recruited to be degraded and recently, a function as mRNA storage depots has been assigned to P bodies. By contrast, SGs are stalled 43S translation pre-initiation complexes, mainly composed of mRNA, translation initiation factors and 40S ribosomal proteins. SGs are thought to function as mRNA triage centers during stress.

9. Conclusion

Trypanosoma cruzi, the parasite responsible of Chagas disease has a complex life cycle including intracellular and extracellular forms, which alternate between invertebrate insect vectors and vertebrate mammalian hosts. *T. cruzi* replicate extracellularly within the insect, but have to infect cells to multiply within the mammal which contrasts with the African trypanosome which is extracellular in both hosts. During their life cycles, they alternate between a mammalian host and an insect vector and undergo profound biochemical and morphological transformations in order to adapt to the different environments changes orchestrated by precise gene regulation programs.

Conflict of interest

I have no conflict of interest to declare.

Author details

Kenechukwu C. Onyekwelu
Department of Medical Biochemistry, College of Medicine, University of Nigeria, Enugu Campus, Nigeria

*Address all correspondence to: kenechukwu.onyekwelu@unn.edu.ng

References

[1] Rassi A Jr, Rassi A, Marin-Neto JA. Chagas disease. Lancet. 2010;**375**:1388-1402

[2] Rassi A Jr, Rassi A, Marcondes de Rezende J. American trypanosomiasis (chagas disease). Infectious Disease Clinics of North America. 2012;**26**(2):257-291

[3] Chagas C. Nova tripanozomiase humana. Estudos sobre a morfologia e o ciclo evolutivo do *Schizotrypanum cruzi* n. gen., n. sp., agente etiológico de nova entidade mórbida do homem. Memórias do Instituto Oswaldo Cruz. 1909;**1**:159-218

[4] Chagas CRJ. Lição de abertura dos cursos da Faculdade de Medicina do Rio de Janeiro—1928. In: Prata AR (org) Carlos Chagas. Coletânea de trabalhos científicos. Brasília: Editora Universidade de Brasília; 1981. pp. 861-883

[5] Maguire JH. Trypanosoma. In: Gorbach SL, Bartlett JG, Blacklow NR, editors. Infectious Diseases. 3rd ed. Philadelphia, PA: Lippincott Williams & Wilkins; 2004. pp. 2327-2334

[6] Tyler KM, Engman DM. The life cycle of *Trypanosoma cruzi* revisited. International Journal of Parasitology. 2001;**31**:472-481

[7] Baral TN. Immunobiology of African trypanosomes: Need of alternative interventions. Journal of Biomedicine and Biotechnology. 2010;**1110-7243**:389153

[8] de Souza W. Cell biology of *Trypanosoma cruzi*. International Review of Cytology. 1984;**86**:197-283

[9] Castro DP, Seabra SH, Garcia ES, de Souza W, Azambuja CP. *Trypanosoma cruzi*: Ultrastructural studies of adhesion, lysis and biofilm formation by *Serratia marcescens*. Experimental Parasitology. 2007;**117**(2):201-207

[10] Nogueira NF, Gonzalez MS, Gomes JE, De Souza W, Garcia ES, Azambuja P, et al. *Trypanosoma cruzi*: Involvement of glycoinositolphospholipids in the attachment to the luminal midgut surface of *Rhodnius prolixus*. Experimental Parasitology. 2007;**116**:120-128

[11] Alves CR, Albuquerque-Cunha JM, Mello CB, Garcia ES, Nogueira NF, Bourguingnon SC, et al. *Trypanosoma cruzi*: Attachment to perimicrovillar membrane glycoproteins of *Rhodnius prolixus*. Experimental Parasitology. 2007;**116**(1):44-52

[12] Terra WR. Evolution of digestive system of insects—Review. Annual Review of Entomology. 1990;**35**:181-200

[13] Ruiz RC, Favoreto S Jr, Dorta ML, Oshiro MEM, Ferreira AT, Manque PM, et al. Infectivity of *Trypanosoma cruzi* strains is associated with differential expression of surface glycoproteins with differential Ca^{2+} signaling activity. Biochemical Journal. 1998;**330**:505-511

[14] Ramirez MI, Ruiz RC, Araya JE, Franco da Silveira J, Yoshida N. Involvement of the stage-specific 82-kilodalton adhesion molecule of *Trypanosoma cruzi* metacyclic trypomastigotes in host cell invasion. Infection and Immunity. 1993;**61**:3636-3641

[15] Santori FR, Dorta ML, Juliano L, Juliano MA, Franco da Silveira J, Ruiz RC, et al. Identification of a domain of *Trypanosoma cruzi* metacyclic trypomastigote surface molecule gp62 required for attachment and invasion of mammalian cells. Molecular and Biochemical Parasitology. 1996;**78**:209-216

[16] Dorta ML, Ferreira AT, Oshiro MEM, Yoshida N. Ca^{2+} signal induced by *Trypanosoma cruzi* metacyclic trypomastigote surface molecules implicated in mammalian cell invasion. Molecular and Biochemical Parasitology. 1995;**73**:285-289

[17] Barr SC, Han W, Andrews NW, Lopez JW, Ball B, Pannabecker TL, et al. A factor from *Trypanosoma cruzi* induces repetitive cytosolic free Ca^{2+} transients in isolated primary canine cardiac myocytes. Infection and Immunity. 1996;**64**:1770-1777

[18] Moreno SNJ, Silva J, Vercesi AE, Docampo R. Cytosolic-free calcium elevation in *Trypanosoma cruzi* is required for cell invasion. Journal of Experimental Medicine. 1994;**180**:1535-1540

[19] Yakubu MA, Majumder S, Kierszenbaum F. Changes in *Trypanosoma cruzi* infectivity by treatments that affect calcium ion levels. Molecular and Biochemical Parasitology. 1994;**66**:119-125

[20] Wilkowsky SE, Wainszelbaum MJ, Isola ELD. *Trypanosoma cruzi*: Participation of intracellular Ca^{2+} during metacyclic trypomastigote-macrophage interaction. Biochemical and Biophysical Research Communications. 1996;**222**:386-389

[21] Favoreto S, Dorta ML, Yoshida N. *Trypanosoma cruzi* 175-kDa protein tyrosine phosphorylation is associated with host cell invasion. Experimental Parasitology. 1998;**89**:188-194

[22] Tardieux I, Nathanson MH, Andrews NW. Role in host cell invasion of *Trypanosoma cruzi*-induced cytosolic free Ca^{2+} transients. Journal of Experimental Medicine. 1994;**179**:1017-1022

[23] Lima MF, Villalta F. Host-cell attachment by *Trypanosoma cruzi* identification of an adhesion molecule. Biochemical and Biophysial Research Communication. 1988;**155**:256-262

[24] Ouaissi MA, Cornette J, Capron A. Identification and isolation of *Trypanosoma cruzi* trypomastigote cell surface protein with properties expected of a fibronectin receptor. Molecular and Biochemical Parasitology. 1986;**19**:201-211

[25] Alves MJM, Abuin G, Kuwajima VY, Colli W. Partial inhibition of trypomastigote entry into cultured mammalian cells by monoclonal antibodies against a surface glycoprotein of *Trypanosoma cruzi*. Molecular and Biochemical Parasitology. 1986;**21**:75-82

[26] Villalta F, Zhang Y, Bibb KE, Burns JM, Lima MF. Signal transduction in human macrophages by gp83 ligand of *Trypanosoma cruz*: Trypomastigote gp83 ligand up-regulates trypanosome entry through the MAP kinase pathway. Biochemical and Biophysical Research Communication. 1998;**249**:247-252

[27] Garcia ES, Azambuja P. Development and interactions of *Trypanosoma cruzi* within the insect vector. Parasitology Today. 1991;**7**:240-244

[28] Kleffman T, Schmidt J, Schaub GA. Attachment of *Trypanosoma cruzi* epimastigotes to hydrophobic substrates and use of this property to separate stages and promote metacyclogenesis. Journal of Eukaryotic Microbiology. 1998;**45**:548-555

[29] Gonzales-Perdomo M, Romero P, Goldenberg S. Cyclic AMP and adenylate cyclase activators stimulate *Trypanosoma cruzi* differentiation. Experimental Parasitology. 1988;**66**:205-212

[30] Gerish G, Malchow D. Cyclic AMP receptors and the control of cell aggregation in *Dyctiostelium*. Advances in Cyclic Nucleotide Research. 1975;**7**:49-65

[31] Strickler JE, Patton CL. Adenosine 3′,5′-monophosphate in reproducing and differentiated trypanosomes. Science. 1975;**190**:1110-l 112

[32] Gomes SL, Mennucci L, Maia JCC. Induction of *Blastocladiella emersonii germi*nation by cyclic adenosine-3′,5′-monophosphate. Cell differenciation. 1980;**9**:169-179

[33] Hoppe J, Wagner KG. Cydic AMP-dependent protein kinase I. A unique allosteric enzyme. Trends in Biochemists Science. 1979;**4**:282-285

[34] Smith SB, White HD, Siegel JB, Krebs EG. Cyclic AMP-dependent protein kinase I: Cyclic nucleotide binding, structural changes, and release of the catalytic subunits. Proceedings of the National Academy of Sciences (USA). 1981;**78**:1591-1595

[35] Mehdy MC, Firtel RA. A secreted factor and cyclic AMP jointly regulate celltype-specific gene expression in *Dyctiostelium discoideum*. Molecular and Cellular Biology. 1985;**5**:107-713

[36] Nagamine Y, Reich E. Gene expression and CAMP. Proceedings of the National Academy of Sciences (USA). 1985;**82**:4606-4610

[37] Oyama M, Blumberg DD. Changes during differentiation in requirements for CAMP for expression of cell-type-specific mRNAs in the cellular slime mold. *Dyctiostelium discoideum*. Developmental Biology. 1986;**117**:550-556

[38] Dedman JR, Brinkley BR, Means AR. Regulation of microfilaments and microtubules by calcium and cyclic AMP. Advances in Cyclic Nucleotides Research. 1979;**11**:131-174

[39] Means AR, Dedman JR. Calmodulin-An intracellular calcium receptor. Nature. 1980;**285**:73-77

[40] Tellez-inon MT, Ulloa RM, Tofwjela M, Toflres HN. 1985. Calmodulin and calcium-dependent cyclic AMP phosphodiesterase activity in Trypanosoma cruzi. Molecular and Biochemical Parasitology. 1985;**14**:143-153

[41] Balber AE. The pellicle and the membrane of the flagellum, flagellar adhesion zone, and flagellar pocket: Functionally discrete surface domains of the bloodstream form of African trypanosomes. Critical Reviews in Immunology. 1990;**10**(3):177-201

[42] Gull K. Host-parasite interactions and trypanosome morphogenesis: A flagellar pocketful of goodies. Current Opinion in Microbiology. 2003;**6**(4):365-370

[43] Gadelha C, Rothery S, Morphew M, McIntosh JR, Severs NJ, Gull K. Membrane domains and flagellar pocket boundaries are influenced by the cytoskeleton in African trypanosomes. Proceedings of the National Academy of Sciences of the United States of America. 2009;**106**(41):17425-17430

[44] Landfear SM, Ignatushchenko M. The flagellum and flagellar pocket of trypanosomatids. Molecular and Biochemical Parasitology. 2001;**115**(1):1-17

[45] Villalta F, Kierszenbaum F. Host-cell invasion by *Trypanosoma cruzi*: Role of cell surface galactose residues. Biochemical and Biophysical Research Communications. 1984;**119**(1):228-235

[46] Yoshida N, Mortara RA, Araguth MF, Gonzalez JC, Russo M. Metacyclic neutralizing effect of monoclonal antibody 10D8 directed to the 35- and 50-kilodalton surface glycoconjugates of *Trypanosoma cruzi*. Infection and Immunity. 1989;**57**(6):1663-1667

[47] Cánepa GE, Meśıas AC, Yu H, Chen X, Buscaglia CA. Structural features

affecting trafficking, processing, and secretion of *Trypanosoma cruzi* mucins. Journal of Biological Chemistry. 2012;**287**(31):26365-26376

[48] Buscaglia CA, Campo VA, Frasch ACC, Di Noia JM. *Trypanosoma cruzi* surface mucins: Host-dependent coat diversity. Nature Reviews Microbiology. 2006;**4**(3):229-236

[49] Frasch ACC. Functional diversity in the trans-sialidase and mucin families in *Trypanosoma cruzi*. Parasitology Today. 2000;**16**(7):282-286

[50] Almeida IC, Gazzinelli RT. Proinflammatory activity of glycosylphosphatidylinositol anchors derived from *Trypanosoma cruzi*: Structural and functional analyses. Journal of Leukocyte Biology. 2001;**70**(4):467-477

[51] Acosta-Serrano A, Almeida IC, Freitas-Junior LH, Yoshida N, Schenkman S. The mucin-like glycoprotein super-family of *Trypanosoma cruzi*: Structure and biological roles. Molecular and Biochemical Parasitology. 2001;**114**(2):143-150

[52] Di Noia JM, D'Orso I, Sánchez DO, Frasch ACC. AU- rich elements in the 3′-untranslated region of a new mucin-type gene family of Trypanosoma cruzi confers mRNA instability and modulates translation efficiency. Journal of Biological Chemistry. 2000;**275**(14):10218-10227

[53] Campo VA, Buscaglia CA, Di Noia JM, Frasch ACC. Immunocharacterization of the mucin-type proteins from the intracellular stage of *Trypanosoma cruzi*. Microbes and Infection. 2006;**8**(2):401-409

[54] Lantos BA, Carlevaro G, Araoz B, Ruiz Diaz P, Camara MM, Buscaglia CA, et al. Sialic acid glycobiology unveils *Trypanosoma cruzi* trypomastigote membrane physiology. PLoS Pathogens. 2016;**12**(4):e1005559. DOI: 10.1371/journal.ppat.1005559

[55] Han MV, Zmasek CM. PhyloXML: XML for evolutionary biology and comparative genomics. BMC Bioinformatics. 2009;**10**:356

[56] Di Noia JM, Sanchez DO, Frasch ACC. The protozoan *Trypanosoma cruzi* has a family of genes resembling the mucin genes of mammalian cells. The Journal of Biological Chemistry. 1995;**270**(41):24146-24149

[57] Di Noia JM, Pollevick GD, Xavier MT, Previato JO, Mendoça-Previato L, Sánchez DO, et al. High diversity in mucin genes and mucin molecules in *Trypanosoma cruzi*. The Journal of Biological Chemistry. 1996;**271**(50):32078-32083

[58] Barreto-Bergter E, Vermelho AB. Structures of glycolipids found in trypanosomatids: Contribution to parasite functions. Open Parasitology Journal. 2010;**4**(1):84-97

[59] Di Noia JM, Buscaglia CA, De Marchi CR, Almeida IC, Frasch ACC. A *trypanosoma cruzi* small surface molecule provides the first immunological evidence that Chagas' disease is due to a single parasite lineage. Journal of Experimental Medicine. 2002;**195**(4):401-413

[60] Buscaglia CA, Di Noia JM. *Trypanosoma cruzi* clonal diversity and the epidemiology of Chagas' disease. Microbes and Infection. 2003;**5**(5):419-427

[61] De Marchi CR, Di Noia JM, Frasch ACC, Neto VA, Almeid IC, Buscaglia CA. Evaluation of a recombinant Trypanosoma cruzi mucin-like antigen for serodiagnosis of Chagas' disease. Clinical and Vaccine Immunology. 2011;**18**(11):1850-1855

[62] Nakayasu ES, Yashunsky DV, Nohara LL, Torrecilhas ACT, Nikolaev VA, Almeida IC. GPIomics: Global analysis of glycosylphosphatidylinositol-anchored molecules of *trypanosoma cruzi*. Molecular Systems Biology. 2009;**5**:261

[63] Schenkman S, Ferguson MAJ, Heise N, Cardoso de Almeida ML, Mortara RA, Yoshida N. Mucin-like glycoproteins linked to the membrane by glycosylphosphatidylinositol anchor are the major acceptors of sialic acid in a reaction catalyzed by trans-sialidase in metacyclic forms of *trypanosoma cruzi*. Molecular and Biochemical Parasitology. 1993;**59**(2):293-303

[64] Yoshida N. Molecular basis of mammalian cell invasion by *trypanosoma cruzi*. Anais da Academia Brasileira de Cie^ncias. 2006;**78**(1):87-111

[65] Urban I, Boiani Santurio L, Chidichimo A, Yu H, Chen X, Mucci J, et al. Molecular diversity of the *trypanosoma cruzi* TcSMUG family of mucin genes and proteins. Biochemical Journal. 2011;**438**(2):303-313

[66] De Pablos LM, Osuna A. Conserved regions as markers of different patterns of expression and distribution of the mucin-associated surface proteins of *trypanosoma cruzi*. Infection and Immunity. 2012;**80**(1):169-174

[67] El-Sayed NM, Myler PJ, Bartholomeu DC, et al. The genome sequence of trypanosoma cruzi, etiologic agent of chagas disease. Science. 2005;**309**(5733):409-415

[68] Frevert U, Schenkman S, Nussenzweig V. Stage-specific expression and intracellular shedding of the cell surface trans-sialidase of *trypanosoma cruzi*. Infection and Immunity. 1992;**60**(6):2349-2360

[69] Moraes Barros RR, Marini MM, Antônio CR, Cortez DR, Miyake AM, Lima FM, et al. Anatomy and evolution of telomeric and subtelomeric regions in the human protozoan parasite *trypanosoma cruzi*. BMC Genomics. 2012;**13**(1):229

[70] Pereira MEA, Loures MA, Villalta F, Andrade AFB. Lectin receptors as markers for *Trypanosoma cruzi*. Developmental stages and a study of the interaction of wheat germ agglutinin with sialic acid residues on epimastigote cell. Journal of Experimental Medicine. 1980;**152**(5):1375-1392

[71] Zingales B, Carniol C, de Lederkremer RM, Colli W. Direct sialic acid transfer from a protein donor to glycolipids of trypomastigote forms of *Trypanosoma cruzi*. Molecular and Biochemical Parasitology. 1987;**26**(1-2):135-144

[72] Tomlinson S, Pontes de Carvalho LC, Vandekerckhove F, Nussenzweig V. Role of sialic acid in the resistance of *Trypanosoma cruzi* trypomastigotes to complement. Journal of Immunology. 1994;**153**(7):3141-3147

[73] Jacobs T, Erdmann H, Fleischer B. Molecular interaction of Siglecs (sialic acid-binding Ig-like lectins) with sialylated ligands on *Trypanosoma cruzi*. European Journal of Cell Biology. 2010;**89**(1):113-116

[74] Schenkman S, Jiang MS, Hart GW, Nussenzweig V. A novel cell surface trans-sialidase of *Trypanosoma cruzi* generates a stage-specific epitope required for invasion of mammalian cells. Cell. 1991;**65**(7):1117-1125

[75] Prioli RP, Rosenberg I, Pereira MEA. High- and low-density lipoproteins enhance infection of *Trypanosoma cruzi* in vitro. Molecular and Biochemical Parasitology. 1990;**38**(2):191-198

[76] Yoshida N, Dorta ML, Ferreira AT, Oshiro MEM, Mortara RA,

Acosta-Serrano A, et al. Removal of sialic acid from mucin-like surface molecules of *Trypanosoma cruzi* metacyclic trypomastigotes enhances parasite-host cell interaction. Molecular and Biochemical Parasitology. 1997;**84**(1):57-67

[77] Schenkman S, De Carvalho LP, Nussenzweig V. *Trypanosoma cruzi* trans-sialidase and neuraminidase activities can be mediated by the same enzymes. Journal of Experimental Medicine. 1992;**175**(2):567-575

[78] Buscaglia CA, Alfonso J, Campetella O, Frasch AC. Tandem amino acid repeats from *Trypanosoma cruzi* shed antigens increase the half-life of proteins in blood. Blood. 1999;**93**(6):2025-2032

[79] Correa PRC, Cordero EM, Gentil LG, Bayer-Santos E, da Silveira JF. Genetic structure and expression of the surface glycoprotein GP82, the main adhesin of *Trypanosoma cruzi* metacyclic trypomastigotes. The Scientific World Journal. 2013;**2013**:156734. DOI: 10.1155/2013/156734

[80] Matsumoto TK, Cotrim PC, Da Silveira JF, Stolf AMS, Umezawa ES. *Trypanosoma cruzi*: Isolation of an immunodominant peptide of TESA (trypomastigote excreted-secreted antigens) by gene cloning. Diagnostic Microbiology and Infectious Disease. 2002;**42**(3):187-192

[81] Campetella O, Sánchez D, Cazzulo JJ, Frasch ACC. A superfamily of *trypanosoma cruzi* surface antigens. Parasitology Today. 1992;**8**(11):378-381

[82] Norris KA, Schrimpf JE, Szabo MJ. Identification of the gene family encoding the 160-kilodalton *Trypanosoma cruzi* complement regulatory protein. Infection and Immunity. 1997;**65**(2):349-357

[83] Cetron MS, Hoff R, Kahn S, Eisen H, Van Voorhis WC. Evaluation of recombinant trypomastigote surface antigens of *Trypanosoma cruzi* in screening sera from a population in rural Northeastern Brazil endemic for chagas' disease. Acta Tropica. 1992;**50**(3):259-266

[84] Freire-De-Lima L, Fonseca LM, Oeltmann T, Mendoncą-Previato L, Previato JO. The trans-sialidase, the major *Trypanosoma cruzi* virulence factor: Three decades of studies. Glycobiology. 2015;**25**(11):1142-1149

[85] Schenkman S, Eichinger D, Pereira MEA, Nussen-zweig V. Structural and functional properties of *Trypanosoma* trans-sialidase. Annual Review of Microbiology. 1994;**48**:499-523

[86] Freitas LM, dos Santos SL, Rodrigues-Luiz GF, Mendes TAO, Rodrigues TS, Gazzinelli RT, et al. Genomic analyses, gene expression and antigenic profile of the trans-sialidase superfamily of *Trypanosoma cruzi* reveal an undetected level of complexity. PLoS ONE. 2011;**6**(10):e25914

[87] Kulkarni MM, Olson CL, Engman DM, McGwire BS. *Trypanosoma cruzi* GP63 proteins undergo stage-specific differential posttranslational modification and are important for host cell infection. Infection and Immunity. 2009;**77**(5):2193-2200

[88] Cuevas IC, Cazzulo JJ, Sánchez DO. gp63 homologues in *Trypanosoma cruzi*: Surface antigens with metalloprotease activity and a possible role in host cell infection. Infection and Immunity. 2003;**71**(10):5739-5749

[89] Kangussu-Marcolino MM, De Paiva RMC, Araújo PR, de Mendonça-Neto RP, Lemos L, Bartholomeu DC, et al. Distinct genomic organization, mRNA expression and cellular localization of members of two amastin sub-families present in *Trypanosoma cruzi*. BMC Microbiology. 2013;**131**(1):10

[90] Rochette A, McNicoll F, Girard J, Breton M, Leblanc E, Bergeron MG, et al. Characterization and developmental gene regulation of a large gene family encoding amastin surface proteins in Leishmania spp. Molecular and Biochemical Parasitology. 2005;**140**(2):205-220

[91] Teixeira SMR, Russell DG, Kirchhoff LV, Donelson JE. A differentially expressed gene family encoding amastin, a surface protein of *Trypanosoma cruzi* amastigotes. Journal of Biological Chemistry. 1994;**269**(32):20509-20516

[92] García EA, Ziliani M, AgüeroF, Bernabó G, Sánchez DO, Tekiel V. TcTASV: A novel protein family in Trypanosoma cruzi identified from a subtractive trypomastigote cDNA library. PLoS Neglected Tropical Disease. 2010;**4**(10):e841. DOI: 10.1371/journal.pntd.0000841

[93] Garcia-Silva MR, das Neves RF, Cabrera-Cabrera F, Sanguinetti J, Medeiros LC, Robello C, et al. Extracellular vesicles shed by *Trypanosoma cruzi* are linked to small RNA pathways, life cycle regulation, and susceptibility to infection of mammalian cells. Parasitology Research. 2014;**113**:285-304

[94] Nogueira PM, Ribeiro K, Silveira ACO, Campos JH, Martins-Filho OA, Bela SR, et al. Vesicles from different *Trypanosoma cruzi* strains trigger differential innate and chronic immune responses. Journal of Extracellular Vesicles. 2015;**4**:28734

[95] Lamour N, Riviere L, Coustou V, Coombs GH, Barrett MP, Bringaud F. Proline metabolism in procyclic *Trypanosoma brucei* is down-regulated in the presence of glucose. Journal of Biological Chemistry. 2005;**280**:11902-11910

[96] Jimenez V. Dealing with environmental challenges: Mechanisms of adaptation in *Trypanosoma cruzi*. Research in Microbiology. 2014;**165**:155-165

Chapter 3

Cell Culture and Maintenance of the Evolutionary Forms of *Trypanosoma cruzi* for Studies of Parasitic Biology

Cláudia Jassica Gonçalves Moreno, Johny Wysllas de Freitas Oliveira, Joice Castelo Branco, Laura Araújo, Aline Maria Queiroz, Sílvia Tavares Donato, Nilton José da Silva Júnior, Emilly Thays da Silva Rodrigues and Marcelo Sousa Silva

Abstract

This chapter aims to present and discuss the main cell culture techniques used for the growth and maintenance of the different evolutionary forms of the protozoan *Trypanosoma cruzi*, the etiologic agent of the Chagas disease. Chagas disease is a neglected tropical disease endemic in several Latin American countries. Here, we intend to present the main difficulties, advantages, and disadvantages of using *Trypanosoma cruzi* cell culture in parasitic biology. Finally, we present the main research opportunities in the context of *Trypanosoma cruzi* parasitic biology using cell culture techniques, such as drug development, characterization of pharmacological targets, molecular markers for diagnosis, structural biology, and many other biomedical applications.

Keywords: *Trypanosoma cruzi*, Chagas disease, cell culture, in vitro assay, metacyclogenesis

1. Introduction

Carlos Chagas described the American trypanosomiasis through microscope observation of hemoflagellate protozoan in the sample blood and named *Trypanosoma cruzi* [1]. The parasite belongs to the Kinetoplastida order and *Trypanosomatidae* family. This family is characterized by the presence of flagellum and only one mitochondrion, which extends throughout the cell body. The mitochondrion matrix contains a complex array of DNA fibrils structure called kinetoplast (kDNA).

Trypanosoma cruzi (*T. cruzi*) is a digenetic parasite that develops through a complex life cycle involving an insect vector and a mammalian host. Trypanosomatidae family shows, during development stages, several forms, which can be easily identified by light microscopy in Giemsa-stained preparations. In the case of *T. cruzi*, three main morphogenetic forms are recognized, and the transition between each form

IntechOpen

	Evolutionary forms			
	Epimastigote	Trypomastigote metacyclic	Trypomastigote	Amastigote
Localization	Intestine of the vector and axenic culture	Rectal ampoule of vector axenic culture	Blood and intracellular spaces of vertebrate-host and cells culture	Inside the cells of vertebrate-host and cells culture
Cell morphology	Elongate	Elongate	Elongate	Spherical
Position of the kinetoplast	Near the nucleus	After the nucleus	After the nucleus	Near the nucleus

These evolutionary forms are identified by the relative position of the kinetoplast in relation to the nucleus.

Table 1.
View in optical microscope of the various developmental stages of Trypanosoma cruzi: epimastigote, intracellular amastigote, trypomastigote metacyclic, and bloodstream forms.

involves morphologic modifications in gene expression and cellular cycle (**Table 1**). Therefore, the evolutionary forms, trypomastigote and amastigote, are found in the vertebrate host, and epimastigote and metacyclic trypomastigotes are found in the invertebrate host [2].

Due to its genetic diversity, the parasite *T. cruzi* represents one of the best model organisms for studies of clonal evolution, with rare events of recombination and/or genetic hybridization. Molecular markers, like the mini-exon genes and the ribosomal DNA 24S, allow the characterization of *T. cruzi* lineages and the subdivision in seven discrete typing units (DTUs, from TcI to TcVI) [3, 4]. Parasite of *T. cruzi* diversity could be associated with adaptation and survives inside different hosts. Therefore, this genotypic variation among the different strains of *T. cruzi* from different geographical regions may explain the frequency of variability in different characteristics of parasitism, such as virulence, subclinical infection rate, clinical symptoms of infection, and susceptibility to antiparasitic drugs [5].

A century of its discovery, Chagas disease remains a cause of early mortality and morbidity in individuals at productive age in endemic areas, generating significant socioeconomic impact. The importance of parasite biology studies is undeniable. In this context, in vitro studies of *T. cruzi* offers the opportunity to understand the coordination of metabolic and cellular change as a response to environmental stimuli in a simple and well-characterized system.

2. How to maintain and cultivate *Trypanosoma cruzi*?

During life cycle, *T. cruzi* alternates among different morphological and functional types and different hosts (mammals and triatomine bugs). Epimastigote replicative form can result in non-replicative and infective trypomastigote forms. Amastigote forms could be achieved by infected mammal cells. The adaptation mechanism of *T. cruzi* to environmental changes and diverse physicochemical conditions represents an important means of survival.

Substantial interest in understanding the cell biology of *T. cruzi* made possible its maintenance and growth in in vitro systems. Several methods have already been described for obtaining the different stages of the parasite, and these can be cultured in non-defined axenic media, defined axenic media, and in various cell types.

The axenic cultures of *T. cruzi* were primarily established in different cell types and tissues with appropriate conditions that allowed the infection process and its survival [6]. During this period, metacyclogenesis process was observed and characterized through detailing all the cellular and/or behavioral transformations undergone by the parasite. By observing the biological changes undergone, key points during the process became possible to be determined that were later mimicked to facilitate the differentiation process [7].

Epimastigote forms are maintained in axenic conditions at 28°C, in the nutritive medium liver infusion tryptose (LIT), supplemented with 10% (v/v) heat-inactivated fetal bovine serum (FBS) as described by Camargo [8]. Under chemically defined conditions and suppression of nutrients, the transformation of epimastigotes into metacyclic trypomastigotes is possible in vitro using a process named metacyclogenesis [9].

To obtain intracellular amastigote forms, usually mammal cells, Vero lineage, murine peritoneal macrophages, fibroblast (mouse L929), and primary heart or skeletal muscles cells are commonly incubated with specific medium supplemented with 10% inactivated FBS (v/v) and allowed to adhere usually for 24 h at 37 ± 2°C in 5% CO_2. After this period, adherent cells can infect with trypomastigote metacyclic using a 10:1 ratio of parasites per cell and incubated at 37°C in 5% CO_2, without FBS. Amastigote can be observed in microscope at 4–24 h after infection. Intracellular amastigote can release within 72 h trypomastigote metacyclic from culture derived. Infected Vero cells can be in continuous cultivation during 6–8 weeks [10].

Infected cells with *T. cruzi* can spontaneously release amastigote forms into the supernatant medium. The viable amastigote forms can be collected and produce the axenic amastigote. Firstly, wash cells with phosphate buffer saline (PBS, pH 7.2) through centrifugation, collect, and then resuspend in appropriate medium, usually Grace medium supplemented with nicotinamide and ATP. This addition can neutralize differentiation of amastigote forms, and the culture can be maintained at 37°C in a humid atmosphere containing 5% CO_2. Every 4 days, the media have to be partially removed by centrifugation to remove the supernatant medium and add the fresh medium to maintain the amastigote forms [11].

3. In vitro differentiation of *Trypanosoma cruzi* epimastigote to trypomastigote forms

Metacyclic trypomastigotes (non-replicative) originating from the insect vector or trypomastigotes from infected cells, such as culture derived and bloodstream/tissue, are infective forms of the parasite. Metacyclogenesis is observed in the stationary phase of culture, and metacyclic trypomastigotes are largely absent during logarithmic growth. However, in vitro differentiation of *T. cruzi* is possible according to biologically active factors and essential nutrient depletion or accumulation of certain metabolites [12]. Generally, metacyclogenesis process can be mimicked using chemical differentiation medium or axenic-defined medium and nutrient suppression for the differentiation of the epimastigote forms to occur in trypomastigotes. The omission of medium supplements, such as glucose, was able to lead to flagellar elongation, as observe by Tyler and Engman [13]. The epimastigote-extending flagellum occurs in order to undergo metacyclogenesis.

By contrast, chemical modification is commonly used for differentiation of epimastigote to trypomastigote forms. Epimastigote forms can be collected from the LIT culture, during the stationary phase of the growth curve, and then resuspended in artificial triatomine urine (TAU) supplemented with proline (TAUP medium) or TAU supplemented with L-proline, L-sodium glutamate, L-sodium aspartate, and D-glucose (TAU3AAG medium) [14]. In the first medium, parasite culture can differentiate after 10 days and 72–96 h using TAU3AAG medium. Recent study demonstrated that reductive environment using urate, a salt or ester of uric acid, could promote epimastigote differentiation with significant increment of trypomastigotes [15]. In addition, Grace medium supplemented with *Triatoma infestans* intestinal homogenate was used for in vitro metacyclogenesis of *T. cruzi* [16].

Metacyclic trypomastigotes are essential for the understanding of host-pathogen interaction and it is remarkable genetic variability between strains. Although, these biological stages in axenic medium are difficult to purify, culture remains with epimastigote stages without showing a high efficiency in the purification of metacyclic trypomastigotes [8].

Complement-mediated lysis of epimastigotes followed by separation of the trypomastigotes by gradient centrifugation through a dense albumin column represents a selective method that exploited to purify viable trypomastigotes from cells culture medium. Alternate pathway activation seems capable by itself generating epimastigote lysis and consequent activation of the classical pathway just in the presence of serum [17].

Other several techniques such as density separation with Percoll [18, 19] and extensive use of chromatography have permitted purifying metacyclic trypomastigotes for the association of specific molecules expressed on this parasite stage membrane separation. Chromatography based on the differential plasma membrane charge between epimastigotes and trypomastigotes using ion exchange chromatography with resins, such as cellulose-DEAE and sephadex, have been developed for the purification of metacyclic trypomastigotes [20, 21]. These previous chromatography techniques presented a major disadvantage such as the decrease in the infective capacity of the obtained trypomastigote [22].

Recently, a new protocol in LIT medium cultures was implemented using sepharose-DEAE as a resin during purification process, which could recover metacyclic trypomastigotes for two different DTUs (I and II). DTU TcI was the one recovered a greater amount of these forms, and parasite infectivity in vitro and in vivo was maintained [20]. Metacyclogenesis with different DTUs exhibited significantly different morphologies and metacyclogenesis pattern [23].

Still it remains crucial to have easy, fast, and reliable tools to obtain purified metacyclic trypomastigote forms of *T. cruzi*. Thus, results acquired with these purified parasites should be interpreted with caution in order to validated biological, chemical, or immunological studies.

4. Advantages and limitation of in vitro *Trypanosoma cruzi* culture

The entire *T. cruzi* cycle can be studied in vitro, and it is one of the easiest to keep in the laboratory. A crucial point, which is considered to be a basic requirement for all the biological assays in vitro system using *T. cruzi* is the selection of methods and adoption of biosafety procedures, including good laboratory facilities and laboratory practices and the use of collective and individual protection tools. Therefore, the training of personnel involved in the manipulation and cultivation of *T. cruzi* strains is required, especially on the handling of drug-resistant strains.

Probably, the main advantages of the use of in vitro systems for maintenance and cultivation of *T. cruzi* are related to costs and biosafety. The in vitro system of *T. cruzi* cultivation presents some characteristics that deserve to be highlighted: (i) reduction of the number of animals used in experimental models of infection, (ii) ethical aspects in animal replacement, (iii) less need for laboratory infrastructure, (iv) reduced time of experiments compared to in vivo methods, and (iv) lower risk of human infection during culture procedures [15].

As mentioned above, the use of *T. cruzi* in in vitro assays allows the researcher to access a variety of strains with different biological and genetic profiles, in addition to the different stages of parasite differentiation. Furthermore, *T. cruzi* clones exhibit a greater rate of multiplication in culture medium.

One of the possibilities of in vitro system is to obtain large amounts of metacyclic trypomastigotes (derived culture or chemical modification medium) with biological properties similar to those of the insect-derived forms, which facilitates the study of the biology of *T. cruzi*. On the other hand, there is a controversy that the strains are kept in artificial culture media, transformed into true "laboratory strains," with total or partial loss of their original characteristics, despite their morphological similarities. Contreras and colleagues demonstrated that culture-derived metacyclic forms of *T. cruzi* can decrease their capacity to invade cells [16].

The great potential of the in vitro models is the study of the interaction *T. cruzi*-host cell throughout amastigote culture. It has been demonstrated the in vitro capacity of invasion in all cell types was tested [10, 24]. As yet, in vivo models, infected animals, as well as in human, show evidence of a parasite tropism by cells of the mononuclear phagocytic system and by muscle and nerve cells, contrary to what is observed in vitro as mentioned before. On the other hand, maintained trypomastigote metacyclic in vivo (mice) condition generally increased virulence [14].

However, the in vitro assay does not provide full information on the behavior and characteristics about *T. cruzi*. The complexity of in vivo system is bigger than in parasite cultivation system. The mechanisms of parasite-host interaction are controlled by cells and innumerable proteins that trigger numerous genetic and physiological signals in the parasite. These interactions cannot be accurately evaluated in in vitro systems.

5. Applications in the cell culture of *Trypanosoma cruzi*

Despite decades of research in *T. cruzi* biology, many questions remain to be solved. The cultivation of *T. cruzi* has been explored in several biomedical researches, mainly as a strategy for the understanding of parasite biology and parasite-host interaction. Thus, these in vitro methodologies have been used to understand mechanisms of parasite invasion, mechanisms of resistance and susceptibility to drugs, identification of virulence factors, signaling, cellular metabolism, etc. Our research group has been using the *T. cruzi* cultures to evaluate the potential of new compounds with antiparasitic activity for the treatment of Chagas disease [25]. The cultivation methods also allow to isolate and characterize the circulating *T. cruzi* genotypes, present in triatomines captured in the natural environment of transmission of Chagas disease [26]. In addition, other robust molecular and structural biology tools can be applied to the study of the biology of *T. cruzi* using cell culture, namely, functional genomic and proteomic studies [27, 28] and structural cryo-microscopy [29].

6. Conclusion

In this chapter, we have shown the importance of cell culture techniques as a complementary tool for studies of parasitic biology. Through the cultivation of different evolutionary forms of *T. cruzi*, we can answer important scientific questions in the context of parasitic diseases, more specifically in Chagas disease. In conclusion, this work shows a reflection about the advantages and limitations of *T. cruzi* cultivation techniques in studies of parasitic biology.

Acknowledgements

CJGM, JWFO, and AMQ are grateful for the financial support provided by Capes/Brazil through postgraduate scholarships. MSS thanks Global Health and Tropical Medicine, Lisbon-Portugal (GHTM-UID/multi/04413/2013), and to Berenice Project supported by FP7 European Union (grant number 305937). We are also grateful to Paulo Fanado for editing this manuscript.

Conflict of interest

The authors declare no conflict of interest.

Author details

Cláudia Jassica Gonçalves Moreno[1,2], Johny Wysllas de Freitas Oliveira[1,2], Joice Castelo Branco[1,3], Laura Araújo[1,3], Aline Maria Queiroz[1,3], Sílvia Tavares Donato[1,4], Nilton José da Silva Júnior[1], Emilly Thays da Silva Rodrigues[1] and Marcelo Sousa Silva[1,2,3,4*]

1 Immunoparasitology Laboratory, Health Sciences Centre, Federal University of the Rio Grande do Norte, Natal, Brazil

2 Biosciences Centre, Federal University of the Rio Grande do Norte, Natal, Brazil

3 Health Sciences Centre, Federal University of the Rio Grande do Norte, Natal, Brazil

4 Global Health and Tropical Medicine, Institute of Hygiene and Tropical Medicine, Universidade Nova de Lisboa, Lisbon, Portugal

*Address all correspondence to: mssilva@ihmt.unl.pt

References

[1] Chagas C. Nova tripanomiaze humana. Estudos sobre a morfolojía e o ciclo evolutivo de *Schizotrypanum cruzi* n. gen., n. sp., ajente etiolójico de nova entidade morbida do homen. Memórias do Instituto Oswaldo Cruz. 1909. DOI: 10.1590/S0074-02761909000200008

[2] De Souza W. Cell biology of *Trypanosoma cruzi*. International Review of Cytology. 1984. DOI: 10.1016/S0074-7696(08)60180-1

[3] Zingales B et al. The revised *Trypanosoma cruzi* subspecific nomenclature: Rationale, epidemiological relevance and research applications. Infection, Genetics and Evolution. 2012. DOI: 10.1016/j.meegid.2011.12.009

[4] Zingales B et al. A new consensus for *Trypanosoma cruzi* intraspecific nomenclature: Second revision meeting recommends TcI to TcVI. Memórias do Instituto Oswaldo Cruz. 2009. DOI: 10.1590/S0074-02762009000700021

[5] Zingales B et al. Drug discovery for chagas disease should consider *Trypanosoma cruzi* strain diversity. Memórias do Instituto Oswaldo Cruz. 2014. DOI: 10.1590/0074-0276140156

[6] Romana C et al. Study of the developmental cycle of *Trypanosoma cruzi* in fowl embryo tissue culture. Memórias do Instituto Oswaldo Cruz. 1942;**37**(1):19-27

[7] Meyer H, de Oliveira MX. Cultivation of *Trypanosoma cruzi* in tissue cultures: A four-year study. Parasitology. 1948;**39**(1-2):91-94

[8] Camargo E. Growth and differentiation in *Trypanosoma cruzi*. I. Origin of metacyclic trypanosomes in liquid media. Revista do Instituto de Medicina Tropical de São Paulo. 1964;**6**:93-100

[9] Contreras VT, Salles JM, Thomas N, Morel CM, Goldenberg S. *In vitro* differentiation of *Trypanosoma cruzi* under chemically defined conditions. Molecular and Biochemical Parasitology. 1985. DOI: 10.1016/0166-6851(85)90073-8

[10] Jorge TCA, Barbosa HS, Moreira AL, De Souza W, Meirelles MNL. The interaction of myotropic and macrophagotropic strains of *Trypanosoma cruzi* with myoblasts and fibers of skeletal muscle. Parasitology Research. 1986. DOI: 10.1007/BF00925477

[11] Velasco JR, Adroher FJ, Osuna A. A survey of culture media of *Trypanosoma cruzi* amastigote forms from infected Vero cells. Chemosphere. 1989. DOI: 10.1016/0045-6535(89)90518-3

[12] Tyler KM, Engman DM. Flagellar elongation induced by glucose limitation is preadaptive for *Trypanosoma cruzi* differentiation. Cell Motility and the Cytoskeleton. 2000. DOI: 10.1002/1097-0169(200008)46:4<269::AID-CM4>3.0.CO;2-V

[13] Tyler KM, Engman DM. Flagellar elongation induced by glucose limitation is preadaptive for *Trypanosoma cruzi* differentiation. Cell Motility and the Cytoskeleton. 2000;**46**(4):269-278

[14] Contreras VT et al. Biological aspects of the Dm 28c clone of *Trypanosoma cruzi* after metacyclogenesis in chemically defined media. Memórias do Instituto Oswaldo Cruz. 1988. DOI: 10.1590/S0074-02761988000100016

[15] Nogueira NP et al. Proliferation and differentiation of *Trypanosoma cruzi* inside its vector have a new trigger:

Redox status. PLoS One. 2015. DOI: 10.1371/journal.pone.0116712

[16] Isola ELD, Lammel EM, Cappa SMG. *Trypanosoma cruzi*: Differentiation after interaction of epimastigotes and *Triatoma infestans* intestinal homogenate. Experimental Parasitology. 1986. DOI: 10.1016/0014-4894(86)90039-1

[17] Nogueira N, Bianco C, Cohn Z. Studies on the selective lysis and purification of *Trypanosoma cruzi*. The Journal of Experimental Medicine. 1975;**142**(1):224-229

[18] Castanys S, Osuna A, Gamarro F, Ruiz-Pérez LM. Purification of metacyclic forms of *Trypanosoma cruzi* by Percoll discontinuous gradient centrifugation. Parasite. 1984;**70**(4):443-449

[19] Rimoldi MT, Katzin VJ, Gonzalez-Cappa SM, de Bracco MME. Isolation by Percoll gradient centrifugation and radiolabeling of bloodstream forms of *Trypanosoma cruzi*. Revista Argentina de Microbiología. 1986;**18**(1):41-44

[20] Cruz-Saavedra L, Muñoz M, León C, Patarroyo MA, Arevalo G, Pavia P, et al. Purification of *Trypanosoma cruzi* metacyclic trypomastigotes by ion exchange chromatography in sepharose-DEAE, a novel methodology for host-pathogen interaction studies. Journal of Microbiological Methods. 2017;**142**:27-32

[21] de Carvalho TU, de Souza W. Separation of amastigotes and trypomastigotes of *Trypanosoma cruzi* from cultured cells. Zeitschrift für Parasitenkunde. 1983;**69**(5):571-575

[22] Villalta FV, Leon W. Effect of purification by DEAE-cellulose column on infectivity of *Trypanosoma cruzi* blood forms. The Journal of Parasitology. 1979;**65**(1):188-189

[23] Abegg CP, de Abreu AP, da Silva JL, de Araujo SM, Gomes ML, Ferreira EC, et al. Polymorphisms of blood forms and in vitro metacyclogenesis of *Trypanosoma cruzi* I, II, and IV. Experimental Parasitology. 2017;**176**:8-15

[24] Romanha AJ et al. *In vitro* and *in vivo* experimental models for drug screening and development for Chagas disease. Memórias do Instituto Oswaldo Cruz. 2010. DOI: 10.1590/S0074-02762010000200022

[25] Parente AMS et al. Analogs of the scorpion venom peptide Stigmurin: Structural assessment, toxicity, and increased antimicrobial activity. Toxins (Basel). 2018. DOI: 10.3390/toxins10040161

[26] Ribeiro AR et al. Biological and molecular characterization of *Trypanosoma cruzi* strains from four states of Brazil. The American Journal of Tropical Medicine and Hygiene. 2018. DOI: 10.4269/ajtmh.16-0200

[27] Avila CC et al. Proteome-wide analysis of *Trypanosoma cruzi* exponential and stationary growth phases reveals a subcellular compartment-specific regulation. Genes (Basel). 2018. DOI: 10.3390/genes9080413

[28] Romagnoli BAA et al. Improvements in the CRISPR/Cas9 system for high efficiency gene disruption in *Trypanosoma cruzi*. Acta Tropica. 2018. DOI: 10.1016/j.actatropica.2017.11.013

[29] Brito Querido J, Mancera-Martinez E, Vicens Q, Bochler A, Chicher J, Simonetti A, et al. The cryo-EM structure of a novel 40S kinetoplastid-specific ribosomal protein. Structure. 2017;**25**(12):1785-1794. DOI: 10.1016/j.str.2017.09.014

Chapter 4

Amazonian Reservoir Hosts of *Trypanosoma cruzi*

Jocelyn Ginette Pérez Lazo, Pedro Mayor and Andrés G. Lescano

Abstract

The epidemiology of *Trypanosoma cruzi* in the wild is a particular and highly dynamic scenario that needs attention due to the increased alteration of the environment caused by different factors including anthropogenic change. This chapter is an updated summary about the known reservoir hosts of *T. cruzi* identified in the countries that share the Amazon rainforest. This information will provide a better understanding of the ecology of *T. cruzi* in sylvatic environments. This chapter will also contribute to address the potential risks of *T. cruzi* infection in Amazonian communities who are in contact with wild animals through hunting and wild meat consumption.

Keywords: *Trypanosoma cruzi*, Chagas disease, Amazon, parasites, wild animals, zoonosis, host

1. Introduction

The Amazon basin comprises multiple South American countries: Brazil (63.9%), Peru (15.6%), Bolivia (11.7%), Colombia (5.6%), Ecuador (2.1%), Venezuela (0.9%) and Guyana (0.2%). It covers over 1.3 billion hectares with 60% of total forest area (**Figure 1**) [1]. Population density in the Amazon basin is low and more than 70% live in urban areas. However, this region is in transition due to both climate change as well as anthropogenic activities such as the expansion of agriculture, road paving and logging that lead to accelerated population growth [1, 2].

The environmental conditions of the Amazon basin are favorable for the transmission of multiple vector-borne diseases. Well-known endemic diseases such as malaria and Leishmaniasis show the highest incidence in the Americas, and recent data shows the circulation of multiple arboviruses [3], with an increasing incidence of Dengue fever over time. Trypanosomatids exist in nature since millions of years ago, and Chagas disease has been identified in 4000–9000 year-old human mummies from Chile and Peru [4, 5]. Over time, the spread of Chagas disease in the Americas expanded from a wild and peridomiciliary cycle to a domestic cycle. This occurred through the domiciliation and domestication of triatomines, the primary vectors of *T. cruzi*. Thus, it was suggested that triatomines evolved from non-blood sucking insects and became mandatory hematophagous insects after undergoing morphological modifications [6, 7]. Likewise, trypanosomes also evolved and adapted to this new opportunity [6]. The evolution of *T. cruzi* was influenced by factors related to its dependence on the host and its environment and the reliance

Figure 1.
The Amazon Basin location in South America. Source: FAO, 2015.

on vertebrates as a blood source. *Trypanosoma cruzi* can spread geographically by both triatomines flying into different areas and passive carriage in vertebrate hosts moving across broad areas [8].

The deforestation of the Amazon basin, 3.6 million hectares of forest lost per year between 2000 and 2010, has become an important factor in the domestication of triatomines due to the scarcity of blood sources among wildlife. Triatomines and other disease vectors have looked for new blood sources and reached areas closer to human dwellings. Social inequality and poor public health systems exacerbate the impact of these factors, especially in rural areas, and further contribute to the transmission of Chagas disease and the emergence of new pathogens.

2. Chagas disease transmission in the Amazon basin

Trypanosomiasis is an ancient enzootic parasitosis in nature, maintained by wild animals and infected vectors. In the wild, it is presumably transmitted primarily by the oral route through predation of infected vectors or mammals, or by contamination of animal nests or shelters with metacyclic forms of the parasite released in the feces of infected triatomines or from scent glands of marsupials i.e. *Didelphis marsupialis* (**Figure 2**) [9, 10]

Besides the wild enzooty, other scenarios for the transmission of *T. cruzi* in the Amazon have been proposed (**Figure 2**). For instance, antropozoonosis, through the accidental transmission of *T. cruzi* from infected vectors or wild animals to humans. This occurs commonly with the invasion of infected vectors and marsupials in human dwellings or when humans invade the forest for different activities such as hunting, fishing, logging, tourism, etc. [11]. The invasion of vectors in human dwellings may occur during deforestation or when their blood sources such as wild mammals are scarce due to deforestation itself, over hunting or during

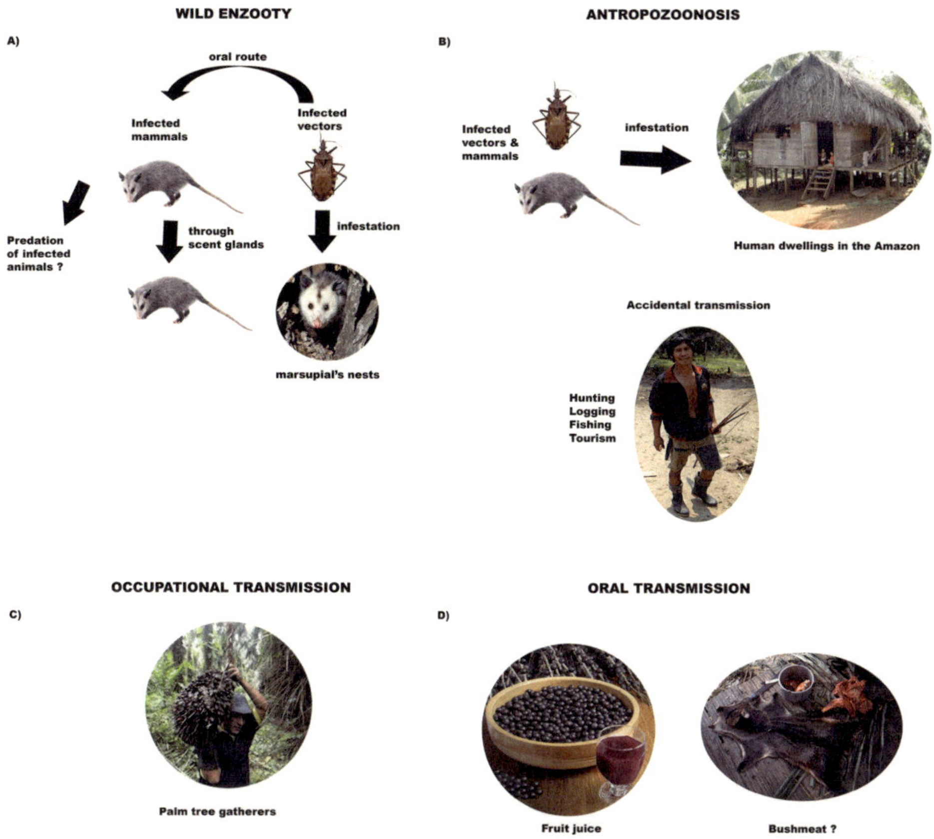

Figure 2.
T. cruzi transmission scenarios in the Amazon Basin. Photo credits: Dr. Pedro Mayor.

the wet season. Thus, vectors fly from the trees at night attracted by the lights and invade the houses in search of blood meals [11, 12]. Nymphs and adults of triatomines infected with *T. cruzi* have been found on palm trees in backyards and inside the houses in Amazonian villages. Blood meals from birds, opossum, rodents, humans, dogs and horses have been found in the gut of triatomines captured in such settings [12].

Trypanosoma cruzi may have different transmission features between areas of different degrees of disturbance. One study carried out in the Brazilian Amazon found that areas with low and intermediate degree of disturbance had higher prevalence of wild mammals with parasitemia detected by hemoculture than in areas with greater environmental disturbance caused by human occupation and agriculture. Areas with intermediate degrees of disturbance also had the highest prevalence of positive domestic mammals detected by IFAT [13]. In contrast, seroprevalence did not differ between areas with difference disturbance levels; one explanation for this finding may be the difference on the ability of reservoirs to infect vectors between geographic areas despite the comparable exposure to the parasite.

Occupational transmission scenarios are well documented among palm tree gatherers, an economically important activity. Palm trees provide shelter for different wild mammals, amphibians and insects. Gatherers of the piassava palm (*Leopoldinia piassaba*) are reportedly bitten by wild triatomines during their camping months in the forest, possibly due to the lack of other blood sources such as wild mammals [14]. Palm gatherers are therefore at higher risk to become infected compared to individuals not involved in this activity [15].

The last scenario is the oral transmission of Chagas disease, often leading to outbreaks of acute Chagas disease (ACD). The Amazon basin is the region where most of such outbreaks have been reported. Actually, Carlos Chagas himself first described ACD in 1909 when a 2-year-old girl from Lassance, Minas Gerais presented with fever, hepatosplenomegaly and Romaña's sign [8]. ACD outbreaks have occurred in both preserved and human disturbed areas [13], but with exposure of humans to the sylvatic cycle as a common factor. Most ACD outbreaks and cases are related to the consumption of food contaminated with trypomastigote forms as reported with the ingestion of fruits from palm trees such as acaí, bacaba, or the ingestion of wild meat [16]. In fact, ACD cases may have increased in the last decade [16]. It was reported an increase of notified suspected cases in Pará state in Brazil from 2010 to 2016, that may be related to the increase of production to acaí juice during those years, but also related to the increase in its consumption since the majority of patients indicated a daily consumption of this juice [16]. Consistently with these findings, juice contaminated with *T. cruzi* was able to experimentally infect mice [17, 18], despite the juice being previously frozen or refrigerated.

In Brazil, *T. cruzi* infection got more attention only after 1969 when ACD cases were reported in Belém, State of Pará [12]. However, sporadic human infections have been reported before this date in the Amazon basin. In a serological study by indirect inmunoflourescence test in a county of 15 villages in the Brazilian Amazon found 0.83% (212/25, 451) of prevalence for *T. cruzi* infection. Children ≤ 10 years had positive serological results indicating recent transmission and acute infection [12]. In Peru, the first ACD case was reported in 1919 in the southeastern Amazonian department of Madre de Dios. Only a few ACD cases have been reported since, although multiple cases have been reported in recent years in Amazonian regions such as Pasco and Loreto [19]. In Ecuador, Chagas disease was first diagnosed in 1929 in Guayas, a coastal province. The first human cases in the Amazon region of Ecuador were yet reported in 1991 in Napo and Sucumbíos provinces [20]. Some studies later demonstrated the exposure and active transmission of *T. cruzi* in more Amazonian provinces such as Orellana, Pastaza, Morona Santiago [20]. Similar to Brazil, ACD cases were also reported in children ranging from 1 to 5 years old in the Ecuadorian Amazon region [21]. In Colombia, the *T. cruzi* haplotype Ia has been isolated among ACD cases from the Amazonian departments of Caquetá and Putumayo, including children and servicemen who were in contact with the forest [22].

Indigenous communities in the Amazon basin are also affected by *T. cruzi* infection. During 2006 to 2010, six ACD cases were reported among children from the Aguaruna, Huambisa and Kandoshi communities in the Peruvian Amazon basin. ACD was attributed to either the biting of infected triatomines found inside the dwellings or oral transmission by drinking "masato" (a fermented beverage made with yuca) or "chapo" (a ripe banana beverage) since no traveling outside the community or blood transfusion was recorded [19]. ACD cases in indigenous communities in Ecuador Amazon region have been reported since 1987; a seroepidemiological study found a 6.0% (61/1011) of positive individuals from 15 of the 18 studied communities. Likewise, the prevalence increased with age, individuals older than 50 showed 18.8% of seroprevalence [21].

3. *Trypanosoma cruzi* reservoir hosts in the wild

The first *T. cruzi* infection in Amazonian wildlife was reported in 1924 among squirrel monkeys (*Saimiri sciureus*) [23]. In mammalian hosts, *T. cruzi* infection could develop depending on factors such as the host health status, the transmission route, the parasite population and co-infection with other parasites. However, reservoir hosts should support the maintenance and dispersion of the parasite in nature [9], and therefore be capable of infecting vectors. Vectors can become infected depending on the parasitemia in the host's blood, which varies among host species and individuals [9]. This is corroborated by previous studies that showed differences in *T. cruzi* infection rates between species of the same genus. For instance, experimental *T. cruzi* infection in the rodent *Thrychomys fosteri* was more severe with high and long-lasting parasitemia compared to *Thrychomys laurentius* [24].

Trypanosoma cruzi can infect different mammal species from all forest strata and canopy levels. **Table 1** lists wild animal species reported positive to *T. cruzi* infection in the Amazon basin. A higher proportion of positive blood culture and serology by immunofluorescence antibody test (IFAT) was observed in mammals from the Amazon compared to specimens from other biomes (Atlantic forest, Caatinga, Cerrado, Pampa and Pantanal) [9]. This study also suggested that 8% of positive animals would be sufficient to maintain the *T. cruzi* cycle in any biome. Different mammalian orders seem to participate in the transmission cycle of *T. cruzi*, but Primates, Didelphimorphia, Chiroptera and Carnivora apparently are the primary taxa due to their high parasitemias and Discrete Typing Units (DTUs) diversity [9].

TcI, TcIII and TcIV are the reported DTUs circulating in the Amazon basin [25]. A very low prevalence of TcIII and TcIV (0.8%) infection was found in 714 *T. cruzi* isolates from five different biomes in Brazil. TcIV was more prevalent than TcIII, and no specific association between genotypes and animal species or geographical distribution was suggested [26]. However, these findings may be biased by the methodology used in this study or the temporal distribution of these DTUs in nature.

The main taxa for the transmission of *T. cruzi* in the Amazon basin are reported below:

3.1 Didelphimorphia

This order includes the hosts most frequently infected by *T. cruzi* in all biomes of Brazil, presenting high rates of positive hemoculture [9, 26]. Didelphimorphia species in the Amazon basin seems to be predominantly infected by TcIII and/or TcIV across different biomes. This order includes the most frequent hosts infected by TcIII, TcIV or mixed trypanosome/genotypes compared to other mammalian taxa [26].

It is suggested that marsupials could act as generalist species due to the diversity of Trypanosoma spp. found simultaneously infecting a single host [9]. For instance, a mixed, triple infection with *Trypanosoma cascavelli* (reported in reptiles), *Trypanosoma dionisii* (reported in bats) and Trypanosoma sp. was reported in *Monodelphis americana* [27]. Other species such as *Philander opossum* show a high prevalence of *T. cruzi* positive hemoculture implying its ability to infect vectors. *Philander opossum* was also the most abundant species present across different areas of the Amazon basin regardless of the season and the degree of land disturbance [13].

Didelphis marsupialis is the specie from this order most commonly reported with *T. cruzi* infection. However, it has been suggested that its presence in a specific area may not be directly related to the endemicity of *T. cruzi*. Low prevalence of the parasite was found in an area despite their great abundance of this mammal, and it may not be a critical specie for the maintenance of the *T. cruzi* sylvatic cycle in the Amazon basin [13]. *Trypanosoma cruzi* infection of *D. marsupialis* could take place orally by ingestion of small mammals or triatomines. However, host-to-host infection could also occur through direct contact with infective metacyclic trypomastigotes released by the scent glands of infected marsupials. Scent glands have demonstrated being suitable for epimastigotes and their differentiation into infective metacyclic forms [10]. Didelphis can also adapt to peridomestic areas, and

Order/Genus	Specie	Methods				References
		H	I	P	DTU	
Artiodactyla						
Sus	*Sus scrofa*		x			[9]
Cingulata						
Cyclopes	*Cyclopes didactylus*	x			Zymodeme 1	[46]
Dasypus	*Dasypus novemcinctus*	x		x	TcIV	[9, 28, 46, 47]
Tamandua	*Tamandua tetradactyla*	x		x	TcI	[9, 48]
Didelphimorphia						
Caluromys	*Caluromys philander*		x			[9]
	Caluromys sp.	x	x	x	TcI	[9, 46, 47]
Didelphis	*Didelphis marsupialis*	x	x	x	TcI, TcII, TcI+ T. rangeli, TcI+ TcII, TcI+ TcIII	[9, 13, 46, 47, 49]
	Didelphis albiventris	x		x		[49]
	Didelphis sp.		x			[9]
Gracilinanus	*Gracilinanus sp.*	x	x	x	TcI	[9, 49]
Marmosa	*Marmosa murina*	x	x	x	TcI	[9, 13]
	Marmosa cinerea	x				[47]
	Marmosa sp.		x			[9]
Marmosops	*Marmosops parvidens*		x			[9]
	Marmosops sp.	x	x	x	TcI	[9]
Metachirus	*Metachirus nudicaudatus*	x			Zymodeme 1	[46]
Micoureus	*Micoureus demerarae*	x	x	x	TcI	[9, 13]
Monodelphis	*Monodelphis domestica*		x			[9]
	Monodelphis brevicaudata	x			Zymodeme 3	[46]
	Monodelphis sp.	x	x	x	TcI, TcI + TcIV	[9]
Philander	*Philander opossum*	x	x	x	TcI, TcI+ TcII	[9, 13, 46, 47, 49]
	Philander sp.	x	x	x	TcI, TcI+ TcIII/TcIV, TcI + T. Rangeli	[9]
Primates						
Alouatta	*Alouatta belzubul*	x	x	x	TcI+ TcIV	[9]
	Alouatta caraya	x	x	x	TcI+ TcIV	[9]

Order/Genus	Specie	Methods				References
		H	I	P	DTU	
Cebuella	*Cebuella pygmaea*	x				[50]
Callicebus	*Callicebus cupreus*			x		[33]
Cebus	*Cebus albifrons*			x		[33]
	Cebus apella			x		[33]
Lagothrix	*Lagothrix poeppigii*			x		[33]
Saguinus	*Saguinus midas niger*	x			Zymodeme 1	[46]
	Saguinus bicolor	x		x	TcI, TcI + T.rangeli	[32]
Saimiri	*Saimiri boliviensis*			x		[33]
	Saimiri sciureus	x		x		[30, 33]
	Saimiri ustus	x				[30]
Sanguinus	*Sanguinus imperator imperator*	x				[50]
	Sanguinus fuscicollis weddelli	x				[50]
Sapajus	*Sapajus libidinosus*	x	x	x	TcI, TcI + T.rangeli	[9]
	Sapajus macrocephalus			x		[33]
Rodentia						
Cuniculus	*Cuniculus paca*	x		x		[28, 47]
Akodon	*Akodon lindberghi*	x		x	TcI	[9]
Cerradomys	*Cerradomys sp.*		x			[9]
Coendou	*Coendou prehensilis*	x	x	x	TcI + T.rangeli	[9]
	Coendou sp.	x			Zymodeme 1	[46, 47]
Dasyprocta	*Dasyprocta prymnolopha*		x			[9]
	Dasyprocta sp.	x		x		[28, 47]
Echymys	*Echymys chrysurus*	x			Zymodeme 1	[46]
Rodentia						
Holochilus	*Holochilus sp.*		x			[9]
Hylaeamys	*Hylaeamys megacephalus*	x	x	x	TcI	[9]
Nectomys	*Nectomys squamipes*					[51]
Oryzomys	*Oryzomys capito*	x				[47]
	Oryzomys sp.		x			[9]

Order/Genus	Specie	Methods				References
		H	I	P	DTU	
Oxymycterus	*Oxymycterus sp.*		x			[9]
	Proechimys roberti		x			[9]
Proechimys	*Proechimys gr. cuvieri*	x	x	x	TcI	[9]
	Proechimys gr. guianensis	x	x			[9, 47]
	Proechimys sp.	x	x	x	TcI+ TcIII/TcIV	[9]
Rattus	*Rattus rattus*	x			Zymodeme 1	[46]
Sciurus	*Sciurus sp.*	x			Zymodeme 1	[46]
Carnivora						
Nasua	*Nasua nasua*	x		x		[28, 47]
Tayra	*Tayra barbara*					[48]
Chiroptera						
Anoura	*Anoura caudifer*	x		x	TcI	[9]
Artibeus	*Artibeus lituratus*	x		x	TcI, TcIV	[9]
	Artibeus cf. fimbriatus	x		x	TcI	[9]
	Artibeus planirostris	x		x	TcI	[9]
Carollia	*Carollia perspicillata*	x		x	TcI	[9]
	Carollia cf. beikeith	x		x	TcI	[9]
	Carollia brevicauda	x		x	TcI	[9]
Choeroniscus	*Choeroniscus minor*					[52]
Chiroptera						
Dermanura	*Dermanura cinereus*	x		x	TcI	[9]
Desmodus	*Desmodus rotundus*			x		[35]
Diaemus	*Diaemus youngi*			x		[35]
Glossophaga	*Glossophaga soricina*	x		x	TcI, TcIV	[9, 52]
Lasiurus	*Lasiurus blossevillii*	x		x	TcI	[9]
Lonchophylla	*Lonchophylla thomasi*	x		x	TcI	[9]
	Lonchophylla mordax					[52]
Molossus	*Molossus major*					[52]
	Molossus ater					[52]

Order/Genus	Specie	Methods				References
		H	I	P	DTU	
Mycronycteris	*Mycronycteris megalotis*					[52]
Noctilio	*Noctilio labialis*					[52]
Phyllostomus	*Phyllostomus discolor*	x		x	TcI	[9]
	Phyllostomus hastatus	x		x	TcIV	[9, 35, 52]
	Phylostomus alongatus					[52]
Plathyrhinus	*Plathyrhinus infuscus*	x		x	TcI	[9]
Saccopterix	*Saccopterix bilineata*					[52]
Tonatia	*Tonatia saurophila*	x		x	TcI	[9]
Trachops	*Trachops cirrhosus*	x		x	TcI	[9, 35]
Uroderma	*Uroderma bilobatum*	x		x	TcI	[9]
Vampyressa	*Vampyressa sp.*	x		x	TcI	[9]

H = hemoculture, I = IFAT, P=PCR.

Table 1.
Free-ranging wild animal's species reported positive to Trypanosoma cruzi infection in the Amazon Basin.

T. cruzi-infected specimens have been found in backyards of houses from different villages in the Brazilian Amazon [12].

Histopathological lesions observed in infected marsupials resemble those presented by Chagas disease patients. Myocarditis with mononuclear infiltrates, cell lysis and inflammatory infiltrates in skeletal muscles, esophagus and small and large intestines are other lesions found in natural infection [12].

3.2 Carnivores

Carnivores are likely to be infected by the oral route, and those whose diet includes insects or flesh present the highest infection rates [9]. The most reported carnivore infected with *T. cruzi* is the coati (*Nasua nasua*) [28]. Coatis are terrestrial but construct their nests in the high canopy level of large trees, which then serve as habitat for triatomines and other insects. Two-thirds of 34 Triatoma and Rhodnius triatomines found in 24 coati nests in the Pantanal state in Brazil were positive to *T. cruzi*. The 62.5% (8/8) of 8 nests infested with triatomines were *T. cruzi*-positive. The triatomines blood meal sources identified by precipitation of the stomach contents included coati and also rodents, birds and marsupials. Moreover, rodents and birds were observed visiting these nests [29].

3.3 Non-human primates

Different species of trypanosomatids have been reported in wild non-human primates from the Amazon. The *T. cruzi* prevalence rates reported in the Brazilian

Amazon range from 10.3% (17/165) using parasitological methods in captured free-ranging non-human primates to 45.5% (45/99) by IFAT in captive and semi-captive primates [30, 31]. However, the *T. cruzi* prevalence in wild non-human primates of the Amazon basin seems to be lower than the prevalence of *T. rangeli*, with the latter showing lasting parasitemias [32]. Several studies corroborate this observation, such as 35.2% versus 10.3% in 165 wild squirrel monkeys captured in the Brazilian Amazon [30], and 75.0% versus 10.4% cultures among 96 tamarins captured in the Amazonas state in Brazil [32]. In a study conducted in the Peruvian Amazon, free-ranging non-human primates had a high prevalence of trypanosomatids (64.3% vs. 27.9%) and *T. cruzi* (8.7% vs. 3.3%) by PCR, compared to captive primates, suggesting that parasite transmission occurs more actively in the sylvatic cycle. Pitheciidae had the highest trypanosomatid prevalence (20/22, 90.9%) and Cebidae had the highest *T. cruzi* prevalence (15/117, 12.8%) [33]. The difference in the prevalence is not well understood but may be related to the route of infection. While *T. cruzi* might be transmitted through either contact with infective forms in the triatomine feces or orally, *T. rangeli* may be primarily transmitted during biting by infected triatomines [32].

There is scarce evidence regarding the physiopathology of *T. cruzi* natural infection in free-ranging non-human primates. Electrocardiography abnormalities were found in a few *T. cruzi*-infected tamarins aged 7 months to 4 years old from an Atlantic Forest reserve in Brazil with infection time spanning 6 months to almost 5 years. It was estimated that infected individuals were 18 times more likely to show detectable electrocardiogram (ECG) abnormalities than those uninfected. Likewise, infected non-human primates were prone to have higher levels of cardiac injury markers such as MBi, a cardiac lesion marker and total protein in serum. Although these findings did not suggest a general health problem, left ventricular hypertrophy was present in some of the infected tamarins similar to the chronic form of Chagas disease in humans [34]. These studies probably underestimate the clinical burden since non-human primates with severe cardiac lesions probably do not survive in the wild and are therefore not included. Despite this limitation, it is important to consider the impact of the physiopathology of natural infection, particularly for at risk species. For instance, *T. cruzi* infection has been reported in *Saguinus bicolor*, a critical endangered monkey in the Amazon basin [32].

3.4 Quiroptera

It was suggested that bats could be bio-accumulator hosts and dispersers of trypanosomatids because of their ability to fly and the great diversity of Trypanosoma species found in bats specially compared to other animal taxon [9]. Out of 1219 Brazilian bats from 76 genera and 94 species, 14% were positive to Trypanosoma sp. by hemoculture and 5% of them were *T. cruzi* positive in single or mixed infections. Although bat is able to be infected with other DTUs most bats were infected with TcI, and TcII was not detected. The highest infection rate was found in the Amazon basin compared to other biomes, suggesting that unaltered areas nurture a high parasite diversity [9]. In the Peruvian Amazon, a 4.1% *T. cruzi* infection prevalence in bats was reported, in both hematophagous (2.7%; 2/73) and non-hematophagous species (6.2%; 3/48) [35]. Interestingly, *T. cruzi* DNA was detected in the salivary glands of *Diaemus youngi*, an hematophagous bat [35]. This highlights the importance of studying the transmission mechanisms of *T. cruzi* in bats and their public health implications for the Amazon basin.

4. Wild meat consumption as a potential risk of oral transmission

Trypanosoma cruzi hosts encompass more than 100 wild and domestic mammalian species, which in turn belong to eight different taxonomic orders distributed in all phytogeography regions of the Neotropics [36]. Since subsistence hunting of wild mammals for consumption is one of the main sources of animal protein in the Amazon Basin, wild meat constitute a potential source for human infection.

In the Amazon region, wild meat represents an important component of household food security, income and a key social and cultural driver. Wild meat is still a key element in Amazon peoples' diet and accounts for a high percentage of daily protein intake. It is estimated that the wild meat consumption rate in rural settlements of the Amazon Basin is 172 g per person per day [37]. A study found that 39% of households in Latin America harvested and consumed wild meat, and dependence was highest among the poorest households [38]. Estimates of the annual wild meat harvest in the Northern Peruvian Amazon are 113,000 animals (1680 tons), and 89,224 tons of meat per year in the Brazilian Amazon [39, 40]. Furthermore, hunting pressure has increased in recent years due to various causes, such as the growth of human populations, access to remaining forests, commercialization of wild meat, increasing use of efficient modern hunting techniques and erosion of traditional hunting institutions due to rapid cultural changes [41].

Notwithstanding its positive nutritional contributions, some serious health concerns may be associated with wild meat consumption in the Amazon basin. Emerging infectious diseases worldwide are increasing over time and are dominated by zoonoses (60%), of which the majority (72%) originates in wildlife [42]. A study conducted in the Peruvian Amazon estimated an annually consumption of 45 animals infected with *T. cruzi*, translating to 0.75 infected animals typically consumed by a large extended family, suggesting recurrent infection opportunities [28].

A few studies from Argentina and Brazil report Chagas disease transmission through consumption of raw, poorly cooked meat, blood or contact with carcasses of wildlife among children [43]. However, some of these studies did not rule out the possibility of vectorial transmission and others have not found proof of exposure to infected triatomines. What it is certainly known is that frequently hunted animals in the Amazon have been reported as *T. cruzi* reservoirs [28, 33], and carnivores might tend to have higher infection rates than non-carnivores suggesting accumulation through wild meat intake. Thus, the consumption of wild meat may be an important risk for human *T. cruzi* infection in the Amazon Basin, which may happen particularly during cleaning of wild meat or contamination of cooking utensils [13]. Other associated risk factors in rural Amazon societies are poor hygienic conditions, unavailability of clean water, inadequate medical care and insufficient knowledge about local diseases [44]. Nevertheless, further studies addressing the relationship between wild meat consumption and Chagas disease are required to better understand the risk of infection in Amazonian communities. Prompt diagnosis, notification of cases and epidemiological studies to assess the risk factors that trigger ACD outbreaks are greatly needed.

5. Domestic animals as reservoir hosts of *Trypanosoma cruzi* in the Amazon basin

Domestic animals should be included in epidemiological studies since they are good sentinels of disease transmission in a geographic area. In the Amazon, several studies report *T. cruzi*-positive domestic animals such as pigs and dogs

[45]. TcI is the primary DTU reported infecting dogs in the Brazilian Amazon [13]. Although a higher prevalence of *T. cruzi* was found in pigs compared to dogs in some anthropogenic disturbed areas, dogs present longer parasitemias. Extended parasitemia suggests that dogs are better able to infect vectors, thereby acting as *T. cruzi* reservoirs. These domestic animals may become infected by different routes, mainly the biting of the infected triatomines since bugs and dogs converge near human dwellings [13]. Other routes of transmission would be the ingestion of the infected bugs or food contaminated with triatomine feces and predation of infected small mammals.

Acknowledgements

This work was supported by the training grant 2D43 TW007393 awarded to Andrés G. Lescano by the Fogarty International Center of the U.S. National Institute of Health, and by ERANet17/HLH-0271.

Conflicts of interest

The authors declare that they have no conflicts of interest regarding the publication of this chapter.

Nomenclature

ACD = acute Chagas disease
DTUs = discrete typing units
ECG = electrocardiogram
IFAT = immunofluorescence antibody test
MBi = cardiac lesion marker
PCR = polymerase chain reaction
TP = total protein

Author details

Jocelyn Ginette Pérez Lazo[1*], Pedro Mayor[2] and Andrés G. Lescano[3]

1 University of Liverpool, Liverpool, United Kingdom

2 Autonomous University of Barcelona, Barcelona, Spain

3 Cayetano Heredia University, Lima, Peru

*Address all correspondence to: jgpl16@liverpool.ac.uk

References

[1] The State of Forests in the Amazon Basin, Congo Basin and Southeast Asia; 2011

[2] Davidson EA et al. The Amazon basin in transition. Nature. 2012;**481**(7381):321-328

[3] Penna G et al. High incidence of diseases endemic to the Amazon region of Brazil, 2001-2006. Emerging Infectious Diseases. 2009;**15**(4):626-632

[4] Guhl F et al. Isolation of *Trypanosoma cruzi* DNA in 4000-year-old mummified human tissue from northern Chile. American Journal of Physical Anthropology. 1999;**108**(4):401-407

[5] Aufderheide AC et al. A 9000-year record of Chagas' disease. Proceedings of the National Academy of Sciences of the United States of America. 2004;**101**(7):2034-2039

[6] Hoare CA. The Trypanosomes of Mammals: A Zoological Monograph. Vol. xvii. Oxford, Edinburgh: Blackwell Scientific Publications; 1972. p. 749

[7] Schofield C. *Trypanosoma cruzi*—The vector-parasite paradox. Memórias do Instituto Oswaldo Cruz. 2000;**95**:535-544

[8] Coura JR, Viñas PA, Junqueira AC. Ecoepidemiology, short history and control of Chagas disease in the endemic countries and the new challenge for non-endemic countries. Memórias do Instituto Oswaldo Cruz. 2014;**109**:856-862

[9] Jansen AM, Xavier SCdC, Roque ALR. *Trypanosoma cruzi* transmission in the wild and its most important reservoir hosts in Brazil. Parasites & Vectors. 2018;**11**:502

[10] Carreira JCA et al. *Trypanosoma cruzi* in the scent glands of *Didelphis marsupialis*: The kinetics of colonization. Experimental Parasitology. 2001;**97**:129-140

[11] Coura JR, Junqueira ACV. Ecological diversity of *Trypanosoma cruzi* transmission in the Amazon basin. The main scenaries in the Brazilian Amazon. Acta Tropica. 2015;**151**:51-57

[12] Teixeira AR et al. Emerging Chagas disease: Trophic network and cycle of transmission of *Trypanosoma cruzi* from palm trees in the Amazon. Emerging Infectious Diseases. 2001;**7**:100-112

[13] Roque ALR et al. *Trypanosoma cruzi* among wild and domestic mammals in different areas of the Abaetetuba municipality (Pará state, Brazil), an endemic Chagas disease transmission area. Veterinary Parasitology. 2013;**193**:71-77

[14] Coura JR, Barrett TV, Naranjo MA. Ataque de populações humanas por triatomíneos silvestres no Amazonas: Uma nova forma de transmissão da infecção chagásica? Revista da Sociedade Brasileira de Medicina Tropical. 1994;**27**:251-253

[15] Brum-Soares LM et al. Morbidade da doença de Chagas em pacientes autóctones da microrregião do Rio Negro, Estado do Amazonas. Revista da Sociedade Brasileira de Medicina Tropical. 2010;**43**:170-177

[16] Santos VRCd et al. Acute Chagas disease in the state of Pará, Amazon region: Is it increasing? Memórias do Instituto Oswaldo Cruz. 2018;**113**:e170298

[17] Passos LAC et al. Sobrevivência e infectividade do *Trypanosoma cruzi* na polpa de açaí: Estudo in vitro e in vivo. Epidemiologia e Serviços de Saúde. 2012;**21**:223-232

[18] Pérez-Gutiérrez E, Agrelo RS, Figueroa R. Technical recommendation on Chagas' disease epidemiology and prevention, focussing its transmission as a disease transmitted by food. Revista da Sociedade Brasileira de Medicina Tropical. 2006;**39**:512-514

[19] Cabrera R et al. New focus of active transmission of Chagas disease in indigenous populations in the Peruvian Amazon basin. Revista da Sociedade Brasileira de Medicina Tropical. 2013;**46**:367-372

[20] Guevara AG et al. Seroepidemiological study of chagas disease in the southern Amazon region of Ecuador. Tropical Medicine and Health. 2013;**41**:21-25

[21] Chico M et al. Chagas disease in Ecuador: Evidence for disease transmission in an indigenous population in the Amazon region. Memórias do Instituto Oswaldo Cruz. 1997;**92**:317-320

[22] Falla A et al. Haplotype identification within *Trypanosoma cruzi* I in Colombian isolates from several reservoirs, vectors and humans. Acta Tropica. 2009;**110**:15-21

[23] Chagas CJSM. Infection naturelle des singes du Pará (Crysotrix sciureus) par *Trypanosoma cruzi*. Comptes rendus des séances de la Société de biologie et de ses filiales. 1924;**2**:75-76

[24] Roque ALR et al. *Trypanosoma cruzi*: Distinct patterns of infection in the sibling caviomorph rodent species *Thrichomys apereoides* laurentius and *Thrichomys pachyurus* (Rodentia, Echimyidae). Experimental Parasitology. 2005;**111**:37-46

[25] Monteiro WM et al. *Trypanosoma cruzi* TcIII/Z3 genotype as agent of an outbreak of Chagas disease in the Brazilian Western Amazonia. Tropical Medicine & International Health. 2010;**15**(9):1049-1051

[26] Barros JHS et al. Identification of novel mammalian hosts and Brazilian biome geographic distribution of *Trypanosoma cruzi* TcIII and TcIV. Acta Tropica. 2017;**172**:173-179

[27] Dario MA et al. High Trypanosoma spp. diversity is maintained by bats and triatomines in Espírito Santo state, Brazil. PLoS One. 2017;**12**:e0188412

[28] Morales EA et al. Prevalence of *Trypanosoma cruzi* and other trypanosomatids in frequently-hunted wild mammals from the Peruvian Amazon. The American Journal of Tropical Medicine and Hygiene. 2017;**97**:1482-1485

[29] de Lima JS et al. Infestation of arboreal nests of coatis by triatomine species, vectors of *Trypanosoma cruzi* , in a large Neotropical wetland. Journal of Vector Ecology. 2015;**40**:379-385

[30] Ziccardi M, Lourenço-de-Oliveira R. The infection rates of trypanosomes in squirrel monkeys at two sites in the Brazilian Amazon. Memórias do Instituto Oswaldo Cruz. 1997;**92**:465-470

[31] Lisboa CV et al. Stable infection of primates with *Trypanosoma cruzi* I and II. Parasitology. 2006;**133**:603

[32] Maia da Silva F et al. Infection rates and genotypes of *Trypanosoma rangeli* and *T. cruzi* infecting free-ranging Saguinus bicolor (Callitrichidae), a critically endangered primate of the Amazon rainforest. Acta Tropica. 2008;**107**:168-173

[33] Aysanoa E et al. Molecular epidemiology of trypanosomatids and *Trypanosoma cruzi* in primates from Peru. EcoHealth. 2017;**14**:732-742

[34] Monteiro RV et al. Clinical, biochemical, and electrocardiographic aspects of *Trypanosoma cruzi* infection in free-ranging golden lion tamarins (*Leontopithecus rosalia*). Journal of Medical Primatology. 2006;**35**:48-55

[35] Villena FE et al. First report of *Trypanosoma cruzi* infection in salivary gland of bats from the Peruvian Amazon. The American Journal of Tropical Medicine and Hygiene. 2018;**99**:723-728

[36] Herrera HM et al. The coati (Nasua nasua, Carnivora, Procyonidae) as a reservoir host for the main lineages of *Trypanosoma cruzi* in the Pantanal region, Brazil. Transactions of the Royal Society of Tropical Medicine and Hygiene. 2008;**102**(11):1133-1139

[37] Nasi R, Taber A, Van Vliet N. Empty forests, empty stomachs? Bushmeat and livelihoods in the Congo and Amazon basins. The International Forestry Review. 2011;**13**(3):355-368

[38] Nielsen MR et al. The importance of wild meat in the global south. Ecological Economics. 2018;**146**:696-705

[39] Bodmer RE, Lozano EP. Rural development and sustainable wildlife use in Peru. Conservation Biology. 2001;**15**(4):1163-1170

[40] Peres CA. Effects of subsistence hunting on vertebrate community structure in Amazonian forests. Conservation Biology. 2000;**14**(1):240-253

[41] Coad L, et al. Toward a Sustainable, Participatory and Inclusive Wild Meat Sector; 2019

[42] Jones KE et al. Global trends in emerging infectious diseases. Nature. 2008;**451**(7181):990-993

[43] Sangenis LHC et al. Transmissão da doença de Chagas por consumo de carne de caça: Revisão sistemática. Revista Brasileira de Epidemiologia. 2016;**19**:803-811

[44] Mayor P et al. Polycystic echinococcosis in Pacas, Amazon region, Peru. Emerging Infectious Diseases. 2015;**21**(3):456-459

[45] Xavier SCdC et al. Distantiae transmission of *Trypanosoma cruzi*: A new epidemiological feature of acute Chagas disease in Brazil. PLoS Neglected Tropical Diseases. 2014;**8**:e2878

[46] Miles MA et al. Chagas's disease in the Amazon Basin: Ii. The distribution of *Trypanosoma cruzi* zymodemes 1 and 3 in Pará state, North Brazil. Transactions of the Royal Society of Tropical Medicine and Hygiene. 1981;**75**:667-674

[47] Lainson R et al. Chagas's disease in the Amazon Basin: 1. *Trypanosoma cruzi* infections in silvatic mammals, triatomine bugs and man in the state of Pará, North Brazil. Transactions of the Royal Society of Tropical Medicine and Hygiene. 1979;**73**:193-204

[48] de Almeida Rodrigues B, de Brito Melo G. Contribuição ao estudo da Tripanosomiase Americana. Memórias do Instituto Oswaldo Cruz. 1942;**37**(1):77-93

[49] da Costa AP et al. *Trypanosoma cruzi* and Leishmania infantum chagasi infection in wild mammals from Maranhão state. Vector Borne and Zoonotic Diseases. 2015;**15**:656-666

[50] Ziccardi M et al. Trypanosomes of non-human primates from the National Centre of primates, Ananindeua, state of Pará, Brazil. Memórias do Instituto Oswaldo Cruz. 2000;**95**:157-159

[51] Deane LM. Sôbre um tripanossomo do tipo cruzi encontrado num rato silvestre, no Estado do Pará. Revista Brasileira de Malariologia e Doenças Tropicais. 1960;**12**:87-102

[52] Dias E et al. Investigações sobre esquizotripanose de morcegos no Estado do Pará; encontro do barbeiro Cavernicola pilosa como transmissor. Revista Brasileira de Biologia. 1942;**2**:103-110

Section 2

Biochemistry and Molecular Biology

Chapter 5

Role of Proteomics in the Study of *Trypanosoma cruzi* Biology

Juan San Francisco, Bessy Gutiérrez and Jorge González

Abstract

Proteomics is the science that studies the proteome, which corresponds to the global expression of proteins at a given time under determined conditions. In the last 20 years, proteomics has emerged as a powerful tool that has allowed the study of proteins that are expressed in the cell under normal or altered conditions as well as post-translational modifications, such as phosphorylation, glycosidation, acetylation, and methylation, among others. In this chapter, we present the main contributions of proteomics to the knowledge of *Trypanosoma cruzi* biology. Proteomes of all *T. cruzi* life cycle stages, secretomes/exoproteomes, post-translational modifications such as phosphorylation or acetylation and immunomes, interactomes, and glycomes are described. The role of proteomics in the identification of new chemotherapeutic targets and potential vaccine candidates will also be discussed.

Keywords: *Trypanosoma cruzi*, proteomics, subcellular proteomics, post-translational modifications, drug resistance, virulence, secretome/exoproteome

1. Introduction

The flagellate protozoan *Trypanosoma cruzi* is the causative agent of Chagas disease, an illness that affects 6–8 million people worldwide [1]. The parasite has a complex life cycle, which starts when the triatomines defecate in the vicinity of the bite, and in the infective stage, the metacyclic trypomastigote (MT) is carried into the wound by scraping and then penetrates across the skin and invades resident cells. Once in the cytosol, MTs transform into amastigotes. This proliferative stage multiply in the cytosol and, after various rounds of binary fission, transform into trypomastigotes, which are released from the cell and disseminated by the bloodstream invading different organs, spreading the parasitic infection [2].

From a biological and biochemical point of view, *T. cruzi* is a genetically heterogeneous species with a wide variety of strains distributed in different vectors and with a great diversity of mammalian hosts. This high genetic variability is related to the variability of virulence and clinical manifestations among strains [3]. Currently, *T. cruzi* is classified into seven discrete typing units (DTUs), TcI to TcVI [4], and more recently, a new one named Tcbat has been included [5].

2. The proteomics and mass spectrometry

Proteomics is a discipline focused on understanding protein expression in biological samples and the possible comprehension of the complex physiological and pathological events using different methods to identify and characterize proteins. Thus, proteomics is the study of the proteome, which corresponds to the global patterns of proteins that determine cell expression under a determinate condition [6].

From 1997, when the first paper on the proteome was published [7], proteomics studies have become an invaluable tool not only for the study of the global cell expression of proteins (the proteome) but also for the dissection of specific processes in cell biology and biochemistry, such as the study of proteins that are expressed in organelles and post-translational modifications [8, 9]. Moreover, by combining proteomics analysis and bioinformatics tools, researchers can report much broader projections, including the identification of new chemotherapeutic targets, new vaccine candidates, new tumor markers with prognostic or diagnostic value [10], and the study of secretomes and immunonomes, among others [11]. Thus, a current PubMed search with the keyword "proteomic" delivers 96,688 published papers.

Initially, Edman sequencing was used, and this method was replaced by mass spectrometry (MS). MS allows the separation of different isotopes, which is possible due to the use of soft ionization procedures, such as matrix-assisted laser desorption/ionization (MALDI) [12] and the electrospray ionization method (ESI) [13]. In both cases, the analytes (peptides) are ionized from the solid phase (MALDI) or liquid phase (ESI) into the gas phase. Then, liquid chromatography (LC) systems can be directly interfaced to mass spectrophotometers. LC tandem mass spectrometry (LC-MS/MS) has been applied to proteomics to separate peptides, and MS/MS records the intact peptides (full MS) before one precursor ion is selected and fragmented. This fragmentation is always induced in a collision cell by nitrogen or argon. Then, fragments are recorded in an MS/MS spectrum, and this fragmentation pattern reveals a specific mass for each amino acid from the peptide [6]. In summary, the amino acids present in the peptide can be determined, as well as the peptide sequence, and finally, the protein containing this peptide is identified.

In kinetoplastids like *T. cruzi*, all protein-encoding genes are organized in large polycistronic transcription units that produce polycistronic precursor RNAs that are then processed to monocistronic mRNAs by the mechanism of trans-splicing [14]. Then, translation of specific mRNAs and/or the stability control regulate the gene expression [15]. Furthermore, post-translational modifications perform a key function in order to modulate the *T. cruzi* protein function. For these reasons, in recent years, proteomics has had a crucial role in evaluating protein expression levels in different *T. cruzi* stages or following specific treatments with drugs, inhibitors, or natural products, and thus, proteomics studies have been transformed in an invaluable tool to reveal adaptive changes in this protozoan [16].

3. The *T. cruzi* proteome

A proteomic analysis of the human pathogen *T. cruzi* using two-dimensional electrophoresis (2-DE) was initially reported by Paba et al. [17]. A *T. cruzi* comparative proteome analysis of different stages was performed. This study showed that few proteins displayed significant differential expression among the three developmental *T. cruzi* stages. Using matrix-assisted laser desorption/ionization time-of-flight (MALDI-TOF) mass spectrometry peptide mass fingerprinting, the researchers identified 26 proteins. Among these, 19 different proteins were identified, mainly heat shock proteins (HSP; chaperones, HSP 60, HSP 70, and HSP 90), elongation factors, glycolytic pathway enzymes (enolase, pyruvate kinase, and 2,3 bisphosphoglycerate mutase), and structural proteins (KMP 11, tubulin, and paraflagellar rod components). In addition, the same researchers reported a comparative proteome analysis of *T. cruzi* developmental stages. In this study, isotope-coded affinity tag technology (ICAT) associated with liquid chromatography-mass spectrometry peptide sequencing (LC-MS/MS) was used. High-confidence sequence information and expression levels for 41 *T. cruzi* polypeptides, including metabolic enzymes, paraflagellar rod components, tubulins, and heat shock proteins, were reported. Twenty-nine proteins displayed similar expression levels in trypomastigotes as well as amastigotes, nine proteins displayed higher levels in trypomastigotes than amastigotes, and three proteins showed higher expression in amastigotes than trypomastigotes [18].

In addition, Parodi-Talice et al. [16] described a standardized and reproducible protocol for analyzing *T. cruzi* proteins by 2-DE, which generated protein reference maps of this parasite. In this report, tubulins, HSP, and prostaglandin F2α synthase were three of the most abundant proteins reported in *T. cruzi*.

A proteomic analysis of all *T. cruzi* stages was performed. Peptidic sequences from 2784 proteins in 1168 protein groups from the annotated *T. cruzi* genome were reported in all parasite stages. Protein products were identified from >1000 genes annotated as "hypothetical" in the parasite genome, including integrants of the mucin-associated surface proteins (MASP). The four parasite stages appear to use distinct energy sources, including histidine for stages present in the insect vectors and fatty acids by intracellular amastigotes [19]. In contrast, a study that identified basic proteins reported different enzymes related to amino acid metabolism in epimastigotes, whereas trans-sialidases and protein paraflagellar rods were found specifically in trypomastigotes [20]. Another proteomics study of *T. cruzi* trypomastigotes identified 1448 nonredundant proteins, 14% of which corresponded to surface proteins, and the majority were anchored via glycophosphatidylinositol (GPI) and related to parasite virulence [21]. According to these authors, the high diversity of expressed proteins on the trypomastigote surface could have implications for both invasion and immune evasion strategies. A comparison of the cell surface proteomes from different *T. cruzi* stages revealed that some enzymes are stage-specific, although the majority were present in more than one stage. Bioinformatics analysis showed that most of the identified proteins were derived from the membrane and were involved in cell invasion, adhesion, cell signaling, and host immune response modulation [22]. However, the biological role of these proteins was not experimentally demonstrated.

The blood trypomastigote (BT) proteome has also been described [23]. In a shotgun analysis, 17,394 peptides were reported, corresponding to 7514 proteins of which 5901 corresponded to *T. cruzi*. Trans-sialidases (TS), bioenergetics-related enzymes, chaperones, and cytoskeletal proteins were found among the top scoring proteins. Gene ontology (GO) studies showed that all *T. cruzi* compartments were evaluated and the most part of proteins were engaged in metabolic processes and/or presented catalytic functions. A comparative study among BT and tissue-culture-derived trypomastigote (TCDT) or metacyclic trypomastigote (MT) proteomic profiles identified 2202 proteins uniquely reported in the BTs. These unique proteins were related to (a) surface proteins, (b) the nonclassical secretion pathway, (c) cytoskeletal dynamics, (d) the cell cycle and transcription, (e) proteolysis, (f) redox metabolism, (g) biosynthetic pathways, (h) bioenergetics, (i) protein folding, (j) cell signaling, (k) vesicular traffic, (l) DNA repair, and (m) cell death [23]. The Brunoro paper identified many proteins belonging to different metabolic parasite pathways and functions that were exclusive to BTs and provided a valuable data set for a better biological understanding of this parasite stage.

The *T. cruzi* cell surface is covered by a dense layer of GPI-anchored molecules. These molecules are involved in a variety of interactions between the parasite and its vertebrate and invertebrate hosts. GPI-anchored rich fractions and other *T. cruzi* membrane proteins were obtained from epimastigotes (EPIs) and MTs and then submitted to two-dimensional liquid chromatography coupled to tandem mass spectrometry (2D-LC-MS/MS). A total of 98 MT proteins and 280 EPI proteins were identified. Of those, approximately 65% (n = 245) had predicted lipid post-translational modification sites (i.e., GPI-anchor, myristoylation, or prenylation), signal-anchor sequences, or transmembrane domains that could explain their solubility in detergent solution. This report showed that MTs, but not EPIs, express a large repertoire of surface glycoproteins, such as GP90 and GP82, which are involved in host cell adhesion and invasion. Taken together, these results showed that *T. cruzi* displays stage-specific protein profiles that are related to the biology of each *T. cruzi* stage [24].

A map of soluble proteins from epimastigotes of *T. cruzi* CL Brener was generated using 2-DE combined with MS. The analysis showed that several proteins, such as tubulin and heat shock proteins, were found in multiple isoforms. The enzyme arginine kinase was also found. This enzyme is absent in mammalian cells and, according to the authors, could represent a potential new chemotherapeutic target [25].

Using conditions optimized for the 6–11 pH range, researchers compared the proteomes of TCDTs and extracellular amastigote-like parasites by MALDI-TOF/TOF. The alkaline 2-DE gels from TCDTs and extracellular amastigotes showed that proteins with a pI over 7.0 were very well depicted (= 65% of proteins detected). Furthermore, the differences in protein expression among the human-hosted *T. cruzi* stages were in agreement with former proteomics reports and were related to the biological properties of each parasite stage [26].

4. The *T. cruzi* secretome/exoproteome

Adhesion and invasion are the first stages of interaction between an obligate intracellular pathogen and its host cell. This process involves cell surface molecules and secreted molecules. Then, the molecules that are released from pathogens may play a critical role during the establishment of infection, especially in immunomodulation or immune evasion, as well as in migration across host tissues, cell adhesion, cell-cell communication, differentiation, proliferation, and morphogenesis [27, 28].

The term secretome was first introduced in a bioinformatics survey of proteins secreted by *Bacillus subtilis* [29]. The authors defined the secretome as the subset of the proteome that is secreted, in addition to the components of cellular machinery used for protein secretion. However, these released proteins are not only secreted proteins but also proteins that arise from other export mechanisms. Thus, the best term to describe the protein content found in the extracellular proximity of a given biological system is the "exoproteome" [30].

According to [31], *T. cruzi* releases proteins associated with two types of vesicles. The large vesicles (LVs), named ectosomes, arise from the plasma membrane, and small vesicles (SVs), similar to exosomes, are released by the flagellar pocket. A third fraction was enriched in soluble proteins and was not associated with extracellular vesicles (EVs) [31]. Label-free quantitative proteomics analysis showed the presence of proteins associated with metabolism, signaling, packaging of nucleic acids, parasite survival, and virulence [31].

The *T. cruzi* exoproteome of TCDTs (Y strain) was recently characterized. NanoLC-MS/MS analysis resulted in the identification of 540 proteins. Bioinformatics analysis predicted most identified proteins as secreted, predominantly by nonclassical pathways, and suggested that these proteins are involved in host cell interaction. Some proteins possess predicted GPI-anchor signals; these proteins are mostly TS, MASP, and surface glycoproteins. The existence of various related proteins to similar functions in the exoproteome likely reflects this parasite's enhanced mechanisms for adhesion, invasion, and internalization of different host cell types and immune evasion [32].

Considering that Chagas disease is a chronic infection in which amastigote forms remain for a long time, the *T. cruzi* laboratory examination using direct observation is tricky, and although antibody detection tests are sensitive, they may give false-positive results. The implementation of novel fast assays aimed for the identification of excretory/secretory parasite-derived molecules in serum would be a useful contribution in the diagnosis of human *T. cruzi* infection. Brossas and coworkers [33] investigated the proteins secreted by *T. cruzi* using MS analyses of conditioned culture media devoid of serum, collected during the emergence of trypomastigotes from infected Vero cells. Indeed, the secretomes of two *T. cruzi* strains from DTU TcVI (VD and CL Brener) were compared. This study led to the identification of 591 *T. cruzi* proteins. Three hundred sixty-three proteins were common to both strains, and most belong to different multigenic super families (e.g., TcS, GP63, MASP, and DGF1). Finally, a report of 94 secreted proteins, present in both DTU TcVI strains that do not correspond to the members of multigene families, was also established. This paper offers the first comparative study of the secretomes from two different *T. cruzi* strains of DTU TcVI. Then, Brossas and coworkers' papers identified a subset of common secreted proteins that could have a potential role as serum biomarkers for *T. cruzi* infection [33].

A novel family of *T. cruzi* surface membrane proteins (TcSMPs), which are conserved among different *T. cruzi* lineages and have orthologs in other *Trypanosoma* species, was described. Several lines of evidence suggest that TcSMP is a membrane-spanning protein located at the cellular surface and is also released into the extracellular milieu. TcSMP displayed a signal anchor and a C-terminal hydrophobic sequence predicted to be a transmembrane domain. In addition, proteomic TcSMP peptides previously found in a membrane-enriched fraction were identified in membrane vesicles as well as in soluble forms in the *T. cruzi* secretome. TcSMP proteins were also located intracellularly, likely associated with membrane-bound structures. These proteins were shown to inhibit metacyclic trypomastigote entry into host cells. TcSMPs could bind to mammalian cells and trigger Ca^{2+} signaling and lysosome exocytosis. This evidence strongly suggests the involvement of these proteins in parasite invasion [34].

5. *T. cruzi* subcellular proteomics

A combination of MS-based proteomics methods and traditional biochemical cell fractionation protocols had been used to characterize subcellular organization. The protein contents of specific subcellular compartments can thus be identified following specific enrichment strategies that concentrate and enrich organelles and subcompartments of the cell [8].

T. cruzi reservosomes show concentrated proteins and lipids obtained from medium together with the main proteolytic enzymes originating from the secretory pathway, being at the same time a storage organelle and the main site of protein degradation and endpoint of the parasite endocytic pathway. By cell fractionation and LC-MS/MS, the EPI reservosome-resident proteins were identified in a total of 709 *T. cruzi*-specific proteins. Among these, 456 had predicted functions and 253 were described as hypothetical proteins. The existence of a high number of previously reported proteins was confirmed. In addition, new different classes of proteins were reported. Among them, transport proteins, proton pumps, and enzymes were also identified. According to the authors, the definition of the reservosome protein profile could be a good tool to assess the molecular signatures, identify molecular markers, and understand the relationship with different parasite organelles [35].

The EPI transition from the exponential growth phase to the stationary phase is a key step that recapitulates the early molecular events of *T. cruzi* metacyclogenesis, opening new possibilities for understanding this process. Thus, a quantitative shotgun proteomics study of the *T. cruzi* EPIs in the exponential and stationary growth phases was performed. More than 3000 proteins were identified and quantified, stressing the regulation of proteins engaged in various subcellular compartments. Ribosomal proteins were positively regulated in the exponential phase, in agreement with the high proliferative rate of this growth phase. Autophagy-related proteins were positively regulated in the stationary growth phase, showing the beginning of the metacyclogenesis. Otherwise, this paper reported the regulation of N-terminally acetylated proteins during growth phase transitioning, adding a new layer of regulation to this process. Thus, this paper reported a proteome-wide rewiring during *T. cruzi* differentiation from the EPIs' proliferative exponential phase to the stationary growth phase, which is the step before cell transformation into the MT stage [36].

In the eukaryotic cell, chromatin-associated proteins are key regulators of many important processes. By high-resolution MS, the *T. cruzi* chromatin proteome was investigated using three different protocols and comparing it between two parasite stages, EPIs and TCDTs. More than 2000 proteins were identified and quantified both in chromatin and nonchromatin extracts. In addition to histones and other known nuclear proteins, trypanosome chromatin also contains metabolic proteins (mainly from carbohydrate pathways), cytoskeleton components, and many other proteins with unknown functions. In addition, EPIs and TCDTs also differ greatly regarding their chromatin-associated factor composition and amount. Although the nucleosome content was the same for both stages, the remaining proteins were rarely detected in TCDTs, suggesting that they have naked chromatin. Proteins associated with DNA proliferation, such as proliferation cell nuclear antigen (PCNA), replication protein A (RPA), and DNA topoisomerases, were exclusively found in the chromatin of EPIs. However, TCDTs showed an enrichment of a histone H2B variant. Moreover, nearly 20% of the EPIs' chromatin-associated proteins were expressed in TCDTs but were placed in the nonchromatin space. Several types of proteins, including phosphatases and a Ran-binding protein, that may be transported among chromatin and the nonchromatin space during differentiation were also identified. These results indicated that the replicative state in trypanosomes involves an increase in chromatin-associated protein content. Given that trypanosomes are early-branching organisms, these data could enhance our understanding of chromatin-associated processes in other cell types [37].

T. cruzi proliferation displays unique features, such as the absence of chromosome condensation and closed mitosis. Thus, *T. cruzi* epimastigotes were lysed, and the nuclear fraction was isolated to study the proteome by LC-MS/MS. A total of 864 proteins were identified, of which 272 were annotated as putative uncharacterized proteins and 275 had not been previously reported in the *T. cruzi* proteome. Bioinformatics analysis using the Database for Annotation, Visualization, and Integrated Discovery (DAVID) was carried out, and proteins were grouped into nuclear proteins in 65 gene clusters, wherein the clusters with the highest enrichment scores harbored members with chromatin organization and DNA-binding functions [38].

Contractile vacuole complexes are critical components of cell volume regulation and have been shown to have other functional roles in several free-living protists.

However, very little was known about the functions on *T. cruzi*. A proteomics and bioinformatics approach to identify proteins localized to the contractile vacuole was performed using the *T. cruzi* epimastigote-enriched fraction of contractile vacuoles, which was analyzed by one-dimensional gel electrophoresis and LC-MS/MS. A total of 109 newly detected proteins were identified, including at least 39 members of the dispersed gene family 1 (DGF-1). This observation suggested that many members of this family are simultaneously expressed in EPIs. In addition, several homologs with known localizations in contractile vacuoles of other organisms were selected. GFP-fusion proteins or specific antibodies were used to determine the localization of each protein. Six of these putative proteins (Rab11, Rab32, AP180, ATPase subunit B, VAMP1, and phosphate transporter) were mostly located in the vacuole bladder. However, TcSNARE2.1, TcSNARE2.2, and calmodulin were located in the spongiome. Calmodulin was a cytoplasmic protein. These observations confirmed the usefulness of merged subcellular fractionation, proteomics assays, and bioinformatics methods in order to know the location of organellar proteins that could be hard to identify when global proteomic approaches are used [39].

Several lines of evidence suggest that the translational system in trypanosomatids shows important differences compared to those of other eukaryotes. A careful data mining quest for ribosomal protein genes in the *T. cruzi* genome database coupled with MS of isolated *T. cruzi* ribosomes was performed. The findings suggested that *T. cruzi* ribosomal proteins have around 50% sequence identity to yeast proteins. In addition, some trypanosomal proteins were longer than yeast proteins because of the existence of many N- or C-terminal extensions, which are unique to trypanosomatids. In detail, L19 and S21 have shown C-terminal extensions of 168 and 164 amino acids, respectively. Even more, two 60S subunit proteins that had not been formerly found in the *T. cruzi* whole proteome, namely, L22 and L42, were detected [40].

6. Parasite biology and reversible post-translational modifications

In eukaryotes, histones are well-conserved proteins that form the basic structure of chromatin and undergo several post-translational modifications, which are important for the control of transcription, replication, DNA damage repair, and chromosome condensation [41].

The processes by which *T. cruzi* acquires infectivity and survives in different hosts involve rapid adaptations to new environments and tight regulation of gene expression, mainly post-transcriptionally. Nevertheless, the chromatin structure/organization of trypanosomatids is similar to that of other eukaryotes, including histone variants and post-translational modifications. Evidence suggests that epigenetic mechanisms also play an important role in the parasite, indicating that new epigenetic targets can be used for chemotherapeutic intervention. From this perspective, MS-based large-scale proteomics has been used to identify *T. cruzi* post-transcriptional modifications in *T. cruzi* canonical and variant histones. A total of 13 distinct modification types were identified, including 9 newly described chemical modifications for trypanosomatids, which included novel and unusual modifications, such as alternative lysine acylations, serine/threonine acetylation, and N-terminal methylation. In the same way, two hundred conserved and unique post-transcriptional sites in the tails and globular regions of all canonical and variant histones were identified. These results provided the first comprehensive map of *T. cruzi* histone marks and provided new evidence that allowed us to gain a deeper understanding of the epigenetic mechanisms that operate in *T. cruzi*. This report could provide information to identify potential candidates for the development of new epigenetic drugs [42].

In trypanosomatids such as *T. cruzi,* the regulation of gene expression is mostly post-transcriptional. The stability of mRNA and the possibility to accede to polysomes are believed to be strongly regulated, enabling *T. cruzi* to accommodate to various environmental situations that occur through its life cycle. Post-transcriptional regulation needs the association among mRNAs and specific proteins to shaping the complexes named ribonucleoproteins (mRNP). The dynamic association among proteins and mRNAs was studied through poly(T) beads to purify and characterize proteins and protein complexes linked to poly-A+ mRNAs. The protein contents of these fractions were studied by LC-MS/MS. A number of 542 protein components of the mRNP complexes associated with mRNAs were detected. Among them, 24 of the retrieved proteins were common to all fractions, whereas others were only found in an individual fraction: EPI polysomal (0.37%) and post-polysomal (2.95%) fractions and stress polysomal (13.8%) and post-polysomal (40.78%) fractions. This proteomic approach provided the first insight into the composition and dynamics of *T. cruzi* mRNPs [43].

Protein acetylation is also a post-translational modification that regulates several cellular pathways. Using proteomics, researchers identified 389 ε-lysine-acetylated sites in 235 proteins in *T. cruzi* epimastigotes. Lysine-acetylated protein sites were enriched in enzymes involved in oxidation/reduction balance, which is required for parasite survival in the host. These observations provide evidence that in *T. cruzi*, protein acetylation participates in the differential regulation of several cellular pathways and helps elucidate the mechanisms involved in parasite infection and survival [44].

In early-branched organisms, histones appear to be less conserved and contain alternative sites for modifications, which could reveal evolutionary unique functions of histone modifications in gene expression and other chromatin-based processes. Using high-resolution MS, researchers studied the *T. cruzi* trypomastigote and epimastigote histone post-translational modifications. A total of 44 new modifications were detected: 18 acetylations, 7 monomethylations, 7 dimethylations, 7 trimethylations, and 4 phosphorylations. Epimastigotes contained more histone modifications than trypomastigotes. Acetylations of lysines at the C-terminus of histone H2A and methylations of lysine 23 of histone H3 were reported to be enriched in MT or TCDT. Contrasting with that, protein phosphorylation at serine 23 of H2B and methylation of lysine 76 of histone H3 predominated in epimastigotes. In parasites undergoing mitosis and cytokinesis, the existence of one or two methylations in lysine 76 was reported. These observations helped reveal the role of histone modifications in gene expression control and cell cycle regulation in *T. cruzi* [45]. However, epigenetic events, such as histone acetylation, affected DNA topology, replication, and gene expression. Histone deacetylases are involved in chromatin compaction and post-translational modifications of cytoplasmic proteins, such as tubulin. Given these facts, histone deacetylase inhibitors, such as trichostatin A, were evaluated by de Oliveira Santos and coworkers to determine their effects on *T. cruzi* epimastigote cell proliferation, viability, cell cycle, and ultrastructure, as well as on histone acetylation and tubulin expression. A quantitative proteomics study showed an increase in histone acetylation after trichostatin A treatment, suggesting that deacetylase inhibitors may represent excellent tools for elucidating trypanosomatid cell biology [46].

Protein N-myristoylation is catalyzed by N-myristoyltransferase (NMT), an essential and potential drug target in *T. cruzi*. Roberts and Fairlamb used a combination of label-free and stable isotope labeling of cells in culture (SILAC)-based proteomics approaches when parasites were treated with or without the N-myristoyltransferase inhibitor DDD85646. A total of 56 proteins present in at least two out of the three experimental studies were identified. Among these,

6 were probably to be false positives, with the remainder 50 starting with the amino acids MG at the N-terminus in one or more of the *T. cruzi* genomes. The greater part of these were proteins of unknown function (32), with the remaining (18) involved in a different spectrum of key cellular and metabolic roles, such as intracellular transport, cell signaling, and protein turnover. In conclusion, the results showed that 0.43–0.46% of the *T. cruzi* proteome corresponded to N-myristoylated proteins, slightly lower than that reported in other eukaryotic organisms (0.5–1.7%) [47].

Glycoproteins are very attractive therapeutic candidate targets because they mediate key processes during the *T. cruzi* life cycle, such as cellular recognition, host cell invasion and adhesion, and immune evasion. Using subcellular fractionation, lectin affinity, and SILAC, Atwood et al. [48] conducted a *T. cruzi* trypomastigote glycoproteomics analysis, which described the identification of organelles and cell surface N-linked glycoproteins of *T. cruzi*. These researchers identified 36 glycosylation sites on 35 glycopeptides that mapped to 29 glycoproteins. They also presented the first evidence for 11 *T. cruzi*-specific glycoproteins and provided experimental data indicating that the MASP and dispersed gene family (DGF-1) were post-translationally modified by N-linked glycans. In another study using lectin-based and hydrophilic interaction liquid chromatography followed by high-resolution LC-MS/MS, a comprehensive glycoproteomics analysis was performed in *T. cruzi* epimastigotes and trypomastigotes. After treatment with glycanases, a number of 1306 N-glycosylation sites in NxS/T/C motifs were reported from 690 *T. cruzi* glycosylated proteins. Among them, 170 and 334 glycoproteins were solely detected in EPI and trypomastigotes, respectively. In addition, global site-specific characterization of the N- and O-linked glycan heterogeneity in the two *T. cruzi* life stages was reached by intact glycopeptide methods, detecting 144/466 unique N-linked and 10/97 unique O-linked intact glycopeptides in epimastigotes/trypomastigotes, respectively. Thus, *T. cruzi* displays a pattern of stage-specific glycoprotein expression that may be the result of parasite interactions with vertebrate and invertebrate hosts during its life cycle [49].

In eukaryotic cells, SUMOylation is an important protein post-translational modification. The C-terminus of proteolytically activated small ubiquitin-like modifier (SUMO) is covalently bound to a lysine residue of the target protein by an isopeptide bond through a pathway that involves an E1-activating enzyme, an E2-conjugating enzyme, and a transfer to the target, in occasions with the support of a protein ligase [50]. The modification is reversed by a protease, which is also responsible for SUMO ripening. Several proteins have been described as SUMO targets and were shown to participate in the regulation of cell cycle progression, transcription, translation, ubiquitination, and DNA repair [50]. Orthologous genes corresponding to the SUMOylation pathway have been reported in *T. cruzi*. Furthermore, the SUMOylation system appears to be functionally active in the parasite. An immunofluorescence analysis showed that *T. cruzi* SUMO (TcSUMO) was predominantly found in the nucleus. To identify SUMOylation targets and further elucidate their physiological roles, researchers generated transfectant *T. cruzi* epimastigote lines expressing a double-tagged *T. cruzi* SUMO, and SUMOylated proteins were enriched by tandem affinity chromatography. By two-dimensional LC-MS/MS, a total of 236 proteins with different biological roles were reported as possible *T. cruzi* SUMO targets. Among these, metacaspase-3 was confirmed as a bona fide SUMOylation substrate by a biochemical approaches. The application of proteomics methods in many different biological systems has allowed to know that orthologs of putative *T. cruzi* SUMOylated proteins are equally modified, suggesting that conserved mechanisms could operate for protein SUMOylation [51].

For the role of phosphorylation, kinase inhibitors have been suggested as novel antiparasitic agents. However, a fundamental understanding of the cell signaling pathways requires a detailed analysis of the involved phosphorylated proteins. A MS-based phosphoproteomic map of *T. cruzi* epimastigotes was performed by LC-MS/MS, dual-stage fragmentation, and multistage activation method. A total of 237 phosphorylated peptides from 119 different proteins were identified. Moreover, 220 phosphorylation sites were unequivocally mapped: 148 on serine, 57 on threonine, and 8 on tyrosine. Even more, immunoprecipitation and immunoblot assays reported the existence of at least seven tyrosine-phosphorylated proteins in *T. cruzi*. The phosphorylated proteins were submitted to GO, InterPro, and BLAST scan and classified according to their roles in cell structure, motility, cellular transport, metabolic pathways, pathogenesis, DNA/RNA/protein turnover, and signaling. These phosphoproteomic information provided novel perspectives into *T. cruzi* phosphorylation pathways and the potential roles of the identified phosphoproteins in *T. cruzi* biology and reinforced the idea that kinases are parasite drug targets [52].

Furthermore, a phosphoproteomic approach was performed to determine the role of transforming growth factor beta (TGF-β) in *T. cruzi* biology. Then, EPI (Y strain) was incubated with TGF-β for 1, 5, 30, and 60 min, which promoted remodeling of the parasite phosphorylation network and protein expression pattern. To identify parasite proteins involved in the TGF-β response, researchers performed 2-DE and MS. The modified proteins were engaged in many parasite functions, such as proteolysis, metabolism, heat shock response, cytoskeleton arrangement, oxidative stress regulation, translation, and signal transduction. After TGF-β treatment, a number of 75 protein spots were found to be positively or negatively regulated more than twofold. Among these, 42 were detected by MS, highlighting, among these, cruzipain, the main *T. cruzi* papain-like cysteine protease. In these experiments, TGF-β addition favored epimastigote proliferation, corroborating 2-DE data in which proteins previously described to be involved in this process were positively stimulated by TGF-β [53].

The MT proteome, which showed differential expression of some proteins during metacyclogenesis [54], has also been determined. Metacyclogenesis is the process by which *T. cruzi* epimastigotes differentiate into metacyclic trypomastigotes and acquire the ability to infect cells and resist complement lysis. This process obviously requires the expression of a new set of proteins associated with parasite virulence. This critical step in parasite biology was first studied by Parodi-Talice and coworkers [55] who analyzed the process by 2-DE coupled to MALDI-TOF MS. A large proportion of unique proteins expressed during metacyclogenesis were observed. Interestingly, 50% of the spots were found to differ between epimastigotes and trypomastigotes. The study provided a 2D map of MT. Sixty-six protein spots were successfully identified, corresponding to 43 different proteins. The expression profiles allowed protein identification during metacyclogenesis and categorize the proteins into three groups in accordance to their maximal level of expression. Various isoforms for a number of proteins, some of them displaying differential expression during metacyclogenesis, were detected. These information suggested that post-translational modifications may be an essential component of the parasite's biological program to regulate gene expression during metacyclogenesis. The Parodi-Talice and coworkers' paper contributed to the identification of relevant proteins implicated in the metacyclogenesis process, providing relevant information about the pathways involved in *T. cruzi* epimastigote differentiation into MTs [55]. However, a quantitative MS-based proteomics study of *T. cruzi* metacyclogenesis was performed, and almost 3000 proteins expressed during the process were quantified. Relevant proteins and pathways involved in the parasite's differentiation and infectivity acquisition were identified, opening new perspectives for further studies that could lead to the identification of new chemotherapeutics targets [54].

Studies of the *T. cruzi* phosphoproteome and the nuclear subproteome were also performed [56], and the results showed that the *T. cruzi* nuclear subproteome was composed of several members of the HSP family and members of the "retrotransposon hot spot" (RHS) protein family as reported by De Castro Moreira Dos Santos et al. (2015).

The differentiation from trypomastigotes into amastigotes, called amastigogenesis, has also been studied *in vitro* by incubating TCDTs in acidic DMEM to perform a comprehensive quantitative proteomics and phosphoproteomics analysis of *T. cruzi* amastigogenesis. Regulated proteins and pathways involved in coordinating amastigogenesis were also identified, indicating that a significant proportion of the regulated proteins were membrane proteins. The authors further reported that modulated phosphorylation events coordinated by protein kinases and phosphatases are involved in signaling pathways and were triggers after incubation in acidic medium [22].

One of the least studied differentiation processes is the passage of infective trypomastigotes to epimastigotes, termed epimastigogenesis [57]. To resolve this issue, researchers established an *in vitro* and *in vivo* model for epimastigogenesis, enabling the characterization of a new infective morphological stage called "recently differentiated epimastigotes." This stage was infective and resistant to complement-mediated lysis both *in vitro* and *in vivo*. Shotgun proteomics of all *T. cruzi* stages revealed that several proteins were upregulated only in recently differentiated epimastigotes (including ERO1, multidrug resistance protein E and ABC transporters). This evidence suggests a role for these proteins in complement-system resistance, as previously observed for other organisms [58, 59]. As indicated by the proteomics analysis, this stage contains a minimal and distinct set of molecules required for *in vitro* and *in vivo* infectivity, suggesting a putative role of epimastigotes in *T. cruzi* infection of mammalian cells [60].

A validated, constraint-based model of the core metabolism of *T. cruzi* strain CL Brener was carried out. The model includes 4 compartments (extracellular space, cytosol, mitochondrion, glycosome), 51 transport reactions, and 93 metabolic reactions covering carbohydrates, amino acids, and energy metabolism. In addition, several replicate high-throughput proteomic data sets were used to specifically examine the metabolism of *T. cruzi* epimastigotes. This paper shows the usefulness of constraint-based models for incorporating several sources of information (e.g., genomics, primary biochemical literature, proteomics) to build testable hypotheses. This model constitutes and focuses on the systematic analysis of *T. cruzi* metabolism under a broad spectrum of conditions and disturbances and should possibly contribute to identify new drug targets [61].

T. cruzi is always subjected to sudden temperature shifts during its life cycle. The adaptive capacity to these changes is critical for parasite survival, reproduction, and transmission. These changes may modify the profile of protein expression that participates in homeostasis during the course of stress treatment. Thus, the proteome of *T. cruzi* EPI exposed to heat shock was analyzed by 2-DE followed by MS for protein identification. A total number of 24 spots differing in abundance were reported. Of the 24 modified spots, 19 showed a higher intensity and 5 showed a smaller intensity in relation to the control. Various functional categories of the reported proteins were determined: metabolism, cell defense, hypothetical proteins, protein fate, protein synthesis, cellular transport, and cell cycle. Proteins engaged in the interaction with the cellular environment were reported too [62].

A proteomics analysis of an organellar cell fraction from *T. cruzi* CL Brener epimastigotes was reported by Ferella et al. [63]. A total number of 396 proteins were determined by LC-MS/MS. Among these, 138 were reported as hypothetical in the genome databases, and the remainder was allocated to many metabolic and biosynthetic pathways, transport, and structural functions. Comparative studies with an entire cell proteome are made possible to validate the expression of 173 additional proteins. Among these, 38 proteins previously detected in other stages were not found in the only large-scale study of the total EPI proteome. A chosen set of identified proteins was further studied in order to know their gene copy number, sequence variation, transmembrane domains, and targeting signals. The genes were cloned, and the recombinant proteins were expressed with a c-myc epitope tag in *T. cruzi* EPIs. Immunofluorescence microscopy showed the location of these proteins in different cellular compartments, such as the ER, acidocalcisome, mitochondrion, and putative cytoplasmic transport or delivery vesicles. These pieces of evidences reinforce the idea that the use of enriched subcellular fractions allowed the identification of *T. cruzi* proteins that were unnoticed by whole cell proteomics approaches [63].

7. Proteomics in experimental chemotherapy and drug resistance

Chagas disease chemotherapy is based mainly on benznidazole (Bz) and nifurtimox (Nx), developed more than four decades ago, which are very efficient nitroderivatives against the acute stage but present limited efficacy during the chronic phase. In fact, both are far from ideal due to substantial secondary side effects, limited efficacy against different parasite isolates, the need for long-term therapy, and their well-known poor activity in the late chronic phase. Thus, identification of improved drugs to treat *T. cruzi*-infected people is urgently needed. Even though many different kinds of natural and synthetic compounds have been described to have *in vitro* and *in vivo* activity against *T. cruzi*, since the introduction of Bz and Nf, only scarce chemotherapeutic compounds, such as allopurinol and some sterol inhibitors, have moved to clinical assays.

The proteome of two *T. cruzi* strains after piplartine treatment was reported [64]. Considerable changes in the expression of enzymes involved in parasite protection against oxidative stress, such as tryparedoxin peroxidase (TXNPx) and methionine sulfoxide reductase (MSR), were observed in both strains. These findings suggested that inhibition of these enzymes could be potential chemotherapeutic targets.

The trypanocidal effects of naturally occurring quinones and their derivatives and naphthoimidazoles derived from β-lapachone N1, N2, and N3 were the strongest. To assess the molecular mechanisms of action of these compounds, researchers studied treated bloodstream trypomastigotes by 2D-DIGE followed by MALDI-TOF/TOF. A total of 61 differentially abundant protein spots were identified when comparing the control with the N1, N2, or N3 treatment for 34 identified spots. The differentially abundant proteins included activated protein kinase C receptor, tubulin isoforms, asparagine synthetase, arginine kinase, elongation factor 2, enolase, guanine deaminase, heat shock proteins, hypothetical proteins, paraflagellar rod components, RAB GDP dissociation inhibitor, succinyl-CoA ligase, ATP synthase subunit B, and methionine sulfoxide reductase. These information allowed the identification of different modes of action for N1, N2, and N3, which indicate a great variety of metabolic pathways involved and open new avenues for the development of new trypanocidal drugs [65].

In addition, proteomics was used to identify proteins involved in N1, N2, and N3 trypanocidal activity. In epimastigotes, naphthoimidazoles are involved in multiple mechanisms: (a) redox metabolism, (b) energy production, (c) ergosterol biosynthesis, (d) cytoskeleton assembly, (e) protein metabolism and biosynthesis, and

(f) chaperone modulation. These compounds were able to induce an imbalance in crucial pathways of the parasite, leading to the loss of metabolic homeostasis and *T. cruzi* death [66].

A proteomic analysis of *T. cruzi* resistance to Bz was reported initially by Andrade et al. [67]. The differential proteomes of *T. cruzi* with selected *in vivo* resistance to Bz (BZR and Clone27R), its susceptible pairs (BZS and Clone9S), and a pair from a population with benznidazole *in vitro*-induced resistance (17LER) and the susceptible pair 17WTS were studied by 2-DE followed by MS. Among 137 spots studied through MS, 110 were detected as 56 distinct proteins. Among the 56 different proteins, 36 were detected in resistant, 9 in susceptible, and 11 in both phenotypes. In the group of the proteins identified in resistant samples, five were found in Cl 27R and in BZR (calpain-like cysteine peptidase, hypothetical protein conserved 26 kDa, putative peptidase, peroxiredoxin, and tyrosine amino transferase) and four in Cl 27R and 17LER (cyclophilin A, glutamate dehydrogenase, iron superoxide dismutase, and nucleoside diphosphate kinase). However, the proteins detected in Bz-susceptible samples, PGF-2a, were found in BZS and 17WTS. A functional category study has shown that proteins engaged in transcription and protein destination were overexpressed in the Bz-resistant phenotype. This report provided large-scale, proteomic information to obtain a better knowledge of the mechanism involved in *T. cruzi* resistance to Bz [67].

8. Laboratory diagnostic and protein identification

Immune complexes (ICs) are the direct result of humoral immune responses. Therefore, the identification of foreign constituents or autoantigens within ICs might provide new insights into the immunopathology of infectious diseases. Ohyama and coworkers [68] studied the immune complexome of plasma samples obtained from *T. cruzi*-infected patients. Then, 20 seropositive plasma samples, including cardiac and/or megacolon chronic patients (n = 11) and asymptomatic patients (n = 9), were analyzed together with 10 seronegative healthy people to investigate the antigens linked to circulating ICs. In this study, 39 *T. cruzi* antigens and 114 human autoantigens particular to patients with Chagas disease were identified. In this report, two *T. cruzi* antigens (surface protease GP63 and glucose-6-isomerase) and six human autoantigens (CD180 antigen, ceruloplasmin, fibrinogen beta chain, fibrinogen beta chain isoform 2 preprotein, isoform gamma-A of fibrinogen γ-chain, serum paraoxonase) were reported in more than 50% of the patients evaluated. Human isoforms short of complement factor H-related protein 2 and *T. cruzi* TS were most often found in the asymptomatic forms of Chagas disease (5/9 for both) compared with the determinate Chagas (0/11, P = 0·046 for human, 1/11, P = 0·0498 for *T. cruzi*). The authors conclude that the Chagas disease immune complexome may reflect the difference in immune status among different clinical stages of chronic form of *T. cruzi* infection [68].

In a successful translation of proteomic-based studies into accessible tools for bench diagnosis, Ruiz-Lancheros and coworkers [69] reported the value of apolipoprotein A-1 and fibronectin fragments as markers of parasitological cure for congenital Chagas disease in Bz-treated children [69].

Using a proteomics approach, researchers identified a *T. cruzi* protein phosphatase 2A in a cytosolic extract previously purified in a microcystin-Sepharose affinity column. This strategy enables the cloning and expression of this protein phosphatase, which is involved in *T. cruzi* differentiation from trypomastigotes to amastigotes [70]. A proteomics approach was also used to identify a *T. cruzi* alfa 5 proteasome subunit, previously recognized by the monoclonal antibody 7E5 [71].

9. Host-parasite interaction

To investigate changes in the cells' phosphoprotein pattern after infection, researchers used proteome profiler human phospho-kinase arrays to investigate the molecular mechanism of *T. cruzi*-induced alteration of the colon during the early infection phase of primary human colonic epithelial cells (HCoEpiCs) infected with *T. cruzi* trypomastigotes at different time points. Significant changes in the phosphorylation pattern that could be related to cellular deregulation in colonic epithelial cells after parasite infection were observed, including a significant increase in the levels of phosphorylated heat shock protein (p-HSP) 27 and transcription factors such as c-Jun and CREB. A significant upregulation of phospho (p-) Akt S473 and p-JNK, which may directly or indirectly modulate CREB and c-Jun phosphorylation, respectively, was also observed together with increased levels of phosphorylated CREB and c-Jun in the nucleus. In addition, p-c-Jun and p-CREB colocalized in the nucleus after 180 min of *T. cruzi* infection. A greater level of p-c-Jun and p-CREB has been associated with inflammatory and profibrotic responses. *T. cruzi* infection of HCoEpiCs provokes an enhanced expression of thrombospondin-1 (TSP-1), which is fibrogenic at higher levels. The authors also found that *T. cruzi* infection influenced the expression of some transcription factors, such as NF-kB and JAK2-STAT1, which can enhance proinflammatory flux. Bioinformatics studies of the phosphoprotein networks derived using the phosphoprotein information serve as a blueprint for *T. cruzi*-mediated cellular differentiation of HCoEpiCs in the lapse of the acute phase of parasite infection.

T. cruzi was shown to release vesicles containing a wide range of surface molecules to affect the host immune responses and cell invasion [72]. A proteomics study reported different compositions of extracellular vesicles released by two *T. cruzi* strains associated with their distinct interaction with host cells [73]. The secretomes of two distinct *T. cruzi* strains (Y and YuYu), which were previously shown to differentially modulate host immune responses, were compared. Extracellular vesicles (EVs) obtained from tissue culture-derived trypomastigotes of both strains were purified and quantitated. Proteomics studies confirmed the higher concentrations of proteins encoded by the trans-sialidase (TS) gene superfamily, MASPs, and several well-known exosomal proteins in the YuYu strain, which also have shown a significant difference among purified EVs and the vesicle-free fraction compared to the Y strain. In order to know if those differences were associated with *T. cruzi* infectivity, J774 phagocytic cells and LLC-MK2 non-phagocytic cells were previously incubated with isolated EVs from both strains and then infected with TCDTs from Y strain. EVs liberated by the YuYu strain provoked smaller levels of infection but an increased intracellular proliferation in J774 macrophages than EVs from the Y strain. On the other hand, YuYu strain-derived EVs induced greater levels of infection of LLC-MK2 cells than Y strain-derived EVs. In summary, the quantitative and qualitative differences in EVs and released proteins from distinct *T. cruzi* strains may be correlated with the parasite capacity to be infective and virulent [73].

The involvement of host cell-derived microvesicles (MVs) during the early interaction between *T. cruzi* MTs and THP-1 cells was previously demonstrated. To determine the contribution of different parasite stages and their extracellular vesicles in the interaction with host cells, researchers assessed TCDTs, MTs, and EPIs, which were shown to provoke distinct levels of MV liberated from THP-1 cells. Nevertheless, only MTs and TCDTs could enhance parasite cell invasion. Fluorescence resonance energy transfer microscopy showed that THP-1-derived MVs can fuse with parasite-derived MVs. Additionally, MVs released from the TCT-THP-1 interaction displayed a bigger fusogenic ability than those from the META- or EPI-THP-1 interaction. However, by proteomics, a greater concentration of proteins from MTs (25%) than TCDTs (12%) or EPIs (5%) was detected in MVs from the parasite-THP-1 interaction. Lastly, sera from patients with chronic Chagas disease at the indeterminate or cardiac phase were able to identify antigens in THP-1-derived MVs resulting only from interaction with META or TCDTs forms. Elucidation of the cellular traffic pathways and the role of MVs in modulation of host immune system could supply significant hints with respect to Chagas disease pathophysiology [74].

T. cruzi has two tryparedoxins (TcTXNI and TcTXNI) pertaining to the thioredoxin superfamily. TXNs are oxidoreductases that have a role in electron transfer between trypanothione and peroxiredoxins [75, 76]. This represents a difference with the mammalian cells, in which these roles are performed by thioredoxins. These differences make TXNs a new and interesting chemotherapeutic target. *T. cruzi* tryparedoxin 1 (TcTXN1) is a cytoplasmic and constitutively expressed protein in *T. cruzi*. To elucidate the *T. cruzi* redox interactome, an active-site mutant protein lacking the resolving cysteine was designed. Then, the *in vitro* complex formation among the mutated TcTXN1 and its natural partner, the cytoplasmic peroxiredoxin, was validated. By the expression of this mutant protein in parasites with an additional 6xHis-tag, heterodisulfide complexes were purified and identified by 2-DE/MS. This allowed us to identify 15 TcTXN1 proteins that were involved in 2 main processes: oxidative metabolism and protein synthesis and degradation. This type of experiment allowed the discovery of several putative TcTXN1-interacting proteins and contributed to a better understanding of the redox *T. cruzi* interactome [77]. In another work, the study was extended to TcTXNII. TcTXNII is a transmembrane protein anchored to the mitochondria surface and endoplasmic reticulum, with a cytoplasmic orientation of the redox domain. This protein is expressed during the metacyclogenesis process. To further characterize the redox interactome of *T. cruzi*, an active-site mutant TcTXNII lacking the resolving cysteine was designed, and through the expression of this mutant protein and incubation with *T. cruzi* proteins, heterodisulfide complexes were isolated and identified by mass spectrometry. This approach allowed the identification of 16 TcTXNII-interacting proteins that are involved in a wide range of cellular processes. Thus, this report indicated the relevance of TcTXNII and contributed to a better understanding of the *T. cruzi* redox interactome [77].

TcI isolates of *T. cruzi* exploit the antioxidant network for enhanced intracellular survival in macrophages and virulence in mice. In these reports, the TcI strains Colombiana (COL) and SylvioX10/4 (SYL) and a cultured clone (TCC) displayed distinct biological features in an experimental murine infection, including high parasitemia and symptomatic cardiomyopathy (SYL), low parasitemia and high tissue tropism (COL), and no pathogenicity (TCC). A proteomic study of the EPI and TCDT stages by 2-DE and MALDI-TOF MS, followed by functional annotation of the differential proteome data sets (≥twofold change, $P < 0.05$) demonstrated that (i) many proteins were engaged in cytoskeletal assembly and remodeling essential for flagellar wave frequency and amplitude and forward motility of the parasite and (ii) the parasite-specific antioxidant network was improved in COL and SYL (versus TCC) trypomastigotes. Immunoblots reveal the increased protein levels of cytoplasmic and mitochondrial tryparedoxin peroxidases and their substrate (tryparedoxin) and iron superoxide dismutase in COL and SYL (versus TCC) trypomastigotes. In addition, COL and SYL (but not TCC) were resistant to experimental treatment with oxidants such as H_2O_2 and peroxynitrite [ONOO(−)]. These parasite isolates were able to resist the intracellular superoxide and nitric oxide response in macrophages and were not killed by them. These observations indicate that protein expression focused on increased motility and control of macrophage-derived free radicals may represent a survival and persistence strategy developed by TcI isolates of *T. cruzi* [78].

10. *T. cruzi* typing

T. cruzi strains have been allocated to seven genetic groups (TcI-TcVI and TcBat), called discrete typing units (DTUs), which constitute groups of parasite isolates that differ in geographical origins, virulence, pathogenicity, and immunological properties [79]. In addition, several clinical manifestations (from asymptomatic to extremely severe illness) have been intended to be linked to *T. cruzi* genetic variability. For this reason, many DTU typing approaches have been proposed. Recently, a *T. cruzi* strain typing assay using MS2 peptide spectral libraries (Tc-STAMS2) was developed [80]. The Tc-STAMS2 methodology used shotgun proteomics associated with a spectral library quest to assign and discriminate *T. cruzi* strains independent of genome information. This approach was based on the built of a library of MS/MS peptide spectra constructed using genotyped *T. cruzi* reference strains. For identification, the MS/MS peptide spectra of unheard *T. cruzi* isolated were identified by the spectral matching algorithm SpectraST. The Tc-STAMS2 approach enabled right identification of all DTUs with high reliance. The approach was robust toward distinct sample preparations, length of chromatographic gradients, and fragmentation techniques.

The Tc-STAMS2 procedure allowed *T. cruzi* strain typing using MS/MS spectra as discriminatory functions and allowed the differentiation of TcI-TcVI DTUs. Similar to genomic-based procedures, the Tc-STAMS2 method permits identification of strains within DTUs. Its robustness toward different experimental and biological variables makes it an invaluable complementary methodology to the existing *T. cruzi* genotyping assays. Furthermore, this approach may be applied to identify DTU-specific properties correlated with the strain phenotype [73].

A proteomics analysis of two *T. cruzi* zymodeme 3 strains was also performed. In fact, two *T. cruzi* Z3 strains, designated 3663 and 4167, isolated from Brazilian Amazon triatomines were studied. These strains exhibited different infection patterns for Vero cells, in which 3663 trypomastigotes were much less infective than 4167 trypomastigotes. By proteomics, the differences in the global protein expression profiles of these two Z3 strains were investigated. Two-dimensional (2D) protein maps were obtained, and some spots were identified by MS. The study showed a meaningful difference in the expression profiles of several proteins in both strains. One of them was cruzipain, a key virulence factor. These data were corroborated by flow cytometric analysis using an anti-cruzipain antibody. This difference could contribute to the infectivity profiles observed for each strain by *in vitro* assays using different cell lines [81].

A phylogenetic character mapping of 26 stocks of *T. cruzi* and 2 stocks of the sister taxon *T. cruzi marinkellei* was performed to test for possible associations between *T. cruzi*-subspecific phylogenetic diversity and levels of protein expression, as examined by proteomics analysis and MS. A strong correlation ($P < 10(-4)$) was observed between genetic distance, as established by multilocus enzyme electrophoresis, and proteomic dissimilarities estimated by proteomic Euclidian distances. Several proteins were found to be specifically associated with *T. cruzi* DTUs. This study explored the previously uncharacterized links between *T. cruzi* infraspecific phylogenetic diversity and gene expression [82].

11. Immune response

One major problem in developing a rational search for *T. cruzi* vaccine candidates is the restricted information with respect to the proteins that are expressed by different strains, clones, and isolates included in the different parasite DTUs.

A proteomics study of *T. cruzi* trypomastigotes identified 1448 nonredundant proteins, of which close to 14% of the found peptidic sequences corresponded to surface proteins, mostly of which were GPI-anchored and associated with *T. cruzi* virulence. Immunoinformatics approaches reported a high number of peptides with predicted high-binding capacity for class I and class II molecules of the major histocompatibility complex (MHC). The many different types of proteins exposed on the trypomastigote surface membrane could have several consequences for parasite cell invasion and evasion of the immune response. Finally, a rational survey to identify potential T-cell epitopes that may be evaluated and validated to develop a prophylactic vaccine to protect against *T. cruzi* infection was performed [21].

T. cruzi BT proteins were immunoprecipitated using serum antibodies obtained from asymptomatic or cardiac (stages B1 and C) patients with different forms of the disease and from healthy people used as controls. Proteins recognized by sera from Chagas disease patients were identified and quantified by MS, and changes in the recognition patterns were further evaluated. Compared to asymptomatic samples, IgG from stage C patients mainly identify the I/6 autoantigen, whereas IgG from B1 patients resulted in a greater recognition of dihydrolipoamide acetyltransferase precursor, calpain cysteine peptidase, and two variants of CAP5.5. In this paper, CAP5.5 identification by human serum immunoglobulin from patients with early myocarditis resulted in a 23-fold abundance variation when compared to sera obtained from asymptomatic patients. These results strongly suggest the role of CAP5.5 in Chagas disease cardiomyopathy.

The lack of appropriate biomarkers to assess treatments and evolution of this disease is the main limitation for clinical trials and patient medical evaluation. Accordingly, these observations may contribute to a better comprehension of *T. cruzi* pathogenesis and to evaluating potential candidates for developing vaccine and diagnostic tests, in addition to the clinical applicability of suitable biomarkers for patient follow-up and prognosis [83].

The major protein antigens targeted by the diverse antibodies of *T. cruzi*-infected mice were studied. To detect global IgG antibody specificities, sera from infected mice were immunoblotted against whole *T. cruzi* extracts. The most immunogenic *T. cruzi* proteins were identified by proteomics using MALDI-TOF-TOF. Among these proteins, pyruvate phosphate dikinase, Hsp-85, and β-tubulin were recognized by mouse IgG as the major protein bands. The *T. cruzi* β-tubulin gene was cloned and expressed in *Escherichia coli*, and recombinant *T. cruzi* β-tubulin was used to immunize mice. Immunized animals increased their specific IgG reactivity and were protected against *T. cruzi* infection. These results strongly suggest that proteomics, used for repertoire analysis, is a valid tool to identify protective antigens that could be vaccine candidates to protect against *T. cruzi* infection [84].

12. Concluding remarks and future directions

T. cruzi infection is a major health issue in America and also in the rest of the world, where around 8 million people are infected by *T. cruzi* and the other 28 million are exposed to the parasite infection. Even more, ~20,000 people die each year, mainly as a result of cardiac problem. Currently, immunoprophylactic strategies have not been developed, and chemotherapy is so far from optimal, based on two old drugs, Nf and Bz. However, these drugs are more active in the acute form of Chagas disease and display several side effects.

In the last 15 years, a considerable number of proteomics studies have contributed to a better understanding of the biology and biochemistry of *T. cruzi*. In fact, the proteomes of the different life cycle stages of the parasite have been described, including the proteomes of different zymodemes and different parasite DTUs. Using a typical experimental workflow as described in **Figure 1**, many different proteomics strategies have been reported. These allowed to determine the composition of different organelles, such as ribosomes, contractile vacuoles, reservosomes, and nuclear protein contents. These proteomics analyses have contributed to the knowledge of the products excreted/secreted by the parasite and have also allowed us to identify the contents of the microvesicles released by different parasite stages and their interaction with host cells. The study of post-translational modifications has also been possible and has contributed to the knowledge of parasite phosphorylation, glycosidation, myristoylation, SUMOylation, histone acetylation, and methylation. Proteomics has also helped elucidate critical biological processes, such as metacyclogenesis, amastigogenesis, and epimastigogenesis. The pathways involved in resistance to Bz and in the action mode of napthoimidazoles isolated from β-lapachone have been studied by proteomics approaches. Interactomics studies that demonstrated typing using M2 peptide spectral libraries and identification of cure markers are also among the major achievements of proteomics in the research field of *T. cruzi* and Chagas disease. In all of these cases, special care was taken in sample preparation, and it was always cleaned using "Stage tip" (or other close related system) in order to remove sample contaminants. In some experiments quantitative proteomics was performed using label-free methods, while in others labeling was performed using iTRAQ, ICAT, SILAC, or others. In all cases, digested proteins were submitted to liquid chromatography systems directly interfaced to mass spectrophotometers (**Figure 1**).

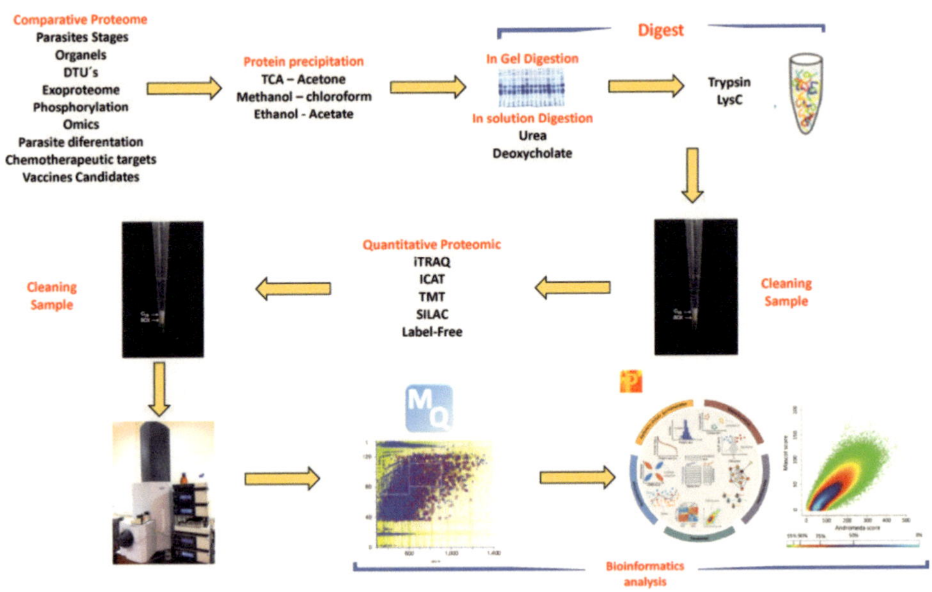

Figure 1.
Flow diagram of the main events in Trypanosoma cruzi proteomics. Qualitative and quantitative analyses using different parasite stages can be performed. Then, the proteins can be precipitated or separated by SDS-PAGE. Proteins are digested, in the gel or in solution using proteases, such as trypsin and Lys C. The peptides are subjected to a cleaning phase, and the peptides can be labeled using techniques such as ICAT, ITRAQ, and SILAC or not be labeled, using label-free quantification (LFQ). Finally, the samples are subjected to a ultrahigh-pressure liquid chromatography (UHPLC) associated to mass spectrometry (MS). The information obtained is submitted to bioinformatics analysis using software such as MaxQuant (or others), for quantitative proteomics analysis, and then by Andromeda to obtain the sequence of the peptides in the different databases and Perseus (or others) to obtain classifications, proteomics interactions, post-transductional modifications, and metabolic pathways.

However, we still expect major contributions of proteomics in many aspects of parasite biology and in various clinical aspects. The subproteomes of acidocalcisomes and the kinetoplast-mitochondrion complex could reveal important information about parasite biology.

In addition, the identification of new biomarkers with diagnostic and prognostic value and for established cure criteria is needed, and proteomics would contribute to this field. In the same way, the identification and validation of new chemotherapeutic targets to develop new and more efficient drugs and the identification of new potential vaccine candidates could also be an expected contribution.

Furthermore, in the future, considering the concept of "precision medicine" or "personalized medicine," proteomics will offer the opportunity to characterize strains isolated from infected patients, identify its DTU, and define appropriate treatments and perhaps appropriate immunoprophylaxis. Moreover, the identification of human sera biomarkers that allow us to predict which patients with the undetermined forms of Chagas disease show cardiomyopathy or digestive tract involvement would be a relevant contribution of clinical proteomics.

Recently, San Francisco et al. [85] reported that the *in vitro* invasion capacity and *in vivo* Balb/c mouse infectiveness of a highly virulent *T. cruzi* cell line were strongly reduced by parasite pretreatment with antisense oligonucleotides targeting trans-sialidase or complement regulatory proteins or with E64d to inhibit cruzipain activity. However, the parasite infectivity was not completely blocked, suggesting that other players, different from these proteins, could be involved in *T. cruzi* virulence. Thus, for comparison of the proteomic profiles of highly virulent and low-virulent cell lines of *T. cruzi*, the molecular bases of *T. cruzi* virulence will be established. For the *T. cruzi* secretome, different interactions among *T. cruzi* secreted proteins and host cell signaling pathways have been proposed [86]. However, the role of these proteins in virulence and modulation of the host immune system has not been shown experimentally. *T. cruzi* secretome studies and immunomics analyses using acute sera of animals or infected patients in the acute phase of the infection would help elucidate the mechanisms that *T. cruzi* uses to subvert the host immune response.

Support

We thank the Seedlings Grant (J.G) and Bridge Found for Research of Excellence Grant (J.G.) and Research Seedling Project SEM-17-02 (J.G.), University of Antofagasta.

Author details

Juan San Francisco, Bessy Gutiérrez and Jorge González*
Molecular Parasitology Unit, Medical Technology Department,
University of Antofagasta, Antofagasta, Chile

*Address all correspondence to: jorge.gonzalez@uantof.cl

References

[1] WHO. WHO. 2018. Available from: http://www.who.int/mediacentre/factsheets/fs340/es/

[2] Tyler KM, Engman DM. The life cycle of *Trypanosoma cruzi* revisited. International Journal for Parasitology. 2001;**31**:472-481

[3] Macedo AM, Machado CR, Oliveira RP, Pena SDJ. *Trypanosoma cruzi*: Genetic structure of populations and relevance of genetic variability to the pathogenesis of chagas disease. Memórias do Instituto Oswaldo Cruz. 2004;**99**:1-12

[4] Zingales B, Andrade SG, Briones MRS, Campbell DA, Chiari E, Fernandes O, et al. A new consensus for *Trypanosoma cruzi* intraspecific nomenclature: Second revision meeting recommends TcI to TcVI. Memórias do Instituto Oswaldo Cruz. 2009;**104**:1051-1054

[5] Zingales B. *Trypanosoma cruzi* genetic diversity: Something new for something knownabout Chagas disease manifestations, serodiagnosis and drug sensitivity. Acta Tropica. 2018;**184**:38-52

[6] Zhao YY, Lin RC. UPLC-MSE application in disease biomarker discovery: The discoveries in proteomics to metabolomics. Chemico-Biological Interactions. 2014;**215**:7-16. DOI: 10.1016/j.cbi.2014.02.014

[7] Humphery-Smith L, Blackstock W. Proteome analysis: Genomics via the output rather than the input code. Journal of Protein Chemistry. 1997;**6**:537-544. DOI: 10.1023/A:1026330015280

[8] Drissi R, Dubois ML, Boisvert FM. Proteomics methods for subcellular proteome analysis. The FEBS Journal. 2013;**280**:5626-5634. DOI: 10.1111/febs.12502

[9] Larance M, Lamond AI. Multidimensional proteomics for cell biology. Nature Reviews. Molecular Cell Biology. 2015;**16**:269-280. DOI: 10.1038/nrm3970

[10] Hristova VA, Chan DW. Cancer biomarker discovery and translation: Proteomics and beyond. Expert Review of Proteomics. 2019;**16**:93-103. Epub ahead of print

[11] Monti C, Zilocchi M, Colugnat I, Alberio T. Proteomics turns functional. Journal of Proteomics. 2019;**198**:36-44. DOI: 10.1016. Epub ahead of print

[12] Karas M, Hillenkamp F. Laser desorption ionization of proteins with molecular masses exceeding 10 000 Daltons. Analytical Chemistry. 1988;**60**:2299-2301. DOI: 10.1021/ac00171a028

[13] Fenn JB, Mann M, Meng CK, Wong SF, Whitehouse CM. Electrospray ionization for mass spectrometry of large biomolecules. Science. 1989;**246**:64-71. DOI: 10.1126/science.2675315

[14] Araújo PR, Teixeira SM. Regulatory elements involved in the post-transcriptional control of stage-specific gene expression in *Trypanosoma cruzi*—A review. Memórias do Instituto Oswaldo Cruz. 2011;**106**:257-266. DOI: 10.1590/S0074-02762011000300002

[15] Clayton CE. Gene expression in Kinetoplastids. Current Opinion in Microbiology. 2016;**32**:46-51. DOI: 10.1016/j.mib.2016.04.018

[16] Parodi-Talice A, Durán R, Arrambide N, Prieto V, Piñeyro MD, Pritsch O, et al. Proteome analysis of the causative agent of Chagas disease: *Trypanosoma cruzi*. International Journal for Parasitology. 2004;**34**:881-886

[17] Paba J, Santana JM, Teixeira ARL, Fontes W, Sousa MV, Ricart CAO. Proteomic analysis of the human pathogen *Trypanosoma cruzi*. Proteomics. 2004;**4**:1052-1059

[18] Paba J, Ricart CAO, Fontes W, Santana JM, Teixeira ARL, Marchese J, et al. Proteomic analysis of *Trypanosoma cruzi* developmental stages using isotope-coded affinity tag reagents. Journal of Proteome Research. 2004;**3**:517-524

[19] Atwood JA, Weatherly DB, Minning TA, Bundy B, Cavola C, Opperdoes FR, et al. The *Trypanosoma cruzi* proteome. Science. 2005;**309**:473-476

[20] Magalhaes A, Charneau S, Paba J, Guércio R, Teixeira A, Santana J, et al. *Trypanosoma cruzi* alkaline 2-DE: Optimization and application to comparative proteome analysis of flagellate life stages. Proteome Science. 2008;**6**:24

[21] Nakayasu ES, Sobreira TJP, Torres R, Ganiko L, Oliveira PSL, Marques AF, et al. Improved proteomic approach for the discovery of potential vaccine targets in *Trypanosoma cruzi*. Journal of Proteome Research. 2012;**11**:237-246

[22] Queiroz RML, Charneau S, Mandacaru SC, Schwämmle V, Lima BD, Roepstorff P, et al. Quantitative proteomic and phosphoproteomic analysis of *Trypanosoma cruzi* amastigogenesis. Molecular & Cellular Proteomics. 2014;**13**:3457-3472

[23] Brunoro GVF, Caminha MA, Ferreira AT d S, Leprevost F d V, Carvalho PC, Perales J, et al. Reevaluating the *Trypanosoma cruzi* proteomic map: The shotgun description of bloodstream trypomastigotes. Journal of Proteomics. 2015;**115**:58-65

[24] Cordero EM, Nakayasu ES, Gentil LG, Yoshida N, Almeida IC, Da Silveira JF. Proteomic analysis of detergent-solubilized membrane proteins from insect-developmental forms of *Trypanosoma cruzi*. Journal of Proteome Research. 2009;**7**:3642-3652

[25] Sodré CL, Chapeaurouge AD, Kalume DE, De Mendonça Lima L, Perales J, Fernandes O. Proteomic map of *Trypanosoma cruzi* CL Brener: The reference strain of the genome project. Archives of Microbiology. 2009;**19**:117-184. DOI: 10.1007/s00203-008-0439-6

[26] Magalhaes AD, Queiroz RML, Bastos IMD, Santana JM, Sousa MV, Ricart CAO, et al. Comparative two-dimensional gel electrophoresis of *Trypanosoma cruzi* mammalian-stage forms in an alkaline pH range. Protein and Peptide Letters. 2015;**22**:1066-1075

[27] Soblik H, Younis AE, Mitreva M, Renard BY, Kirchner M, Geisinger F, et al. Life cycle stage-resolved proteomic analysis of the excretome/secretome from *Strongyloides ratti*—Identification of stage-specific proteases. Molecular & Cellular Proteomics. 2011;**12**:M111.010157. DOI. 10.107/M111.010157

[28] Harnett W. Secretory products of helminth parasites as immunomodulators. Molecular and Biochemical Parasitology. 2014;**195**:130-136. DOI: 10.1016/j.molbiopara.2014.03.007

[29] Tjalsma H, Antelmann H, Jongbloed JDH, Braun PG, Darmon E, Dorenbos R, et al. Proteomics of protein secretion by *Bacillus subtilis*: Separating the "secrets" of the Secretome. Microbiology and Molecular Biology Reviews. 2004;**68**:207-233. DOI: 10.1128/MMBR.68.2.207-233.2004

[30] Armengaud J, Christie-Oleza JA, Clair G, Malard V, Duport C. Exoproteomics: Exploring the world around biological systems. Expert Review of Proteomics. 2012;**9**:561-575

[31] Bayer-Santos E, Aguilar-Bonavides C, Rodrigues SP, Cordero EM, Marques AF, Varela-Ramirez A, et al. Proteomic analysis of *Trypanosoma cruzi* secretome: Characterization of two populations of extracellular vesicles and soluble proteins. Journal of Proteome Research. 2013;**12**:883-897

[32] Queiroz RML, Ricart CAO, Machado MO, Bastos IMD, Santana JM d, Sousa MV d, et al. Insight into the exoproteome of the tissue-derived Trypomastigote form of *Trypanosoma cruzi*. Frontiers in Chemistry. 2016;**4**:1-10

[33] Brossas JY, Gulin JEN, Bisio MMC, Chapelle M, Marinach-Patrice C, Bordessoules M, et al. Secretome analysis of *Trypanosoma cruzi* by proteomics studies. PLoS One. 2017;**12**:e0185504. DOI: 10.1371/journal.pone.0185504

[34] Martins NO, Souza RT d, Cordero EM, Maldonado DC, Cortez C, Marini MM, et al. Molecular characterization of a novel family of *Trypanosoma cruzi* surface membrane proteins (TcSMP) involved in mammalian host cell invasion. PLoS Neglected Tropical Diseases. 2015;**9**(11):e0004216. DOI: 10.1371/journal.pntd.0004216

[35] Sant'Anna C, Nakayasu ES, Pereira MG, Lourenço D, De Souza W, Almeida IC, et al. Subcellular proteomics of *Trypanosoma cruzi* reservosomes. Proteomics. 2009;**9**:1782-1794. DOI: 10.1002/pmic.200800730

[36] Avila CC, Mule SN, Rosa-Fernandes L, Viner R, Barisón MJ, Costa-Martins AG, et al. Proteome-wide analysis of *Trypanosoma cruzi* exponential and stationary growth phases reveals a subcellular compartment-specific regulation. Genes (Basel). 2018;**9**(8). pii: E413. DOI: 10.3390/genes9080413

[37] Leandro de Jesus TC, Calderano SG, Vitorino FN d L, Llanos RP, Lopes M d C, de Araújo CB, et al. Quantitative proteomic analysis of replicative and nonreplicative forms reveals important insights into chromatin biology of *Trypanosoma cruzi*. Molecular & Cellular Proteomics. 2017;**16**:23-38. DOI: 10.1074/mcp.M116.061200

[38] dos Santos Júnior Ade C, Kalume DE, Camargo R, Gómez-Mendoza DP, Correa JR, Charneau S, et al. Unveiling the *Trypanosoma cruzi* nuclear proteome. PLoS One. 2015;**10**(9):e0138667. DOI: 10.1371/journal.pone.0138667

[39] Ulrich PN, Jimenez V, Park M, Martins VP, Atwood J, Moles K, et al. Identification of contractile vacuole proteins in *Trypanosoma cruzi*. PLoS One. 2011;**6**:e18013. DOI: 10.1371/journal.pone.0018013

[40] Ayub MJ, Atwood J, Nuccio A, Tarleton R, Levin MJ. Proteomic analysis of the *Trypanosoma cruzi* ribosomal proteins. Biochemical and Biophysical Research Communications. 2009;**382**:30-34. DOI: 10.1016/j.bbrc.2009.02.095

[41] Koyama M, Kurumizaka H. Structural diversity of the nucleosome. Journal of Biochemistry. 2018;**163**:85-95. DOI: 10.1093/jb/mvx081

[42] Picchi GFA, Zulkievicz V, Krieger MA, Zanchin NT, Goldenberg S, De Godoy LMF. Post-translational modifications of *Trypanosoma cruzi* canonical and variant histones. Journal of Proteome Research. 2017;**16**:1167-1179. DOI: 10.1021/acs.jproteome.6b00655

[43] Alves LR, Ávila AR, Correa A, Holetz FB, Mansur FCB, Manque PA, et al. Proteomic analysis reveals the dynamic association of proteins with translated mRNAs in *Trypanosoma cruzi*. Gene. 2010;**452**(2):72-78. DOI: 10.1016/j.gene.2009.12.009

[44] Moretti NS, Cestari I, Anupama A, Stuart K, Schenkman S. Comparative proteomic analysis of lysine acetylation in trypanosomes. Journal of Proteome Research. 2018;**17**:374-385. DOI: 10.1021/acs.jproteome.7b00603

[45] De Jesus TCL, Nunes VS, Lopes MDC, Martil DE, Iwai LK, Moretti NS, et al. Chromatin proteomics reveals variable histone modifications during the life cycle of *Trypanosoma cruzi*. Journal of Proteome Research. 2016;**15**:2039-2051. DOI: 10.1021/acs.jproteome.6b00208

[46] de Oliveira Santos J, Zuma AA, de Luna Vitorino FN, da Cunha JPC, de Souza W, Motta MCM. Trichostatin A induces *Trypanosoma cruzi* histone and tubulin acetylation: Effects on cell division and microtubule cytoskeleton remodelling. Parasitology. 2018;**146**:543-552. DOI: 10.1017/S0031182018001828. Epub ahead of print

[47] Roberts AJ, Fairlamb AH. The N-myristoylome of *Trypanosoma cruzi*. Scientific Reports. 2016;**6**:31078. DOI: 10.1038/srep31078

[48] Atwood JA 3rd, Minning T, Ludolf F, Nuccio A, Weatherly DB, Alvarez-Manilla G, et al. Glycoproteomics of *Trypanosoma cruzi* trypomastigotes using subcellular fractionation, lectin affinity, and stable isotope labeling. Journal of Proteome Research. 2006;**5**(12):3376-3384. DOI: 10.1021/pr060364b

[49] Alves MJ, Kawahara R, Viner R, Colli W, Mattos EC, Thaysen-Andersen M, et al. Comprehensive glycoprofiling of the epimastigote and trypomastigote stages of *Trypanosoma cruzi*. Journal of Proteomics. 2017;**151**:182-192. DOI: 10.1016/j.jprot.2016.05.034

[50] Ovaa H, Vertegaal ACO. Probing ubiquitin and SUMO conjugation and deconjugation. Biochemical Society Transactions. 2018;**46**:423-436. DOI: 10.1042/BST20170086

[51] Bayona JC, Nakayasu ES, Laverrière M, Aguilar C, Sobreira TJP, Choi H, et al. SUMOylation pathway in *Trypanosoma cruzi*: Functional characterization and proteomic analysis of target proteins. Molecular & Cellular Proteomics. 2011;**10**(12):M110.007369. DOI: 10.1074/mcp.M110.007369

[52] Nakayasu ES, Gaynor MR, Sobreira TJP, Ross JA, Almeida IC. Phosphoproteomic analysis of the human pathogen *Trypanosoma cruzi* at the epimastigote stage. Proteomics. 2009;**9**(13):3489-3506. DOI: 10.1002/pmic.200800874

[53] Ferrão PM, de Oliveira FL, Degrave WM, Araujo-Jorge TC, Mendonça-Lima L, Waghabi MC. A phosphoproteomic approach towards the understanding of the role of TGF-β in *Trypanosoma cruzi* biology. PLoS One. 2012;**7**:e38736. DOI: 10.1371/journal.pone.0038736

[54] de Godoy LMF, Marchini FK, Pavoni DP, Rampazzo R d CP, Probst CM, Goldenberg S, et al. Quantitative proteomics of *Trypanosoma cruzi* during metacyclogenesis. Proteomics. 2012;**12**:2694-2703

[55] Parodi-Talice A, Monteiro-Goes V, Arrambide N, Avila AR, Duran R, Correa A, et al. Proteomic analysis of metacyclic trypomastigotes undergoing *Trypanosoma cruzi* metacyclogenesis. Journal of Mass Spectrometry. 2007;**42**:1422-1432

[56] Marchini FK, de Godoy LMF, Rampazzo RCP, Pavoni DP, Probst CM, Gnad F, et al. Profiling the *Trypanosoma cruzi* phosphoproteome. PLoS One. 2011;**6**(9):e25381. DOI: 10.1371/journal.pone.0025381

[57] Rondinelli E, Silva R, de Oliveira Carvalho J, de Almeida

Soares CM, de Carvalho EF, de Castro FT. *Trypanosoma cruzi*: An *in vitro* cycle of cell differentiation in axenic culture. Experimental Parasitology. 1988;**66**:197-202. DOI: 10.1016/0014-4894(88)90091-4

[58] Weisburg JH, Curcio M, Caron PC, Raghu G, Mechetner EB, Roepe PD, et al. The multidrug resistance phenotype confers immunological resistance. The Journal of Experimental Medicine. 1996;**183**:2699-2704. DOI: 10.1084/jem.183.6.2699

[59] Nakamura S, Shchepetov M, Dalia AB, Clark SE, Murphy TF, Sethi S, et al. Molecular basis of increased serum resistance among pulmonary isolates of non-typeable *Haemophilus influenzae*. PLoS Pathogens. 2011;7(1):e1001247. DOI: 10.1371/journal.ppat.1001247

[60] Kessler RL, Contreras VT, Marliére NP, Aparecida Guarneri A, Villamizar Silva LH, Mazzarotto GACA, et al. Recently differentiated epimastigotes from *Trypanosoma cruzi* are infective to the mammalian host. Molecular Microbiology. 2017;**104**:712-736. DOI: 10.1111/mmi.13653

[61] Roberts SB, Robichaux JL, Chavali AK, Manque PA, Lee V, Lara AM, et al. Proteomic and network analysis characterize stage-specific metabolism in *Trypanosoma cruzi*. BMC Systems Biology. 2009;**16**(3):52. DOI: 10.1186/1752-0509-3-52

[62] Pérez-Morales D, Lanz-Mendoza H, Hurtado G, Martínez-Espinosa R, Espinoza B. Proteomic analysis of *Trypanosoma cruzi* epimastigotes subjected to heat shock. Journal of Biomedicine & Biotechnology. 2012;**2012**:902803. DOI: 10.1155/2012/902803

[63] Ferella M, Nilsson D, Darban H, Rodrigues C, Bontempi EJ, Docampo R, et al. Proteomics in *Trypanosoma cruzi*-localization of novel proteins to various organelles. Proteomics. 2008;**8**:2735-2749. DOI: 10.1002/pmic.200700940

[64] Vieira GAL, Silva MTA d, Regasini LO, Cotinguiba F, Laure HJ, Rosa JC, et al. *Trypanosoma cruzi*: Analysis of two different strains after piplartine treatment. Brazilian Journal of Infectious Diseases. 2018;**22**:208-218. DOI: 10.1016/j.bjid.2018.02.009

[65] Brunoro G, Faça V, Caminha M, Ferreira A, Trugilho M, de Moura K, et al. Differential gel electrophoresis (DIGE) evaluation of naphthoimidazoles mode of action: A study in *Trypanosoma cruzi* bloodstream Trypomastigotes. PLoS Neglected Tropical Diseases. 2016;**8**:e0004951

[66] Menna-Barreto RFS, Beghini DG, Ferreira ATS, Pinto AV, De Castro SL, Perales J. A proteomic analysis of the mechanism of action of naphthoimidazoles in *Trypanosoma cruzi* epimastigotes *in vitro*. Journal of Proteomics. 2010;**73**:2306-2315. DOI: 10.1016/j.jprot.2010.07.002

[67] Andrade HM, Murta SMF, Chapeaurouge A, Perales J, Nirdé P, Romanha AJ. Proteomic analysis of *Trypanosoma cruzi* resistance to benznidazole. Journal of Proteome Research. 2008;**7**:2357-2367. DOI: 10.1021/pr700659m

[68] Ohyama K, Huy NT, Yoshimi H, Kishikawa N, Nishizawa JE, Roca Y, et al. Proteomic profile of circulating immune complexes in chronic Chagas disease. Parasite Immunology. 2016;**38**:609-617. DOI: 10.1111/pim.12341

[69] Ruiz-Lancheros E, Rasoolizadeh A, Chatelain E, Garcia-Bournissen F, Moroni S, Moscatelli G, et al. Validation of apolipoprotein A-1 and fibronectin fragments as markers of parasitological

cure for congenital chagas disease in children treated with benznidazole. Open Forum Infectious Diseases. 2018;**5**:ofy236. DOI: 10.1093/ofid/ofy236

[70] González J, Cornejo A, Santos MRM, Cordero EM, Gutiérrez B, Porcile P, et al. A novel protein phosphatase 2A (PP2A) is involved in the transformation of human protozoan parasite *Trypanosoma cruzi*. The Biochemical Journal. 2003;**374**:647-656

[71] Gutiérrez B, Osorio L, Motta MCM, Huima-Byron T, Erdjument-Bromage H, Muñoz C, et al. Molecular characterization and intracellular distribution of the alpha 5 subunit of *Trypanosoma cruzi* 20S proteasome. Parasitology International. 2009;**58**:367-374

[72] de Pablos Torró LM, Moreira LR, Osuna A. Extracellular vesicles in chagas disease: A new passenger for an old disease. Frontiers in Microbiology. 2018;**9**:1140. DOI: 10.3389/fmicb.2018.01190

[73] Ribeiro KS, Vasconcellos CI, Soares RP, Mendes MT, Ellis CC, Aguilera-Flores M, et al. Proteomic analysis reveals different composition of extracellular vesicles released by two *Trypanosoma cruzi* strains associated with their distinct interaction with host cells. Journal of Extracellular Vesicles. 2018;**7**(1):1463779. DOI: 10.1080/20013078.2018.1463779

[74] Ramirez MI, Deolindo P, de Messias-Reason IJ, Arigi EA, Choi H, Almeida IC, et al. Dynamic flux of microvesicles modulate parasite–host cell interaction of *Trypanosoma cruzi* in eukaryotic cells. Cellular Microbiology. 2017;**19**(4). DOI: 10.1111/cmi.12672

[75] Piacenza L, Peluffo G, Alvarez MN, Kelly JM, Wilkinson SR, Radi R. Peroxiredoxins play a major role in protecting *Trypanosoma cruzi* against macrophage- and endogenously-derived peroxynitrite. The Biochemical Journal. 2008;**410**:359-368

[76] Piacenza L, Zago MP, Peluffo G, Alvarez MN, Basombrio MA, Radi R. Enzymes of the antioxidant network as novel determiners of *Trypanosoma cruzi* virulence. International Journal for Parasitology. 2009;**39**:1455-1464

[77] Arias DG, Piñeyro MD, Iglesias AA, Guerrero SA, Robello C. Molecular characterization and interactome analysis of *Trypanosoma cruzi* tryparedoxin II. Journal of Proteomics. 2015;**120**:95-104. DOI: 10.1016/j.jprot.2015.03.001

[78] Zago M, Hosakote Y, Koo S, Dhiman M, Piñeyro M, Parodi-Talice A, et al. TcI isolates of *Trypanosoma cruzi* exploit the antioxidant network for enhanced intracellular survival in macrophages and virulence in mice. Infection and Immunity. 2016;**6**:1842-1856

[79] Zingales B, Miles MA, Campbell DA, Tibayrenc M, Macedo AM, Teixeira MMG, et al. The revised *Trypanosoma cruzi* subspecific nomenclature: Rationale, epidemiological relevance and research applications. Infection, Genetics and Evolution. 2012;**12**:240-253

[80] de Oliveira GS, Kawahara R, Rosa-Fernandes L, Mule SN, Avila CC, Teixeira MMG, et al. Development of a *Trypanosoma cruzi* strain typing assay using MS2 peptide spectral libraries (Tc-STAMS2). PLoS Neglected Tropical Diseases. 2018;**12**:00006351. DOI: 10.1371/journal.pntd.0006351

[81] Kikuchi SA, Sodré CL, Kalume DE, Elias CGR, Santos ALS, de Nazaré Soeiro M, et al. Proteomic analysis of two *Trypanosoma cruzi* zymodeme 3 strains. Experimental Parasitology. 2010;**126**:540-551

[82] Telleria J, Biron DG, Brizard J-P, Demettre E, Seveno M, Barnabe C, et al. Phylogenetic character mapping of proteomic diversity shows high correlation with subspecific phylogenetic diversity in *Trypanosoma cruzi*. Proceedings of the National Academy of Sciences. 2010;**107**:20411-20416. DOI: 10.1073/pnas.1015496107

[83] Caminha MA, de Lorena VMB, de Oliveira W, Perales J, Carvalho PC, Lima DB, et al. *Trypanosoma cruzi* immunoproteome: Calpain-like CAP5.5 differentially detected throughout distinct stages of human Chagas disease cardiomyopathy. Journal of Proteomics. 2018;**194**:179-190

[84] Montalvão F, Nascimento DO, Nunes MP, Koeller CM, Morrot A, Lery LMS, et al. Antibody repertoires identify β-tubulin as a host protective parasite antigen in mice infected with *Trypanosoma cruzi*. Frontiers in Immunology. 2018;**9**:671. DOI: 10.3389/fimmu.2018.00671

[85] San Francisco J, Barría I, Gutiérrez B, Neira I, Muñoz C, Sagua H, et al. Decreased cruzipain gp85/trans-sialidase family protein expression contributes to loss of *Trypanosoma cruzi* trypomastigote virulence. Microbes and Infection. 2017;**19**:55-61. DOI: 10.1016

[86] Watanabe Costa R, da Silveira JF, Bahia D. Interactions between *Trypanosoma cruzi* secreted proteins and host cell signaling pathways. Frontiers in Microbiology. 2016;7:338. DOI: 10.3389/fmicb.2016.00388

Chapter 6

Biology of the *Trypanosoma cruzi* Genome

Luisa Berná, Sebastián Pita, María Laura Chiribao, Adriana Parodi-Talice, Fernando Alvarez-Valin and Carlos Robello

Abstract

The genome of *Trypanosoma cruzi* was first made available in 2005, and the intrinsic genome complexity of this parasite has hindered high-quality genome assembly and annotation. Recent technological developments in long read sequencing allowed to circumvent this problem, showing very interesting features in the genome architecture of *T. cruzi*, allowing to accurately estimate gene copy numbers, abundance and distribution of repetitive sequences (including satellites and retroelements), and the complexity of multigene families implied in host-parasite interactions. The genome of *T. cruzi* is composed of a "core compartment" and a "disruptive compartment" which exhibit opposite GC content and gene composition, with high differences on their regulatory regions. The novel tandem and dispersed repetitive sequences identified, in addition to recombination events, allows to conclude that genome plasticity is a key survival strategy during its complex life cycle.

Keywords: genome, *Trypanosoma cruzi*, compartmentalization, core and disruptive compartments

1. Introduction

The complex genome of *Trypanosoma cruzi* reflects its complex life. These parasites are able to invade almost any kind of cell to freely circulate in blood or extracellular matrix, to pass through the digestive tract of its insect vector and survive after being eliminated in feces. This stressful lifestyle strongly requires a fine regulation of gene expression, which in turn is reflected on its genome organization. Although the focus of this chapter is the nuclear genome (hereinafter called generically "genome"), it is worth mentioning that *Trypanosoma*tids have another genome that contained in their single mitochondria called kinetoplast DNA. This exhibits unique architectural and functional features: it consists of a dense network of two types of circular DNA molecules called maxicircles and minicircles. Maxicircles, of several kb in length, are equivalent to regular mtDNA of other eukaryotes, whereas minicircles are much shorter (seldom longer than 2 kb) and encode gRNAs. These are short RNA molecules responsible for guiding RNA editing, a process of posttranscriptional modifications that consists in the addition and deletion of uridines. Although editing is not exclusive of trypanosomatids, only in this group it involves massive changes in several (mitochondrialy encoded) genes.

IntechOpen

Trypanosomes have peculiarities in transcription and genome organization that differentiate them from the majority of eukaryotes. Protein-coding genes are organized in clusters separated by relatively short intergenic regions, located on the same DNA strand [1] and—with a few exceptions—do not contain introns. Clusters are transcribed as long nuclear polycistronic units, and maturation implies 3′ polyadenylation—characteristic of eukaryotes—and trans-splicing, a peculiar mechanism of mRNA maturation. Trans-splicing is the process by which two RNAs encoded in different genome locations (trans) react to form a unique transcript, where the 5′ moiety contains the spliced leader sequence (~40 nt), and the rest contains the transcribed gene [2, 3]. The spliced leader (SL) is transcribed from a tandem array as a precursor of ~140 nt whose 3′ end is removed and SL inserted to an AG splice-acceptor site on a pre-mRNA molecule, through a molecular mechanism that resembles cis-splicing [4–6]. Usually polypyrimidine-rich motifs precede AG splice acceptor. Since SL-RNA is the target of capping, trans-splicing is responsible for the addition of the 7-methylguanosine cap-like (cap4) on RNAs [7]. It has been described decades ago that this process is coupled to the polyadenylation of the 3′ end of the upstream gene, co-transcriptionally. As a consequence, a molecule of mature mRNA (capped, trans-spliced and polyadenylated) is released from the polycistron and exported to the cytoplasm, where it can be translated. Unlike other organisms, where trans-splicing also occurs, in trypanosomatids it affects almost all genes. Therefore, in trypanosomes the 5′ UTR is the sequence segment located between the SL and the start codon, whereas the 3′ UTR is defined in the same way as in other eukaryotes. With the exception of genes tandemly repeated, polycistronic units do not contain functionally related genes, and usually individual genes from the same transcription unit can show markedly different expression patterns along life cycle [1, 3]. Gene expression in trypanosomes is regulated mainly at the posttranscriptional level, and numerous studies have shown the relevance of 3′ UTR regions in regulation, affecting mRNA stability or translation, and hence differential expression [3, 8]. Different elements in the 3′ UTRs together with the presence of a high number of RNA binding proteins could explain, at least in part, differential expression [9–11], although the exact mechanisms allowing gene specificity are still unknown.

An important issue that still is not clear is whether *T. cruzi* constitutes a single species or a complex of species. Initially two groups of *T. cruzi* were described (I and II) based on biological and biochemical criteria as well as molecular techniques [12]. The first study using molecular phylogeny (sequences of coding genes) clearly showed that at least three major lineages (A, B and C) were present in this parasite [13], and that the distances between these groups are equivalent to the distance between different species of *Leishmania*. Currently six groups or discrete typing units (DTUs) named TcI-TcVI were proposed [14], and *T. cruzi* isolates from bats were included as a seventh DTU [15, 16]; where TcV and TcVI are hybrid lineages derived from haplotypes TcII and TcIII [16]. However, the high biological and genetic diversity of the *T. cruzi* strains, even at the intra-DTU level, indicates that DTUs constitute a useful working definition, but not a definitive classification. The new era of genomic studies through next generation sequencing (NGS) is providing new insights on the above-mentioned unsolved questions.

2. Genome organization

2.1 Chromosomes

In *T. cruzi* mitosis occurs without a complete disruption of nuclear envelope. In addition, although nucleosomes are present, chromatin does not condense up to chromosomes, so they cannot be visualized by microscopy. This feature has made classic cytogenetic studies unsuitable for these parasites. Instead, *T. cruzi* karyotype has been determined by molecular biology techniques, mainly pulsed field gel electrophoresis (PFGE) in combination with Southern blot [17–19]. Early studies revealed complex chromosomal patterns, evidenced by different PFGE profiles among strains, and allowed to infer that *T. cruzi* was at minimum diploid [20]. Size of chromosomal bands ranges from 0.45 to 4 Mb, without minichromosomes, and the number of chromosomes was estimated mainly through probes used as genetic markers. Depending on the probes and PFGE conditions, chromosomes ranged between 19 to 40 per haploid genome, showing that *T. cruzi* is mainly diploid, although the sizes of homologous chromosomes can differ significantly [17–19, 21–23].

A milestone was achieved in 2005 when the draft genomes of *L. major*, *T. brucei*, and *T. cruzi* were simultaneously published and referred as to the "TriTryps" [24–26]. This opened a new era in biology research on these parasites. A distinctive feature in *T. cruzi* was the already known highly repetitive nature of its genome (50%): in fact 5–10% of the genome is composed by the 195 bp satellite, and the rest of the repetitive DNA is composed of multigene families, tandem repeats and retrotransposable elements [27]. This feature gave rise to a highly fragmented assembly, resulting in that chromosome number and structure or, at least large contigs, could not be obtained. Attempts to recover full length chromosome sequencing, used a combined strategy based on synteny maps with *T. brucei* chromosomes and BAC ends sequencing. By this means 41 virtual chromosomes were obtained for the hybrid CL-Brener strain. Although this strategy represented a substantial improvement in comparison to previous versions of the genome, the issue of assembly fragmentation remained as a limitation for diverse types of analyses that require high precision. A recent milestone in the area was the first publication of long read sequencing of two *T. cruzi* genomes (Dm28c and TCC strains), which allowed to circumvent the limitation of high fragmentation imposed by the Sanger method [28], as well as by short reads NGS methods. Using this approach, also described for Bug strain [29], contigs of more than 1 Mb were obtained, probably covering whole chromosomes, but fragmentation still persists in some regions of the genomes. The exact number of chromosomes and their organization will be finally achieved through the combination of long read sequencing methods, optical maps techniques and polymer-based modeling, a field that has undergone a dramatic acceleration in the last decade [30].

2.2 Ploidy

Although PFGE and fluorescence cytophotometry were useful methods to depict the complex variability of *T. cruzi* karyotypes, it was not until the advent of next generation sequencing techonologies (NGS) that ploidy—or chromosomal copy number variation (CCNV)—analyses could be studied more in detail. Aneuploidy, the gain or loss of chromosomal copies, is of particular importance since it gives clues about the relevance of genome plasticity in the context of parasite fitness. This phenomenon has been detailed studied in *Leishmania spp.*, whose "mosaic" aneuploidies—ploidy variations within isolates from a strain and even between individual cells from a population – were related to drug resistance, regulation on gene expression, or host adaptation [31–33]. On the contrary, PFGE, fluorescence cytophotometry and high-throughput sequencing data analyses agreed on the ploidy stability of *T. brucei* and its subspecies: *T. b. brucei*, *T. b. gambiense* and *T. b. rhodesiense* [34–36]. Remarkably, a field isolated *T. congolense* triploid was reported, suggesting that Salivarian evolutionary lineage species, such as *T. brucei* and *T. congolense*, can sustain euploidies but not massive aneuploidies [37].

Strains	DTU	Size (Mb)	Contigs	N50	L50	GC%	Genes	Proteins	Sequencing plataform	References
Trypanosoma cruzi										
Dm28c	TcI	53,3	636	317.638	47	51,6	18759	15319	PacBio + illumina	[28]
TCC	TcVI (hybrid)	87,1	1.236	264.196	92	51,7	29109	24191	PacBio + illumina	[28]
Bug2148	TcV (hybrid)	55,2	929	200.364	64	51,3	-	-	PacBio	[29]
CL Brener	TcVI (hybrid)	89,9	29.495	88.624	212	51,7	23696	19607	Sanger	[25]
Esmeraldo-like*		32,5	41	--	--	--	11106	10338	--	[49]
Non-Esmeraldo-like*		32,5	41	--	--	--	11398	10831	--	[49]
Dm28c	TcI	27,3	1.210	78.389	86	50,6	11398	11348	454	[81]
G	TcI	25,2	1.450	74.655	91	47,4	13488	12708	454	[82]
CL	TcVI (hybrid)	65,0	7.764	73.547	95	39,8	34248	32278	454	[82]
S23b	TcII	28,1	7.145	20.992	332	45,2	-	-	Illumina	[38]
S92a	TcII	27,1	7.134	20.493	310	46,4	-	-	Illumina	[38]
S11	TcII	28,5	7.855	18.630	346	45,1	-	-	Illumina	[38]
S44a	TcII	17,2	4.971	17.818	232	45,0	-	-	Illumina	[38]
S15	TcII	27,5	9.197	17.779	370	46,2	-	-	Illumina	[38]
231	TcIII	35,4	8.469	14.202	586	48,6	-	-	Illumina	[83]
S162a	TcII	27,3	8.588	12.390	448	45,3	-	-	Illumina	[38]
Y	TcII	39,0	9.821	11.962	561	49,8	-	-	Illumina	[29]
Ycl4	TcII	26,1	6.664	10.716	560	46,6	-	-	Illumina	[38]
Ycl2	TcII	25,9	6.884	10.600	563	46,6	-	-	Illumina	[38]
Ycl6	TcII	25,8	6.967	10.394	549	46,6	-	-	Illumina	[38]
S154a	TcII	19,3	6.946	5.877	859	49,6	-	-	Illumina	[38]
Y	TcII	30,0	8.952	5.474	1305	50,6	-	-	454	[39]
Colombiana	TcI	30,9	9.338	5.189	1394	50,8	-	-	454	[39]
Sylvio X10/1	TcI	38,6	27.019	2.307	2599	51,2	10861	10847	454	[53, 84]
Arequipa	TcI	19,1	10.224	1.932	3156	50,9	-	-	454	[39]
Trypanosoma cruzi marinkellei										
B7	--	34,2	23154	2846	2511	50,9	10117	10104	454 + Illumina	[84]

Table 1.
Genomes of Trypanosoma cruzi

In *T. cruzi*, since CCNV analysis deeply depends on high quality, chromosome-level assembled reference genomes, it was extremely difficult to implement. However, in spite of this limitation, some approaches were done using CLBrener genome as reference [38]. Taking into account the poorly assemble reference genome at that moment, and the repetitive nature of *T. cruzi* genome, only reads with high mapping quality were used in CCNV estimations. The single-copy genes ploidy estimation (SCoPE) was the methodology utilized by the authors. In this methodology, estimation of chromosomal somy is based on the ratio between the mean coverage of all single-copy genes (unique genomic sequences) in a given chromosome and the genome coverage. After including several *T. cruzi* strains from different DTUs, authors proposed that—as was observed in *Leishmania*—the aneuploidy pattern varies among and within *T. cruzi* lineages. In addition, as observed with PFGE, CCNV is considerably frequent between *T. cruzi* strains, including those within a same DTU. Authors propose that TcI appears to be more stable, and TcII had large differences between strains, suggesting that this mechanism is widely used by the parasite to expand groups of genes [39]. Nevertheless, unlike *L. donovani*, CCNV on *T. cruzi* seems to be stable on parasite population, at least for TcII analysis on Y strain and derived clones [38].

2.3 Genome size

The genome size of *T. cruzi* has been estimated by different methodologies such as flow cytometry, renaturation kinetic analysis, microfluorometry, chemical analysis, molecular karyotyping and genome sequencing. Every approach agreed on that *T. cruzi* genome size is variable. Polymorphism has been shown between DTUs, between strains within the same DTU, and even between isolates from the same strain [40–47]. From a wide genome size quantification and analysis including more than fifty strains from DTUs TcI to TcVI [46] it was found that: (i) maximum difference observed between strains was 47.5%; (ii) TcI was the smallest genome, (iii) TcV and TcVI were the least variable, (iv) parental genomes mean gene content (TcI: 88.4 Mb, TcII: 106.5 Mb, TcIII: 119.2 Mb), and similar results on the reduced size of TcI, with few exceptions was further observed [47].

Genome size estimation by bioinformatic analysis of NGS data, as was mentioned before, is hampered due to the massive presence of repetitive sequence regions, which reach up to 50% of the genome [25, 48]. This generates assembly fragmentation and collapse—gene and repetitive sequences, leading to copy number underestimation—which represents a challenge to the correct genome size estimation. In fact, as reflected on **Table 1**, the assembly size is far below the estimations made by DNA measurements methods. Only third generation sequenced genomes appear to represent more accurate figures [28, 29].

3. Genome architecture and composition

The publication the of first *T. cruzi* genome in 2005 [25] was a cornerstone of the study of its genome complexity. Although the CL-Brener sequenced strain turned out to be a hybrid that made the analyzes more arduous, at that time it was corroborated that more than 50% of the genome of *T. cruzi* corresponds to repetitive sequences—mainly retrotrasposons, multigenic families and tandem repeats—including the discovery of the new gene family of a new family of mucin associated surface proteins (MASP). Around 12,500 genes could be identified, but the assembly was fragmented into more than 5400 scaffolds (ordered contigs usually joined with unknown sequences filled as "N"), and the complete sequence of the genome was not obtained, being the total genome size about 67 Mb (half of it corresponding to each haplotype). Later on, based on the scaffolds already defined [25], BAC ends sequencing and synteny maps with *T. brucei*, it was possible to recover full length pseudo-chromosomes [49], although still maintaining thousands of sequences as "unassigned contigs." Since these initial publications, several *T. cruzi* genomes have been reported to be sequenced by NGS, although massive sequencing could not improve the low resolution in complex and highly fragmented regions (**Table 1** and cites therein).

The advent of long read sequencing technologies helped to tackle part of the assembly fragmentation issue, and to better understand *T. cruzi* genome complexity. In 2018 the genomes of two *T. cruzi* strains (Dm28c and TCC, belonging to TcI and TcVI respectively) were sequenced by using Pacbio technology, showing substantial improvements: assemblies of Dm28c and TCC were of 53.2 and 86.7 Mb distributed in 599 and 1142 contigs, respectively, which implied a high reduction of fragmentation [28] (see N50 stats, **Table 1**). Completeness of these genomes was achieved, obtaining for the case of Dm28c all its haploid genome, totaling 53.3 Mb. This size is consistent with the most precise estimations made by fluorescent nucleic acid dye [47]. For the hybrid strain TCC, composed of two relatively divergent parental lineages, it is assumed that the diploid size that includes both parental haplotypes should be recovered, i.e., 106–122 Mb for TCC [46, 47], which compared with the 86.7 Mb indicates that segregation cannot be achieved in those regions with high identity. The ability to separate haplotypes opens new possibilities for the study of the evolutionary processes that occurred in *T. cruzi* and can be useful to provide insights on how hybrids were generated and evolved. Moreover, recombination events can be identified and studied [28]. The hybrid strain Bug 2148 (TcV) was recently long-read sequenced and assembled in 934 contigs, also resolving the fragmentation in a large degree; although the expected genome size is 106–135 Mb [46, 47], the total assembly size is 55.2 Mb and it is striking that there is no evidence of haplotype separation [29] as would be expected for a hybrid strain.

Even using this new technology, these assemblies still have some fragmentation mainly due to the size of the tandem repeats. In particular, the well-characterized 195 bp satellite that can reach clusters of 50 kb, contributes as a major factor to assembly fragmentation avoiding its complete resolution [50–52]. In fact, these genomes contain several contigs entirely composed of this repeat, which together encompasses more than 5% of the genome (see below).

3.1 Genome compartments and gene composition

Since genomic annotation, especially in *T. cruzi*, is arduous and often the goal of genomic sequencing escapes the annotation, it has not been performed in all genomes. For those genomic projects of *T. cruzi* that have the annotation (see **Table 1**) quite similar number of coding genes per haplotype was determined, a minimum of ~10,800 for Sylvio [53] and a maximum of 15,300 for Dm28c [28]. These genes can be divided into two large groups, those of well conserved core genes, and those coding for the multigenic surface families, several of which are unique for *T. cruzi* (see below). In fact, the improvements in the assemblies allowed us to determine that the genome of *T. cruzi* is composed of two compartments. These compartments, called "core" and "disruptive" [28] vary in gene content and nucleotide composition. The "core compartment" is composed of conserved and hypothetical conserved genes, it has a lower GC content (48%) and exhibits synteny conservation with *T. brucei* and *L. major*, whereas the "disruptive compartment" is mainly composed by the surface multigene families trans-sialidase, MASP, and mucins, and exhibits a higher GC content (53%).

3.2 Gene organization

As mentioned, genes in trypanosomatids are organized into non-overlapping clusters on the same DNA strand with unrelated predicted functions. Genes are transcribed as polycistrons and subsequently trans-spliced and polyadenylated. In *T. cruzi* gene clusters can range from ~30 to 500 kb separated by divergent or convergent strand-switch regions (SSR) [54]. Although no evidence of shared consensus motive or patterns has been found among them, the SSR are functionally active. For instance, transcription initiation and termination take place [2, 55, 56], but it is also observed that they are involved in the origin of DNA replication [57], and centromeric function [58, 59]. The SSRs exhibit some properties such as a different composition in comparison to the rest of the genome and higher intrinsic curvature [60, 61], associated in turn with transcriptional regulation. Indeed SSRs from the disruptive compartment are longer than those from the core compartment (mean length ~4.5kb and ~1.5kb respectively).

4. *Trypanosoma cruzi* repetitive genome

One of the outstanding features of the *T. cruzi* genome is its repetitive nature. Three types of sequences contribute to this characteristic: multigenic families, retrotransposons and satellite DNA (tandem repeat sequences).

4.1 Multigene Families

A main characteristic of *T. cruzi* genome is the large number of multigene families, many of them having hundreds of members. The largest families in *T. cruzi* genome are shown in **Table 2**. TS, Mucins and MASP are located in the disruptive compartment of the genome, whereas GP63, DGF-1 and RHS are distributed in both compartments [28]. We will focus on families from the disruptive compartment (MASP, Mucins and TS), and GP63 as an example of a very expanded family in *T. cruzi*. It is noteworthy that these families code for proteins directly involved in interaction with the host, both at the cellular level (adhesion, invasion, infection) and in immune modulation responses, mainly because most of TS, Mucins, MASP and GP63 proteins are GPI anchored, i.e., they are constitutive part of the functionally relevant cell surface of *T. cruzi*.

Family*	Dm28c	TCC
trans-sialidase (TS)	1491	1734
MASP	1045	1332
RHS	784	1222
Mucins	574	1018
GP63	378	710
DGF-1	215	491
UDP-Gal or UDP-GlcNAc-dependent glycosyltransferase	115	118
Elongation factor (1-alpha, 1-gamma and 2)	81	167
Glutamamyl carboxypeptidase	71	86
Protein Associated with Differentiation	61	60
Kinesin	58	81
TASV	45	92
Syntaxin binding protein	45	85
Heat shock protein 70	35	43

*including pseudogenes

Table 2.
Gene families groups in T. cruzi.

4.1.1 Trans-sialidases

Trans-sialidases and trans-sialidase-like proteins (TS) constitute a large and polymorphic superfamily [25, 28, 29] whose name comes from the ability to transfer sialic acid from host glycoconjugates to parasite's mucins [70, 62]. This activity is highly relevant since *T. cruzi* is unable to synthesizes sialic acid *de novo*, and sialic acid containing glycoproteins are demonstrated to be relevant for infection [70, 62, 63]. However, only a very few members of TS family are predicted to be enzymatically active [29], whereas the rest of them have other relevant roles such as binding to host molecules, immunomodulation, apoptosis or invasion [64]. It should be very important to rename this family since its current denomination leads to confusion. The hallmark of the family is the presence of the canonical amino acid motif VTVXNVXLYNR, although some members have a degenerated version of it [64]. TS proteins can be secreted or membrane anchored, in which case they exhibit an *N*-terminal signal peptide and GPI signal sequence at the *C*-terminal region of the protein. Genomic analysis of TS gene family in CL-Brener revealed that TS family was clustered in eight groups, which were classified by the presence or absence of additional motifs like FRIP, Asp box and the SAPA [65, 66]. In this classification the Group I is defined as those sequences with a predicted enzymatic activity, which corresponds to 4% of the total TS genes [67]. By long read sequencing, a more precise gene copy number could be determined on TCC and Dm28c strains: 1734 and 1491 TS genes respectively; with these new protein sequences the classification should be updated. Draws the attention that both strains exhibit a substantially high percentage of pseudogenes: 41.6% in TCC and 38% in Dm28c [28], which suggest that they could not constitute "inert material." This point deserves further studies to determine if pseudogenes are expressed, and/or can constitute a source of variability, among their possible functions. Most of TS genes are overexpressed in trypomastigotes, but a small percentage are upregulated in amastigotes or epimastigotes at the transcriptional level [68].

4.1.2 Mucins

Mucins and mucin like glycoproteins are the main acceptors of sialic acid through the trans-sialidase TS activity [69], and participate in adhesion, protection against lysis, invasion and immune evasion [70]. The first mucin-like gene cloned and the predicted protein exhibited an internal tandem repeat with the canonical sequence T_8LP_2, flanked by an *N*-terminal signal peptide and a *C*-terminal GPI anchor signal sequence. Further studies revealed the presence of a complex family with genes coding for proteins with similar N and C termini but with non-repetitive, variable and serine and threonine rich domains, also classified as mucins. Those groups with repetitive domains and without repetitive domains were designated TcMUCI and TcMUCII [10], and the presence of a mosaic sequences between both groups led to the proposal of a common ancestor and further diversification [70]. Another group of smaller mucin genes, TcSMUG [71], are expressed in the insect stages, and were subclassified in large and small TcSMUG (L and S) [70]. Due to the complexity of this family manual curation is needed for annotation of these genes. Our group used the following criteria: genes exhibiting an *N*-terminal signal peptide, a C-terminal GPI anchor signaling, and T rich sequences such as T_8KP_2, $T_{6\text{-}8}KAP$ or $T_{6\text{-}8}QAP$, finding 1018 and 574 mucin genes in TCC and Dm28c respectively [28], and around 20% were classified as pseudogenes in both strains. Regarding the expression of TcMUC and TcSMUG in life stages of *Trypanosoma cruzi*, trypomastigotes presented higher expression levels of both TcMUC groups, and in contrast with biochemical reports [70], in amastigotes the highest expressed mucins belong to TcMUCII instead of TcMUCI [68].

4.1.3 MASPs

One of the most surprising result after assembly and annotation of the first *T. cruzi* genome [25], was the discovery of a new gene family composed of approximately 1300 genes, and named as mucin associated surface protein (MASP), because of their location in proximity or clustered with mucin genes. MASP family is characterized by conserved *N*-terminal signal peptide, a conserved *C*-terminal domain containing a GPI anchor addition site, and a variable central region [25]. One of the proposed roles of this gene family is the immune system evasion during the acute phase of Chagas disease [72]. CL-Brener clone contains 1377 *masp* genes, among which 771 appear to be intact genes and 433 (31%) are pseudogenes [25], and analysis in Dm28c and TCC yield similar results: 1045 and 1332 genes where 36 and 33% respectively are pseudogenes [28]. Regarding the expression of this gene family, 97% of *masp* genes are upregulated in trypomastigotes, and a discrete number of genes are expressed specifically in amastigotes or epimastigotes [68].

4.1.4 GP63

GP63 are GPI anchored metalloproteases present in the Tritryps. However, whereas *L. major* contains six *gp63* genes and *T. brucei* has thirteen copies, in *T. cruzi* this family is widely expanded: 400 genes or pseudogenes were identified in CL-Brener [25] and Dm28c [28], and more than 700 in TCC [28]. Strikingly, more than 60% of these genes on the three strains are annotated as pseudogenes. Although the role of this family in innate immune evasion and invasion, has been extensively studied in *Leishmania* [73, 74], little is known about its role in *T. cruzi*. The reason of the expansion of this gene family in the *T. cruzi* genome remains to be elucidated as well as its role on this parasite. Transcriptomic analysis revealed that most of the members are highly expressed in trypomastigotes, whereas a few genes are expressed almost exclusively in amastigotes. Interestingly, phylogenetic analysis using 3′ UTR sequences of *gp63* genes showed three groups of sequences clearly distinguished; one group associated with genes highly expressed in trypomastigotes, another one with genes highly expressed in amastigotes, and a third group of genes with almost no expression in any stage of the parasite [68]. This result strongly supports a major role of the 3′ UTR in posttranscriptional regulation of this family that deserves further studies.

4.2 Transposable elements

Transposable elements (TEs) are repeated DNA sequences, which have the ability to move from one to another *locus* in the genome. This was why they have been referred to as "junk" DNA, selfish sequences or genomic parasites. However, growing evidence is indicating the great importance that TEs play in the evolution of genes and genomes in a wide range of organisms, including trypanosomatids [75, 76]. *T. cruzi* genome lacks class II elements (DNA transposons), bearing only class I retroelements. Within them—according to Wicker [77] TEs classification—*T. cruzi* presents three autonomous families: VIPER, a tyrosine recombinase (YR) element which belongs to the DIRS order; L1Tc, a non-LTR element of the *ingi* clade; and CZAR, also a non-LTR element from the CRE clade which is site-specific, inserting only on the SL gene [25, 76, 78]. On the other hand, non-autonomous elements have been also identified. SIRE, have similarity with the VIPER 5′ and 3′ ends, resembling what nowadays are called solo-LTR. NARTc is the non-autonomous couple of L1Tc elements, as has been classically described for LINE/SINE-like couples. Finally, TcTREZO has been described as another site-specific retroelement, inserted within *masp* genes [79]. Although it has been characterized as a non-LTR retroelement due to the presence of a poly-A tail and a secondary structure which will be promoting its retrotranscription, no conserved domains have been detected on this element. Hence, TcTREZO could be an ancient non-autonomous retroelement. All of the VIPER, CZAR and TcTREZO copies are defective (no complete domains where found), whereas L1Tc was the only one which showed putative active copies [28].

4.3 Tandem repeats

Although NGS platforms implied an enormous progress for our knowledge about genomes composition and evolution, tandem repeats were not that benefited. Tandem repeats are commonly classified in micro, mini and macro-satellite, depending on their monomer or cluster length. Microsatellites are those whose monomers are from 2 to 5 bp, minisatellites from 15 to 100 bp, and finally macrosatellites or just called satellites involves repeats greater than 100 bp [80]. Even with very deep genome coverage, short read lengths cause problems for *de novo* assemblies, especially in tandem repeat rich regions. Due to this trouble, tandem repeats can be considered as neglected sequences in the majority of genome analyses. Although great efforts were done, fragmentation of the genome assembly occurs frequently where repeated sequences are located. In fact, the massive major 195 bp satellite (TcSAT1 named in repbase) described for the first time by Sloof et al. [50], represents a huge challenge for contig assembly. Although PacBio reads enable to develop an improved assembly and characterization of tandem repeats characterization and assembly, the size of some clusters exceeds that of the reads. In fact, some small-size contigs (50 kb) are composed entirely by the 195 bp satellite sequence.

In summary, genomic studies are essential for understanding *Trypanosoma cruzi* biology, and the new technologies will give responses to still unanswered questions: Which molecular mechanisms allow to regulate specific genes, without consensus sequences? Is *Trypanosoma cruzi* a unique species? How many chromosomes do they have? How are chromosomes organized? Which role plays the highly repetitive genome on its plasticity? And we can continue, to reinforce the idea that much remains to be done.

Financial support

This work was supported by Institut Pasteur de Montevideo (S.P. postdoctoral fellowship) from UK Research and Innovation via the Global Challenges Research Fund under grant agreement 'A Global Network for Neglected Tropical Diseases' grant number MR/P027989/1. LB, APT, FAV and CR are members of the Sistema Nacional de Investigadores (SNI-ANII, Uruguay).

Author details

Luisa Berná[1], Sebastián Pita[1,2], María Laura Chiribao[1,3], Adriana Parodi-Talice[1,2], Fernando Alvarez-Valin[4] and Carlos Robello[1,3]*

1 Laboratory of Host Pathogen Interactions-UBM, Institut Pasteur de Montevideo, Uruguay

2 Sección Genética, Facultad de Ciencias, Universidad de la República, Uruguay

3 Departamento de Bioquímica, Facultad de Medicina, Universidad de la República, Uruguay

4 Sección Biomatemática, Facultad de Ciencias-Universidad de la República, Uruguay

*Address all correspondence to: robello@pasteur.edu.uy

References

[1] Vanhamme L, Pays E. Control of gene expression in trypanosomes. Microbiology and Molecular Biology Reviews. 1995;**59**(2):223-240

[2] Martínez-Calvillo S et al. Gene expression in trypanosomatid parasites. BioMed Research International. 2010;**2010**

[3] Kramer S. Developmental regulation of gene expression in the absence of transcriptional control: The case of kinetoplastids. Molecular and Biochemical Parasitology. 2012;**181**(2):61-72

[4] Sutton RE, Boothroyd JC. Evidence for trans splicing in trypanosomes. Cell. 1986;**47**(4):527-535

[5] Michaeli S. Trans-splicing in trypanosomes: Machinery and its impact on the parasite transcriptome. Future Microbiology. 2011;**6**(4):459-474

[6] Campbell DA, Thomas S, Sturm NR. Transcription in kinetoplastid protozoa: Why be normal? Microbes and Infection. 2003;**5**(13):1231-1240

[7] Freistadt MS et al. Direct analysis of the mini-exon donor RNA of *Trypanosoma brucei*: Detection of a novel cap structure also present in messenger RNA. Nucleic Acids Research. 1987;**15**(23):9861-9879

[8] Clayton CE. Gene expression in kinetoplastids. Current Opinion in Microbiology. 2016;**32**:46-51

[9] da Silva RA, Bartholomeu DC, Teixeira SM. Control mechanisms of tubulin gene expression in *Trypanosoma cruzi*. International Journal for Parasitology. 2006;**36**(1):87-96

[10] Di Noia JM et al. AU-rich elements in the 3′-untranslated region of a new mucin-type gene family of *Trypanosoma cruzi* confers mRNA instability and modulates translation efficiency. The Journal of Biological Chemistry. 2000;**275**(14):10218-10227

[11] Jager AV, Muia RP, Campetella O. Stage-specific expression of *Trypanosoma cruzi* trans-sialidase involves highly conserved 3′ untranslated regions. FEMS Microbiology Letters. 2008;**283**(2):182-188

[12] Tibayrenc M et al. Genetic characterization of six parasitic protozoa: Parity between random-primer DNA typing and multilocus enzyme electrophoresis. Proceedings of the National Academy of Sciences. 1993;**90**(4):1335-1339

[13] Robello C et al. Evolutionary relationships in *Trypanosoma cruzi*: Molecular phylogenetics supports the existence of a new major lineage of strains. Gene. 2000;**246**(1-2):331-338

[14] Zingales B et al. A new consensus for *Trypanosoma cruzi* intraspecific nomenclature: Second revision meeting recommends TcI to TcVI. Memórias do Instituto Oswaldo Cruz. 2009;**104**(7):1051-1054

[15] Marcili A et al. A new genotype of *Trypanosoma cruzi* associated with bats evidenced by phylogenetic analyses using SSU rDNA, cytochrome b and Histone H2B genes and genotyping based on ITS1 rDNA. Parasitology. 2009;**136**(6):641-655

[16] Zingales B et al. The revised *Trypanosoma cruzi* subspecific nomenclature: Rationale, epidemiological relevance and research applications. Infection, Genetics and Evolution. 2012;**12**(2):240-253

[17] Henriksson J, Åslund L, Pettersson U. Karyotype variability in

Trypanosoma cruzi. Parasitology Today. 1996;**12**(3):108-114

[18] Santos MR et al. The *Trypanosoma cruzi* genome project: Nuclear karyotype and gene mapping of clone CL Brener. Memórias do Instituto Oswaldo Cruz. 1997;**92**(6):821-828

[19] Henriksson J et al. Chromosome specific markers reveal conserved linkage groups in spite of extensive chromosomal size variation in *Trypanosoma cruzi*. Molecular and Biochemical Parasitology. 1995;**73**(1-2):63-74

[20] Gibson WC, Miles MA. The karyotype and ploidy of *Trypanosoma cruzi*. The EMBO Journal. 1986;**5**(6):1299-1305

[21] Henriksson J et al. Chromosomal size variation in *Trypanosoma cruzi* is mainly progressive and is evolutionarily informative. Parasitology. 2002;**124**(3):277-286

[22] Vargas N, Pedroso A, Zingales B. Chromosomal polymorphism, gene synteny and genome size in *T. cruzi* I and *T. cruzi* II groups. Molecular and Biochemical Parasitology. 2004;**138**(1):131-141

[23] Cano MI et al. Molecular karyotype of clone CL Brener chosen for the *Trypanosoma cruzi* genome project. Molecular and Biochemical Parasitology. 1995;**71**(2):273-278

[24] Berriman M et al. The genome of the African trypanosome *Trypanosoma brucei*. Science. 2005;**309**(5733):416-422

[25] El-Sayed NM et al. The genome sequence of *Trypanosoma cruzi*, etiologic agent of Chagas disease. Science. 2005;**309**(5733):409-415

[26] Ivens AC et al. The genome of the kinetoplastid parasite, *Leishmania* major. Science. 2005;**309**(5733):436-442

[27] El-Sayed NM et al. Comparative genomics of trypanosomatid parasitic protozoa. Science. 2005;**309**(5733):404-409

[28] Berná L et al. Expanding an expanded genome: Long-read sequencing of *Trypanosoma cruzi*. Microbial Genomics. 2018;**4**(5)

[29] Callejas-Hernández F, Gironès N, Fresno M. Genome Sequence of *Trypanosoma cruzi* Strain Bug2148. Genome Announcements. 2018;**6**(3):e01497-e01417

[30] Parmar JJ, Woringer M, Zimmer C. How the genome folds: The biophysics of four-dimensional chromatin organization. Annual Review of Biophysics. 2019;**48**

[31] Downing T et al. Whole genome sequencing of multiple *Leishmania donovani* clinical isolates provides insights into population structure and mechanisms of drug resistance. Genome Research. 2011;**21**(12):2143-2156

[32] Dujardin J-C et al. Mosaic aneuploidy in *Leishmania*: The perspective of whole genome sequencing. Trends in Parasitology. 2014;**30**(12):554-555

[33] Mannaert A et al. Adaptive mechanisms in pathogens: Universal aneuploidy in *Leishmania*. Trends in Parasitology. 2012;**28**(9):370-376

[34] Almeida LV et al. Chromosomal copy number variation analysis by next generation sequencing confirms ploidy stability in *Trypanosoma brucei* subspecies. Microbial Genomics. 2018;**4**(10)

[35] Borst P et al. On the DNA content and ploidy of trypanosomes. Molecular and Biochemical Parasitology. 1982;**6**(1):13-23

[36] Hope M et al. Analysis of ploidy (in megabase chromosomes) in

Trypanosoma brucei after genetic exchange. Molecular and Biochemical Parasitology. 1999;**104**(1):1-9

[37] Tihon E et al. Evidence for viable and stable triploid *Trypanosoma* congolense parasites. Parasites & Vectors. 2017;**10**(1):468

[38] Reis-Cunha JL et al. Chromosomal copy number variation reveals differential levels of genomic plasticity in distinct *Trypanosoma cruzi* strains. BMC Genomics. 2015;**16**(1):499

[39] Reis-Cunha JL, Valdivia HO, Bartholomeu DC. Gene and chromosomal copy number variations as an adaptive mechanism towards a parasitic lifestyle in trypanosomatids. Current Genomics. 2018; **19**(2):87-97

[40] Castro C, Craig SP, Castañeda M. Genome organization and ploidy number in *Trypanosoma cruzi*. Molecular and Biochemical Parasitology. 1981;**4**(5-6):273-282

[41] Dvorak JA et al. *Trypanosoma cruzi*: Flow cytometric analysis. I. Analysis of total DNA/organism by means of mithramycin-induced fluorescence 1, 2. The Journal of Protozoology. 1982;**29**(3):430-437

[42] Kooy RF et al. On the DNA content of *Trypanosoma cruzi*. Molecular and Biochemical Parasitology. 1989;**36**(1):73-76

[43] Lanar DE, Levy LS, Manning JE. Complexity and content of the DNA and RNA in *Trypanosoma cruzi*. Molecular and Biochemical Parasitology. 1981;**3**(5):327-341

[44] McDaniel JP, Dvorak JA. Identification, isolation, and characterization of naturally-occurring *Trypanosoma cruzi* variants. Molecular and Biochemical Parasitology. 1993;**57**(2):213-222

[45] Thompson CT, Dvorak JA. Quantitation of total DNA per cell in an exponentially growing population using the diphenylamine reaction and flow cytometry. Analytical Biochemistry. 1989;**177**(2):353-357

[46] Lewis MD et al. Flow cytometric analysis and microsatellite genotyping reveal extensive DNA content variation in *Trypanosoma cruzi* populations and expose contrasts between natural and experimental hybrids. International Journal for Parasitology. 2009;**39**(12):1305-1317

[47] Souza RT et al. Genome size, karyotype polymorphism and chromosomal evolution in *Trypanosoma cruzi*. PLoS One. 2011;**6**(8):e23042

[48] Pita S et al. The Tritryps comparative repeatome: Insights on repetitive element evolution in Trypanosomatid pathogens. Genome Biology and Evolution. 2019;**11**(2):546-551

[49] Weatherly DB, Boehlke C, Tarleton RL. Chromosome level assembly of the hybrid *Trypanosoma cruzi* genome. BMC Genomics. 2009;**10**(1):255

[50] Sloof P et al. Characterization of satellite DNA in *Trypanosoma brucei* and *Trypanosoma cruzi*. Journal of Molecular Biology. 1983;**167**(1):1-21

[51] Gonzalez A et al. Minichromosomal repetitive DNA in *Trypanosoma cruzi*: Its use in a high-sensitivity parasite detection assay. Proceedings of the National Academy of Sciences. 1984;**81**(11):3356-3360

[52] Elias MCQ et al. Organization of satellite DNA in the genome of *Trypanosoma cruzi*. Molecular and Biochemical Parasitology. 2003;**129**(1):1-9

[53] Franzén O et al. Shotgun sequencing analysis of *Trypanosoma cruzi* I Sylvio

X10/1 and comparison with T. cruzi VI CL Brener. PLoS Neglected Tropical Diseases. 2011;**5**(3):e984

[54] Myler PJ et al. *Leishmania* major Friedlin chromosome 1 has an unusual distribution of protein-coding genes. Proceedings of the National Academy of Sciences. 1999;**96**(6):2902-2906

[55] Tosato V et al. Secondary DNA structure analysis of the coding strand switch regions of five *Leishmania* major Friedlin chromosomes. Current Genetics. 2001;**40**(3):186-194

[56] Worthey E et al. *Leishmania* major chromosome 3 contains two long convergent polycistronic gene clusters separated by a tRNA gene. Nucleic Acids Research. 2003;**31**(14):4201-4210

[57] McDonagh PD, Myler PJ, Stuart K. The unusual gene organization of *Leishmania* major chromosome 1 may reflect novel transcription processes. Nucleic Acids Research. 2000;**28**(14):2800-2803

[58] Obado SO et al. Functional mapping of a trypanosome centromere by chromosome fragmentation identifies a 16-kb GC-rich transcriptional "strand-switch" domain as a major feature. Genome Research. 2005;**15**(1):36-43

[59] Obado SO et al. Repetitive DNA is associated with centromeric domains in Trypanosoma brucei but not *Trypanosoma cruzi*. Genome Biology. 2007;**8**(3):R37

[60] El-Sayed NM et al. The sequence and analysis of *Trypanosoma brucei* chromosome II. Nucleic Acids Research. 2003;**31**(16):4856-4863

[61] Smircich P, El-Sayed NM, Garat B. Intrinsic DNA curvature in trypanosomes. BMC Research Notes. 2017;**10**(1):585

[62] Schenkman S et al. A novel cell surface trans-sialidase of *Trypanosoma cruzi* generates a stage-specific epitope required for invasion of mammalian cells. Cell. 1991;**65**(7):1117-1125

[63] Schenkman S et al. Mucin-like glycoproteins linked to the membrane by glycosylphosphatidylinositol anchor are the major acceptors of sialic acid in a reaction catalyzed by trans-sialidase in metacyclic forms of *Trypanosoma cruzi*. Molecular and Biochemical Parasitology. 1993;**59**(2):293-303

[64] Freire-de-Lima L et al. The trans-sialidase, the major *Trypanosoma cruzi* virulence factor: Three decades of studies. Glycobiology. 2015;**25**(11):1142-1149

[65] Buscaglia CA et al. Tandem amino acid repeats from *Trypanosoma cruzi* shed antigens increase the half-life of proteins in blood. Blood. 1999;**93**(6):2025-2032

[66] Cazzulo J, Frasch A. SAPA/trans-sialidase and cruzipain: Two antigens from *Trypanosoma cruzi* contain immunodominant but enzymatically inactive domains. The FASEB Journal. 1992;**6**(14):3259-3264

[67] Freitas LM et al. Genomic analyses, gene expression and antigenic profile of the trans-sialidase Superfamily of *Trypanosoma cruzi* reveal an undetected level of complexity. PLoS One. 2011;**6**(10):e25914

[68] Berna L et al. Transcriptomic analysis reveals metabolic switches and surface remodeling as key processes for stage transition in *Trypanosoma cruzi*. PeerJ. 2017;**5**:e3017

[69] Acosta A, Schenkman RP, Schenkman S. Sialic acid acceptors of different stages of *Trypanosoma cruzi* are mucin-like glycoproteins linked to the parasite membrane by GPI anchors. Brazilian Journal of Medical and Biological Research. 1994;**27**(2):439-442

[70] Buscaglia CA et al. *Trypanosoma cruzi* surface mucins: Host-dependent coat diversity. Nature Reviews Microbiology. 2006;**4**(3):229-236

[71] Urban I et al. Molecular diversity of the *Trypanosoma cruzi* TcSMUG family of mucin genes and proteins. The Biochemical Journal. 2011;**438**(2):303-313

[72] dos Santos SL et al. The MASP Family of *Trypanosoma cruzi*: Changes in gene expression and antigenic profile during the acute phase of experimental infection. PLoS Neglected Tropical Diseases. 2012;**6**(8)

[73] Yao C, Donelson JE, Wilson ME. The major surface protease (MSP or GP63) of *Leishmania* sp. biosynthesis, regulation of expression, and function. Molecular and Biochemical Parasitology. 2003;**132**(1):1-16

[74] Brittingham A et al. Role of the *Leishmania* surface protease gp63 in complement fixation, cell adhesion, and resistance to complement-mediated lysis. The Journal of Immunology. 1995;**155**(6):3102-3111

[75] Bringaud F et al. Members of a large retroposon family are determinants of post-transcriptional gene expression in *Leishmania*. PLoS Pathogens. 2007;**3**(9):e136

[76] Thomas MC et al. The biology and evolution of transposable elements in parasites. Trends in Parasitology. 2010;**26**(7):350-362

[77] Wicker T et al. A unified classification system for eukaryotic transposable elements. Nature Reviews Genetics. 2007;**8**(12):973

[78] Bringaud F et al. Role of transposable elements in trypanosomatids. Microbes and Infection. 2008;**10**(6):575-581

[79] Souza RT et al. New *Trypanosoma cruzi* repeated element that shows site specificity for insertion. Eukaryotic Cell. 2007;**6**(7):1228-1238

[80] Charlesworth B, Sniegowski P, Stephan W. The evolutionary dynamics of repetitive DNA in eukaryotes. Nature. 1994;**371**(6494):215

[81] Grisard EC et al. *Trypanosoma cruzi* clone Dm28c draft genome sequence. Genome Announcements. 2014;**2**(1):pii:e01114-13

[82] Bradwell KR et al. Genomic comparison of *Trypanosoma conorhini* and *Trypanosoma rangeli* to *Trypanosoma cruzi* strains of high and low virulence. BMC Genomics. 2018;**19**(1):770

[83] Baptista RP et al. Assembly of highly repetitive genomes using short reads: The genome of discrete typing unit III *Trypanosoma cruzi* strain 231. Microbial Genomics. 2018;**4**(4)

[84] Franzén O et al. Comparative genomic analysis of human infective *Trypanosoma cruzi* lineages with the bat-restricted subspecies *T. cruzi* marinkellei. BMC Genomics. 2012;**13**:531

Chapter 7

L-arginine Metabolism in the Infection with *Trypanosoma cruzi*

Laila Gutiérrez-Kobeh and Arturo A. Wilkins-Rodríguez

Abstract

Trypanosoma cruzi is the causal agent of Chagas disease that affects 6–7 million people around the world, principally in Latin America. This disease is characterized for the presence of an acute phase in which the host immune response plays a central role in the elimination of the parasite. If the parasite is not efficiently eliminated, patients can remain asymptomatic or develop a chronic infection. One of the cells that are primarily infected with this intracellular parasite is macrophages (Mϕ). Mϕ present a wide array of activation states with classically activated macrophages in one pole (CAMϕ) and alternatively activated macrophages (AAMϕ) in the other. One of the most important differences between these two activation states is the presence of the inducible nitric oxide synthase (iNOS or NOS2) in CAMϕ and arginase 1 (Arg-1) in AAMϕ; both enzymes share the same substrate, L-arginine, and are reciprocally regulated by the action of Th1 cytokines in the case of NOS2 and Th2 cytokines in the case of Arg-1. The activation of CAMϕ permits the production of nitric oxide (NO), highly trypanotoxic, while the activation of AAMϕ allows the synthesis of polyamines, necessary for parasite duplication. L-arginine is a very important metabolite situated in the center between the elimination and perpetuation of *T. cruzi*.

Keywords: arginase-1, L-arginine, inducible nitric oxide synthase, macrophages, *trypanosoma cruzi*

1. Introduction

Trypanosoma cruzi is the causal agent of Chagas disease that affects 6–7 million people around the world, mainly in Latin America [1], although in the last years it has also become a potential public health problem in developed countries due to the constant migrations with cases reported in the USA, Canada, Europe, Japan, and Australia [2].

This intracellular obligate parasite enters the human host in the form of metacyclic promastigotes that are released from the triatomine feces during the blood meal, through damaged skin or mucosae. Alternatively, infection can occur through other routes such as oral, congenital, blood transfusions, or organ transplants. After entering the host, trypomastigotes are phagocytized mainly by macrophages, where they transform to amastigotes, the intracellular form that has the ability to replicate. In order to evade the host immune response and ensure its persistence inside macrophages, *Trypanosoma* has developed multiple strategies. One of these has as a target L-arginine metabolism. Macrophages can eliminate amastigotes or permit their survival depending on the balance of two inducible enzymes nitric oxide synthase (iNOS or NOS2) and arginase-1 (Arg-1) that share the same substrate: L-arginine. During the activation of macrophages in the context known as classical activation, L-arginine is metabolized by iNOS giving rise to the production of nitric oxide (NO), one important trypanotoxic agent that permits these cells to destroy the parasite. On the other hand, during the activation of macrophages in the context known as alternative activation, L-arginine is metabolized by Arg-1 giving rise to the production of polyamines that favor multiplication and persistence of *Trypanosoma* in these cells. Thus, L-arginine is situated as a frontier between the elimination and survival of *Trypanosoma* in macrophages, and its metabolism is a determinant factor for the evolution of the disease.

2. Phases of the infection with *Trypanosoma cruzi*

The infection with *T. cruzi* presents an acute phase that is auto-limiting and can go unnoticed in many infected individuals. During this phase, parasites actively duplicate in different cells and tissues such as macrophages; muscular cells of smooth, striated, and cardiac muscles; adipocytes; and cells of the central nervous system [3]. While some patients succumb during the acute phase of the disease, the development of an adaptive immune generally permits the control of infection with *T. cruzi*. If the parasite is not completely eradicated, individuals remain infected for life, and a dynamic equilibrium is established with the parasite that results in different clinical outcomes. In this way, while many individuals chronically infected remain in an asymptomatic intermediate phase, a significant proportion (30–35%) of patients develop cardiac or digestive manifestations that can drive them to congestive cardiac failure, arrhythmias, and eventually death or develop colon or esophageal megasyndromes. All of these are irreversible pathologic changes that occur even though the presence of the parasite is scarce. One experimental model that recapitulates chagasic myocarditis is present in infected mice for long periods with different *T. cruzi* strains that develop chronic lesion in the myocardium [4, 5].

3. Generalities of L-arginine

L-arginine is one of the most versatile amino acids at the metabolic level. Besides serving as a precursor for protein synthesis, it is also a precursor of multiple compounds of great biologic importance such as urea, nitric oxide, polyamines, L-proline, glutamate, creatinine, and agmatine (**Figure 1**) [6, 7].

In adult mammals, L-arginine is a nonessential amino acid; nevertheless, during childhood and certain physiologic or pathologic conditions (e.g., pregnancy, sepsis, trauma, catabolic stress, intestinal or renal damage), it is considered as a semi-essential amino acid or conditioned nonessential, due to the fact that its consumption exceeds the capacity of being synthesized by the organism and has to be supplied exogenously [8–10]. In mammals, the provision of L-arginine depends on its procurement through the protein diet, endogenous synthesis (de novo synthesis), and its release during the process of protein replacement (**Figure 1**) [6]. Approximately 40% of the L-arginine that is obtained from the protein diet is catabolized in the intestine before entering the circulation [11]. In the absence of the contribution by the protein diet, approximately 80% of the L-arginine that enters the circulation derives from the protein replacement, and the remaining percentage is obtained through the novo synthesis [11]. L-arginine metabolism occurs basically in the liver and kidney; nevertheless, other tissues and cells also possess the required enzymes to metabolize it, including some cells of the immune response [12]. Regarding last point, it is

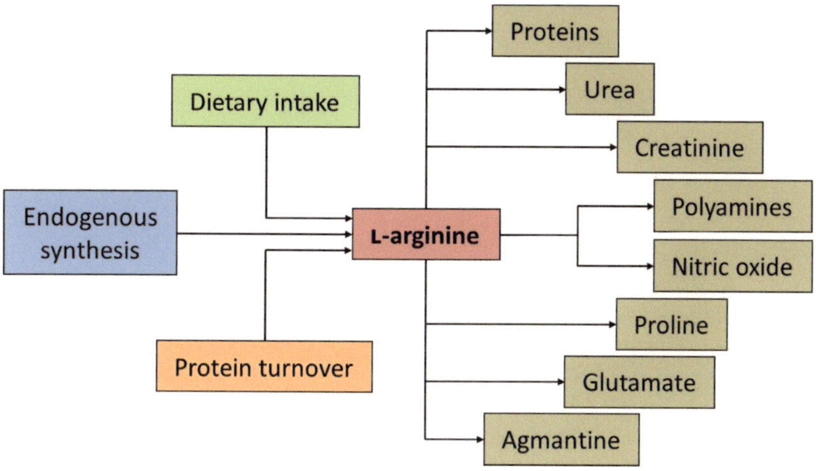

Figure 1.
Sources of L-arginine in mammals and its metabolic products.

interesting to note that a complete urea cycle has been described in macrophages [13]. Although only two enzymes directly involved in L-arginine synthesis have been identified (arginine succinate synthetase and arginine succinate lyase that are the third and fourth enzymes of the urea cycle), four enzymes utilize this amino acid as substrate: arginine decarboxylase, arginine/glycine aminotransferase, different isoforms of arginase (Arg), and the different isoforms of the nitric oxide synthases (NOS), the last two being the most studied and characterized [12]. In mammals two arginase isoforms exist, Arg-1 and Arg-2, that catalyze the same reaction but differ in cellular expression and subcellular localization. Arg-1 is cytosolic and is highly expressed in the liver and some cells of the immune response. Compared to Arg-1, Arg-2 is mitochondrial and is expressed in a great variety of peripheral tissues, mainly in the kidney, prostate, small intestine, and mammary glands during lactation [14]. Regarding NOS, this enzyme is present in three isoforms: neuronal NOS (nNOS or NOS1), inducible nitric oxide synthase (iNOS or NOS2), and endothelial NOS (eNOS or NOS3). NOS 1 is expressed in specific neurons of the central nervous system (CNS), and NOS3 is mostly expressed in endothelial cells [15]. NOS 2 is not usually expressed in cells, but its expression can be induced by bacterial lipopolysaccharide, cytokines, and other agents. Although primarily identified in macrophages, the expression of this enzyme can be stimulated in almost any cell or tissue, provided that the appropriate inducing agents are present [16].

4. L-arginine metabolism in the immune response: special emphasis in macrophages

In the immune response, L-arginine metabolism through NOS2 and Arg-1 has a pivotal role in the regulation of the effector capabilities of macrophages, dendritic cells, and neutrophils [17–20] during infectious processes caused by a great variety of microorganisms: different species of *Mycobacterium*, *Leishmania*, *Trypanosoma*, *Schistosoma*, and *Salmonella*, among others [21, 22].

L-arginine metabolism in the immune response acquired great relevance with the discovery that murine macrophages express both NOS2 and Arg-1 and that their expression is reciprocally regulated by the action of Th1/proinflammatory cytokines (e.g., IFN-γ and TNF-α) and Th2/anti-inflammatory (e.g., IL-4, IL-10, and IL-13) that determine the activation state of macrophages [19, 23–27].

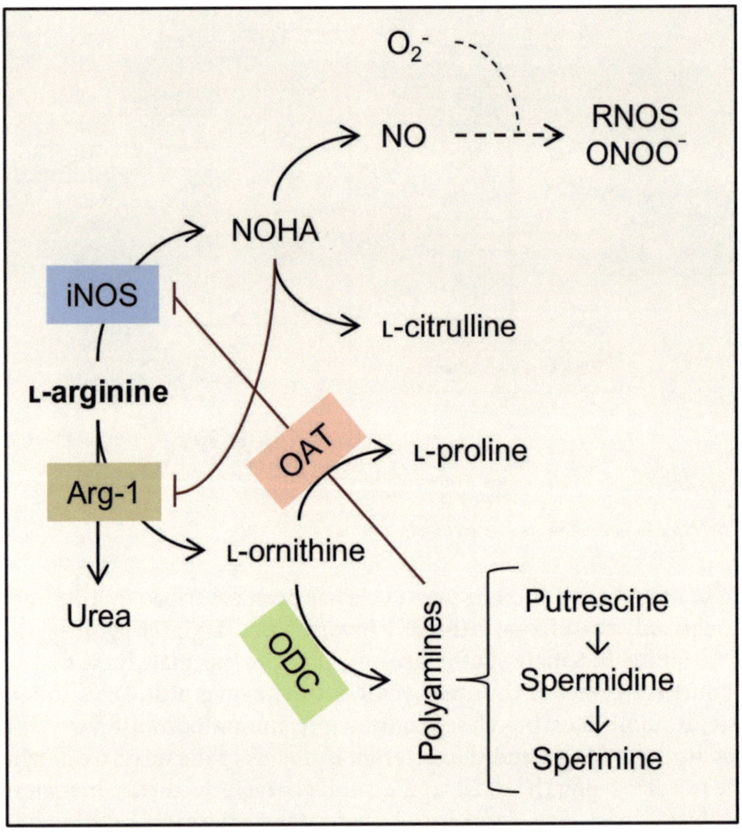

Figure 2.
L-arginine metabolism through iNOS and Arg-1. iNOS, inducible nitric oxide synthase; Arg-1, arginase 1; NOHA, N^{ω}–OH-L-arginine; OAT, ornithine-aminotransferase; ODC, ornithine-decarboxylase; NO•, nitric oxide; O_2^-, superoxide, RNOS, oxygen and nitrogen reactive species; ONOO-, peroxynitrite.

Th1 cytokines activate macrophages in a classical way (CAMΦ) and induce the expression and function of NOS2, while Th2 cytokines activate macrophages in an alternative way (AAMΦ) and induce the expression and function of Arg-1.

NOS2 or iNOS is an oxide-reductase responsible for the synthesis of L-citrulline and nitric oxide (NO•) from L-arginine in the presence of NADPH and oxygen. This reaction occurs through two successive reactions: the monooxygenation of L-arginine that drives to the production of the intermediary N^{ω}-OH- L-arginine (NOHA) and the subsequent hydrolysis of this last compound, thus producing L-citrulline and NO• (**Figure 2**). NOS2 generates both NO• and superoxide (O_2^-) that together react to form the radical peroxynitrite ($ONOO^-$) [28]. This last compound has been identified as a reactive species derived both from oxygen and nitrogen (RONS) that constitutes the principal cytostatic or cytotoxic mechanism of CAMΦ to fight the infections generated by virus, bacteria, fungi, and protozoan parasites [20, 23, 26, 29].

5. Immune response to *Trypanosoma cruzi*

Inside the mammalian host, macrophages represent an important site for the duplication of *T. cruzi*. One of the most important mechanisms in the protective immunity against *T. cruzi* is the activation of macrophages in order to achieve the elimination of parasites (**Figure 3**). CAMϕ are able eliminate *T. cruzi* thanks to

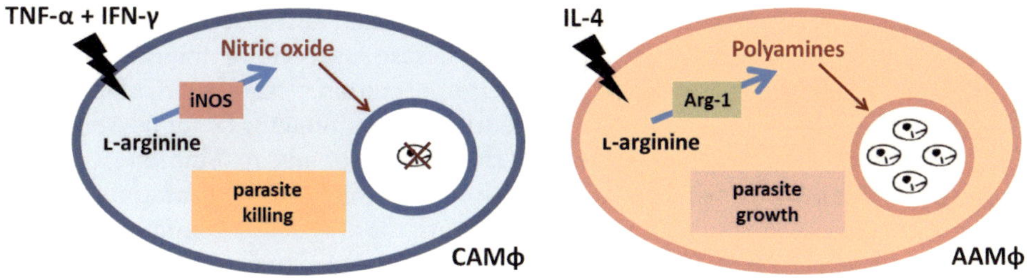

Figure 3.
L-arginine metabolism in macrophages during T. cruzi infection. In classically activated macrophages (CAMΦ), inducible nitric oxide synthase (iNOS) expression and function are induced. L-arginine metabolism through this enzyme entails the production of nitric oxide that possesses great trypanocidal capacity. In alternatively activated macrophages (AAMΦ), arginase 1 (Arg-1) expression and function are induced. L-arginine metabolism through this enzyme entails the production of polyamines that favor T. cruzi multiplication inside macrophages.

NOS2 and RONS that kill intracellular parasites by the modification of structural properties of *T. cruzi* molecules. On the other hand, the different forms of AAMϕ present high levels of mannose receptor (MR) and an overregulation of Arginase 1 [30]. Arg-1 hydrolyzes L-arginine in urea and L-ornithine; the latter is the principal intracellular source for the synthesis of polyamines and trypanothione. Polyamines are small cationic molecules required for cellular proliferation and macrophage homeostatic processes, besides being vital for the intracellular growth of *Trypanosoma* [19, 31]. Both inducible enzymes share L-arginine as substrate, and the expression and function of both enzymes are reciprocally regulated by the action of Th1 and Th2 cytokines. Thus, L-arginine is situated as a frontier between the elimination and survival of *Trypanosoma* in host cells, and its metabolism is a determinant factor in the evolution of the disease.

In response to the defense mechanisms of the host, parasites have developed several strategies in order to escape host immune response and take advantage of some host's molecules. In this way, parasites must reduce the production of toxic molecules, including nitric oxide and its derivatives, that are synthesized by the immune system, in particular by macrophages [32–34]. In addition, internalized parasites of different *T. cruzi* strains are able to escape from the parasitophorous vacuole of resident macrophages [35], a strategy that utilizes a variety of molecules with antioxidant properties [36, 37]. Nevertheless, as the infection progresses, the evasion strategies displayed by *T. cruzi* are widely surpassed by the development of a humoral specific immune response and the activation of macrophages by IFN-γ and other cytokines. As has been previously mentioned, the infection with *T. cruzi* can have an acute or a chronic phase. One of the possible causes of the passage from one phase to another is the fact that the effector immune response against the parasite is insufficient or inappropriate due to a deficient activation of the specific immune response or an excessive regulation of this response.

6. Role of Arg-1 in the infection with *Trypanosoma*

The induction of Arg-1 in macrophages promotes the infection of parasites of the genus *Trypanosoma* by providing nutrients derived from polyamines, since *Trypanosoma* parasites cannot generate their own source of ornithine through the activity of a functional arginase [38, 39]. The increase in arginase activity counteracts the host's immune response and favors parasite growth. It has been shown that in African trypanosomiasis caused by *Trypanosoma gambiense*, there is an increase in the serum level of Arg-1 that returns to basal values after the treatment

[40]. Similarly, in experimental murine trypanosomiasis caused by *Trypanosoma brucei*, macrophage Arg-1 activity represents a disease susceptibility marker [41]. In *T. brucei* Arg-1 activity is induced by excretion/secretion factors, particularly TbKHC1, kinesin H chain, and has been identified as an inductor factor of Arg-1 [38]. Other studies have demonstrated that the addition of an Arg-1 inhibitor reduces parasite growth, which is restored with L-ornithine supplementation. The essential requirement of L-ornithine is related with the absence of a functional arginase in *Trypanosoma* [39], which results in a dependence toward host's arginase for the synthesis of polyamines and trypanothione, which are essential for parasite survival, growth, and differentiation [42]. The difluoromethylornithine, structural analog of L-ornithine, has been used alone or in combination with nifurtimox as an effective drug against African trypanosomiasis [43]. Nevertheless, its administration is difficult and requires large amounts of i.v. injected fluids, which limits its use in remote areas. Thus, it is of utmost importance to find easier ways to select polyamine synthesis as a target against *Trypanosoma.* Alternatively, inhibitors of the route that conducts to arginase activity might reduce parasite loads in infected animals.

7. Conclusion

Trypanosoma cruzi is the causal agent of Chagas disease that affects 6–8 million people primarily in Latin America. It is an intracellular parasite that infects a variety of cells, among which macrophages are a very important target and thus transcendental for the immune response against the parasite. Macrophages can traverse through a gradient of stages of activation with classically activated macrophages in one end and alternatively activated macrophages in the other. These two phases of activation are characterized by the expression of two enzymes that are reciprocally regulated and share the same substrate: L-arginine. Classically activated macrophages express iNOS of NOS2 that is induced by Th1 cytokines and catalyze the conversion of L-arginine to L-citrulline and NO. Contrarily, alternatively activated macrophages express Arg-1 that is induced by Th2 cytokines. Thus, L-arginine metabolism is in the center of *Trypanosoma* elimination of survival. The better knowledge of this route during the different stages of *Trypanosoma* infection is of great importance for the better comprehension of disease progression and design of drugs.

Acknowledgements

This work was funded by project number IN218119 from Papiit, DGAPA, UNAM, to LGK.

Conflict of interest

Authors declare no conflict of interests.

Author details

Laila Gutiérrez-Kobeh* and Arturo A. Wilkins-Rodríguez
Research Unit UNAM-INC, Research Division, School of Medicine, National Autonomous University of Mexico-National Institute of Cardiology "Ignacio Chávez", Mexico City, Mexico

*Address all correspondence to: lgutierr@unam.mx

References

[1] Coura JR, Vias PA. Chagas disease: A new worldwide challenge. Nature. 2010;**465**(7301):S6-S7. DOI: 10.1038/nature09221

[2] Tibayrenc M. Genetic subdivisions within *Trypanosoma cruzi* (discrete typing units) and their relevance for molecular epidemiology and experimental evolution. Kinetoplastid Biology and Disease. 2003;**2**(1):12. DOI: 10.1186/1475-9292-2-12

[3] Andrade SG, Andrade ZA. Pathology of prolonged experimental Chagas disease. Revista do Instituto de Medicina Tropical de Sao Paulo. 1968;**10**(3):180-187

[4] Laguens RP, Meckert PC, Gelpi RJ. Chronic Chagas disease in the mouse. I. Electrocardiographic and morphological patterns of the cardiopathy. Medicina (Buenos Aires). 1981;**41**(1):35-39

[5] Pinto Dias JC. The indeterminate form of human chronic Chagas disease. A clinical epidemiological review. Revista da Sociedade Brasileira de Medicina Tropical. 1989;**22**(3):147-156

[6] Morris SM Jr. Arginine: Beyond protein. American Journal of Clinical Nutrition. 2006;**83**(2):508S-512S. DOI: 10.1093/ajcn/83.2.508S

[7] Morris SM Jr. Arginases and arginine deficiency syndromes. Current Opinion in Clinical Nutrition and Metabolic Care. 2012;**15**(1):64-70. DOI: 10.1097/MCO.0b013e32834d1a08

[8] Bernard AC, Mistry SK, Morris SM Jr, O'Brian WE, Tsuei BJ, Maley ME, et al. Alterations in arginine metabolic enzymes in trauma. Shock. 2001;**15**(3):215-219

[9] Luiking YC, Hallemeesch MM, Vissers YL, Lamers WH, Deutz NE. In vivo whole body and organ arginine metabolism during endotoxemia (sepsis) is dependent on mouse strain and gender. Journal of Nutrition. 2004;**134**(10 Suppl):2768S-2774S; discussion 2796S–2797S. DOI: 10.1093/jn/134.10.2768S

[10] Bronte V, Zanovello P. Regulation of immune responses by L-arginine metabolism. Nature Reviews Immunolology. 2005;**5**(8):641-654. DOI: 10.1038/nri1668

[11] Wu G, Morris SM Jr. Arginine metabolism: Nitric oxide and beyond. Biochemical Journal. 1998;**336**:1-17. DOI: 10.1042/bj3360001

[12] Mori M, Gotoh T. Arginine metabolic enzymes, nitric oxide and infection. Journal of Nutrition. 2004;**134**(10 Suppl):2820S-2825S; discussion 2853S. DOI: 10.1093/jn/134.10.2820S

[13] Hofmann F, Kreusch J, Maier KP, Munder PG, Decker K. The urea cycle in different types of macrophages. Biochemical Society Transactions. 1978;**6**(5):990-993. DOI: 10.1042/bst0060990

[14] Munder M. Arginase: An emerging key player in the mammalian immune system. British Journal of Pharmacology. 2009;**158**(3):638-651. DOI: 10.1111/j.1476-5381.2009.00291.x

[15] Forstermann U, Sessa WC. Nitric oxide synthases: Regulation and function. European Heart Journal. 2012;**33**:829-837, 837a-837d. DOI: 10.1093/eurheartj/ehr304

[16] Forstermann U, Closs EI, Pollock JS, Nakane M, Schwarz P, Gath I, et al. Nitric oxide synthase isozymes. Characterization, purification, molecular cloning, and functions. Hypertension. 1994;**23**:1121-1131

[17] Mayer AK, Bartz H, Fey F, Schmidt LM, Dalpke AH. Airway epithelial cells modify immune responses by inducing an anti-inflammatory microenvironment. European Journal of Immunology. 2008;**38**(6):1689-1699. DOI: 10.1002/eji.200737936

[18] Munder M, Mollinedo F, Calafat J, Canchado J, Gil Lamaignere C, Fuentes JM, et al. Arginase I is constitutively expressed in human granulocytes and participates in fungicidal activity. Blood. 2005;**105**(6):2549-2556. DOI: 10.1182/blood-2004-07-2521

[19] Munder M, Eichmann K, Morán JM, Centeno F, Soler G, Modolell M. Th1/Th2-regulated expression of arginase isoforms in murine macrophages and dendritic cells. Journal of Immunology. 1999;**163**(7):3771-3777

[20] Wilkins-Rodriguez AA, Escalona-Montaño AR, Becker I, Gutiérrez-Kobeh L. Regulation of the expression of nitric oxide synthase by *Leishmania mexicana* amastigotes in murine dendritic cells. Experimental Parasitology. 2010;**126**(3):426-434. DOI: 10.1016/j.exppara.2010.07.014

[21] Wanasen N, Soong L. L-arginine metabolism and its impact on host immunity against *Leishmania* infection. Immunology Research. 2008;**41**(1):15-25. DOI: 10.1007/s12026-007-8012-y

[22] Das PA, Lahiri A, Chakravorty D. Modulation of the arginase pathway in the context of microbial pathogenesis: A metabolic enzyme moonlighting as an immune modulator. PLoS Pathogens. 2010;**6**(6):e1000899. DOI: 10.1371/journal.ppat.1000899

[23] Mosser DM, Edwards JP. Exploring the full spectrum of macrophage activation. Nature Reviews Immunology. 2008;**8**(12):958-969. DOI: 10.1038/nri2448

[24] Modolell M, Corraliza IM, Link F, Soler G, Eichmann K. Reciprocal regulation of the nitric oxide synthase/arginase balance in mouse bone marrow-derived macrophages by TH1 and TH2 cytokines. European Journal of Immunology. 1995;**25**(4):1101-1104. DOI: 10.1002/eji.1830250436

[25] Nathan C, Xie QW. Regulation of biosynthesis of nitric oxide. Journal of Biological Chemistry. 1994;**269**(19):13725-13728

[26] MacMicking J, Xie QW, Nathan C. Nitric oxide and macrophage function. Annual Review of Immunology. 1997;**15**:323-350. DOI: 10.1146/annurev.immunol.15.1.323

[27] Munder M, Eichmann K, Modolell M. Alternative metabolic states in murine macrophages reflected by the nitric oxide synthase/arginase balance: Competitive regulation by $CD4^+$ T cells correlates with Th1/Th2 phenotype. Journal of Immunology. 1998;**160**(11):5347-5354

[28] Xia Y, Dawson VL, Dawson TM, Snyder SH, Zweier JL. Nitric oxide synthase generates superoxide and nitric oxide in arginine-depleted cells leading to peroxynitrite-mediated cellular injury. Proceedings of the National Academy of Sciences USA. 1996;**93**(13):6770-6774

[29] Bogdan C. Nitric oxide and the immune response. Nature Immunology. 2001;**2**(10):907-916. DOI: 10.1038/ni1001-907

[30] Stempin C, Tanos TB, Coso OA, Cerbán FM. Arginase induction promotes *Trypanosoma cruzi* intracellular replication in cruzipain-treated J774 cells through the activation of multiple signaling pathways. European Journal of Immunology. 2004;**34**:200-209. DOI: 10.1002/eji.200324313

[31] Stempin C, Giordanengo L, Gea S, Cerbán F. Alternative activation and

increase of *Trypanosoma cruzi* survival in murine macrophages stimulated by cruzipain, a parasite antigen. Journal of Leukocyte Biology. 2002;**72**:727-734. DOI: 10.1189/jlb.72.4.727

[32] Vincendeau P, Daulouède S. Macrophage cytostatic effect on *Trypanosoma musculi* involves an L-arginine-dependent mechanism. Journal of Immunology. 1991;**146**:4338-4343

[33] Bogdan C. Nitric oxide synthase in innate and adaptive immunity: An update. Trends in Immunology. 2015;**36**:161-178. DOI: 10.1016/j.it.2015.01.003

[34] Nogueira N, Cohn Z. *Trypanosoma cruzi*: Mechanism of entry and intracellular fate in mammalian cells. Journal of Experimental Medicine. 1976;**143**(6):1402-1420. DOI: 10.1084/jem.143.6.1402

[35] Metz G, Carlier Y, Vray B. *Trypanosoma cruzi* upregulates nitric oxide release by IFN-γ-preactivated macrophages, limiting cell infection independently of the respiratory burst. Parasite Immunology. 1993;**15**:693-699. DOI: 10.1111/j.1365-3024.1993.tb00584.x

[36] Pakianathan DR, Kuhn RE. *Trypanosoma cruzi* affects nitric oxide production by murine peritoneal macrophages. Journal of Parasitology. 1994;**80**:432-437

[37] De Muylder G, Daulouède S, Lecordier L, Uzureau P, Morias Y, Van DenAbbeele J, et al. A *Trypanosoma brucei* kinesin heavy chain promotes parasite growth by triggering host arginase activity. PLoS Pathogens. 2013;**9**:e1003731. DOI: 10.1371/journal.ppat.1003731

[38] Hai Y, Kerkhoven EJ, Barrett MP, Christianson DW. Crystal structure of an arginase-like protein from *Trypanosoma brucei* that evolved without a binuclear manganese cluster. Biochemistry. 2015;**54**:458-471. DOI: 10.1021/bi501366a

[39] Namangala B, De Baetselier P, Noël W, Brys L, Beschin A. Alternative versus classical macrophage activation during experimental African trypanosomiasis. Journal of Leukocyte Biology. 2001;**69**:387-396. DOI: 10.1189/jlb.69.3.387

[40] Gobert AP, Daulouede S, Lepoivre M, Boucher JL, Bouteille B, Buguet A, et al. L-Arginine availability modulates local nitric oxide production and parasite killing in experimental trypanosomiasis. Infection and Immunity. 2000;**68**:4653-4657. DOI: 10.1128/IAI.68.8.4653-4657.2000

[41] Raes G, Brys L, Dahal BK, Brandt J, Grooten J, Brombacher FG, et al. Macrophage galactose-type C-type lectins as novel markers for alternatively activated macrophages elicited by parasitic infections and allergic airway inflammation. Journal of Leukocyte Biology. 2005;**77**:321-327. DOI: 10.1189/jlb.0304212

[42] Fairlamb AH, Cerami A. Metabolism and functions of trypanothione in the Kinetoplastida. Annual Reviews Microbiology. 1992;**46**:695-729. DOI: 10.1146/annurev.mi.46.100192.003403

[43] Priotto G, Kasparian S, Mutombo W, Ngouama D, Ghorashian S, Arnold U, et al. Nifurtimoxeflornithine combination therapy for second-stage African *Trypanosoma brucei gambiense* trypanosomiasis: A multicentre, randomised, phase III, non-inferiority trial. Lancet. 2009;**374**:56-64. DOI: 10.1016/S0140-6736(09)61117-X

Chapter 8

The Journey of *Trypanosoma cruzi* under the Redox Baton

Marcia Cristina Paes

Abstract

Trypanosoma cruzi is a protozoan responsible for Chagas disease and has a complex life cycle including vertebrate (mammals) and invertebrate (insects) hosts. The parasite presents proliferative and infective forms that are challenged throughout their cycle as different sources of nutrients, pH, immune system, and levels of reactive oxygen species (ROS). Although ROS cause damage to cells and tissues when their levels are controlled, they are involved in signal transduction pathways involved in cell growth and differentiation. Curiously, the proliferation of epimastigote inside the bug insect is favored by high levels of ROS from the digestion of blood meal, and it is regulated by a cellular signaling mechanism involving heme and CaMKII. On the other hand, the differentiation of epimastigote into metacyclic trypomastigote in the rectum occurs in the reduced state. Interestingly, when the parasite infects the vertebrate, the immune system recognizes this pathogen and macrophages become activated. Thus, NADPH oxidase produces ROS that helps the parasite enter the mammalian cells, improving the infection. The parasite thrives inside the mammalian cells also involving ROS. Thus, the life cycle of *Trypanosoma cruzi* obeys a fine tuning of the redox state, not affecting the host cells and being helpful to the parasite.

Keywords: redox state, reactive oxygen species, redox signaling, host-parasite interaction, iron

1. Introduction

1.1 Chagas disease

Chagas disease was described in 1909 by a Brazilian researcher, Carlos Chagas, who discovered a new trypanosomiasis in Minas Gerais, Brazil, during his work on an anti-malaria campaign [1, 2]. The disease presents three phases: acute, indeterminate, and chronic. The acute phase is asymptomatic and presents nonspecific symptoms and signs, such as inflammatory lesions at the site of entry of the parasite (chagoma) and fever. At this stage, the parasitic load in the blood is high. The indeterminate phase is characterized by the presence of antibodies against *T. cruzi* and absence of clinical manifestations of the disease. The chronic phase of the disease involves the cardiac system, digestive system, or both. The patients at this stage may develop (1) a Chagas' heart disease that compromises the cardiac function by increasing the size of the heart and tissue damage (fibrosis) and (2) chronic inflammation and destruction of parasympathetic neurons leading to progressive enlargement of the esophagus (megaesophagus), sigmoid colon, or rectum (megacolon) [3, 4].

1.2 The biological cycle of *Trypanosoma cruzi*

The causative agent of the disease is a flagellate protozoan that belongs to the order Kinetoplastida and family Trypanosomatidae, the *Trypanosoma cruzi*. Classically, transmission of Chagas disease occurs through insect vectors of the subfamily Triatominae, popularly known as barbers. However, there are other transmission routes such as oral, congenital, blood transfusion, organ transplantation, and laboratory accidents [5].

T. cruzi presents different evolutionary forms that alternate with each other throughout its cycle. Trypomastigote forms present in the blood of infected vertebrate hosts are ingested by the insect vector, where they differentiate into epimastigote replicative forms. These forms undergo another process of differentiation througout the intestine, however this time to infectious and non-replicative forms, the metacyclic trypomastigotes. In turn, these are released with the feces and penetrate into the vertebrate host through the sting of the triatomines or through another portal of entry, such as mucosae. Trypomastigotes are able to invade host cells and differentiate into amastigotes in the intracellular environment. Such forms multiply through binary divisions and are transformed into the infective trypomastigote forms still within the host cell. With the disruption of the plasma membrane of the vertebrate host cell, the trypomastigote forms gain the bloodstream and invade other cells and tissues of the mammal or can be sucked in by new triatomine restarting the cycle [6].

It is known that during its life cycle *T. cruzi* is exposed to different redox environments inside the invertebrate and vertebrate hosts [7, 8] and the ability of *T. cruzi* to adapt to the redox state contributes to the success of the infection [9]. Additionally, in terms of a physiological approach, ROS play a vital role in *T. cruzi*-vector interactions, because heme, a molecule from the insect blood digestion, triggers epimastigote proliferation through a redox-sensitive signaling mechanism [10].

1.3 Redox signaling

Cells generate ROS endogenously and constitutively when oxygen is partially reduced in mitochondria-producing oxidants, the so-called reactive oxygen species (ROS), such as superoxide radicals ($O_2^{\bullet-}$) and hydrogen peroxide (H_2O_2) [11]. To maintain their hemostasis, cells adopt strategies called antioxidant defense. ROS participate in signal transduction pathways involved in cell growth and differentiation [12]. However, when oxidant levels are high, the oxidative/antioxidant balance within the cells disrupts the redox signaling and the redox control, which can lead to cellular damage [13–16]. This exacerbation of the endogenous production of ROS is known as oxidative stress. These oxidant species can lead to lipid peroxidation, affecting membrane integrity, DNA damage, and oxidation of sugars and protein thiols [14, 15]. On the other hand, controlled ROS increase leads to a temporary imbalance that represents the physiological basis for redox regulation [16, 17]. Indeed, redox processes have fundamental implications in biology.

In addition to ROS, other reactive species have notable impacts on redox biology, including the reactive nitrogen species (RNS), such as nitric oxide, nitrogen dioxide (both free radicals), peroxynitrite, and nitrite/nitrate. Besides these, forms of cysteine, methionine, and some low-molecular-mass compounds such as glutathione and trypanothione are called reactive sulfur species (RSS). Another group of reactive species is the reactive carbonyl species (RCS) including various forms of metabolically generated aldehydes and electronically excited (triplet) carbonyls. Finally, reactive selenium species (RSeS) include low molecular mass such as selenocysteine and selenomethionine residues in proteins [17].

2. ROS and *Trypanosoma cruzi*

2.1 The journey inside the bug insect

2.1.1 The epimastigotes and redox environment

Evidence in the literature indicates that the interaction between *T. cruzi* and triatomines is essential for the successful spread of Chagas disease [18] and several factors and molecules have been shown to be important in establishing the infection. After feeding, the insect vector digests the blood in the midgut, where hemoglobin protein degradation occurs and a large amount of heme is released. Heme is a molecule known to increase the formation of reactive oxygen species (ROS) and is able to alter membrane selectivity and permeability [19, 20]. These reactive species can also be generated as a by-product of aerobic metabolism of the parasite [7, 21]. Therefore, the former region of the midgut represents an environment rich in nutrients, but it is potentially an oxidative environment as well. Then, *Trypanosoma cruzi* needs to deal with high concentrations of heme and ROS while inhabiting the midgut of the vector. The epimastigote form, present in this environment, is the replicative and non-infective form that is able to increase its rate of proliferation in the presence of heme in a dose-dependent manner [22], and this heme-induced *T. cruzi* growth is associated with calcium-calmodulin-dependent kinase II (CaMKII) activity [23].

Besides heme, ROS have been shown to trigger proliferation of the epimastigote forms of *Trypanosoma cruzi* [10]. According to these authors, the growth of the parasites in the presence of these molecules is regulated by a cellular signaling mechanism involving CaMKII and the redox status, since the antioxidants, such as urate and GSH, inhibited heme-induced ROS and parasite proliferation. In addition, Myr-AIP, a specific CaMKII inhibitor, extinguishes heme-induced ROS in epimastigotes, decreasing parasite growth. To exclude the possibility of other molecules similar to heme being able to induce a potent proliferative effect on *T. cruzi*, tests were carried out with protoporphyrin IX (PPIX), mesoporphyrin IX (MPIX), Fe mesoporphyrin IX (Fe-MPIX), Sn protoporphyrin IX (SnPPIX), and Zn protoporphyrin IX (ZnPPIX), and only heme showed a potent proliferative effect [10]. These data show that the parasite had to adapt to high concentrations of ROS in order to establish itself in such an oxidizing environment.

Heme and two classical oxidants, H_2O_2 and the well-known superoxide generator, paraquat, are able to promote the growth of epimastigotes in vitro [24]. This effect was reversed in the presence of other reductive molecules (GSH, a thiol-based antioxidant found in the hemolymph of triatomines; urate, an important antioxidant rich in the urine of these insects [25], and n-acetylcysteine (NAC), a classic antioxidant) suggesting a competition between these molecules of antagonistic redox status. An important physiological molecule present in the midgut is hemozoin, a crystal composed of heme dimers [26, 27] that *Rhodnius prolixus*, a Chagas disease vector, uses as an efficient detoxification pathway of heme. The addition of this crystal to an epimastigote culture does not produce an increase in the proliferation of these cells [24].

Thus, the redox environment is considered to be very important for *Trypanosoma cruzi*. Furthermore, the parasite needs ROS for growth inside the vector. If this hypothesis is correct, the disturbance of ROS levels in vivo would lead to differences in the levels of epimastigote proliferation within the intestine. In fact, the heme molecule and ROS are examples of important relationships between parasite and vector because they are capable of promoting the proliferation of the epimastigote forms, but when the insect is fed with blood and antioxidants, such as NAC and urate, the proliferation in vivo decreases as demonstrated in vitro [10, 24].

Observing the effect of the physiological molecules (heme, hemozoin, and urate), it is possible to confirm that there is a modulation between molecules of antagonistic redox status, indicating an inhibitory role of reductive molecules on epimastigote proliferation and confirming the requirement of an oxidant signal to promote the growth of these parasites. Furthermore, in 2017, Nogueira and collaborators showed that heme affects the mitochondrial function of *T. cruzi* epimastigotes and, as a consequence, mitochondrial ROS production is increased, triggering parasite proliferation [28].

2.1.2 Differentiation of epimastigotes into metacyclic trypomastigotes

Still on its journey inside the vector, *Trypanosoma cruzi* reaches the rectum of the bug. This region greatly favors metacyclogenesis, and one important factor is the reductive environment promoted by the high concentration of urate. The levels of metacyclic trypomastigotes are increased in the presence of urate and other antioxidants both in vitro and in vivo. On the other hand, the proliferation of epimastigotes decreases in reductive environments [10, 24].

When the blood meal is supplemented with antioxidants, there is a shift in the redox status of the gut compartments (anterior midgut, posterior midgut, and rectum), increasing differentiation of the parasites in an unusual midgut region and greatly favoring metacyclogenesis in the bug rectum. Notably, contrary to proliferation, the differentiation process appears to be favored by reductive environments [24].

A *Trypanosoma cruzi* eIF2α kinase (TcK2) was characterized by Augusto and collaborators [29] as a transmembrane protein located in organelles that accumulate nutrients in the proliferative forms. The heme molecule has been shown to bind specifically to the catalytic domain of the kinase, inhibiting its activity. On the other hand, in the absence of heme, TcK2 is activated, preventing cell growth and inducing the differentiation of epimastigote forms into infectious and nonproliferative forms. Parasites without TcK2 lose this differentiation ability, and heme is not stored in reserve organelles, as demonstrated by Lara and collaborators [21], remaining in the cytosol. Furthermore, if ROS levels are not controlled in TcK2 null, they cause damage to the parasite, including death. Thus, in wild cells, heme has been shown to be a key factor for growth control and differentiation by regulating an unusual type of eIF2α kinase in *T. cruzi* [29].

As demonstrated by science, the coevolution between parasites and their insect vectors has promoted an elegant strategy for the development and maintenance of the protozoa in the invertebrate vector.

2.2 The transmission of the disease: metacyclic trypomastigotes infect the vertebrate hosts—a new journey

2.2.1 The participation of NADPH oxidase in the infection

The immune system of the higher vertebrates is able to recognize pathogens and respond through their innate immune responses. ROS is an important component of this response produced by phagocytes and can be highly toxic. Macrophages are one of the first lines of defense in mammals, especially against pathogens [30], and become activated facing such challenges.

The $O_2^{\bullet -}$ production after NADPH oxidase activation in macrophages is converted inside the phagosome to H_2O_2 (spontaneously or via superoxide dismutase), and this ROS production, termed the "oxidative burst" of activated phagocytic cells, usually kills the pathogens. In order to infect the vertebrate host, *T. cruzi* metacyclic trypomastigotes invade macrophages and overcome the highly oxidative conditions generated inside the phagosome. Then, biochemical changes occur [9, 31, 32] including antioxidant enzyme activities [33], and, curiously, *Trypanosoma cruzi* depends on ROS involved in this activation process to establish the infection in the vertebrate host [8]. The NADPH oxidase (Phox) activation and this $O_2^{\bullet-}$ production are directly involved in increased infection of macrophages by *T. cruzi* since mice deficient in the $gp91^{phox}$ (Phox KO), subunit of NADPH oxidase, macrophages present reduced parasitism [8, 34].

Peroxynitrite is also highly lethal and used by phagocytes against pathogens. It is formed when nitric oxide (NO) and $O_2^{\bullet-}$ react with each other. Thus, the production of peroxynitrite is decreased by the inhibition of ROS or NO production [35]. Paiva and collaborators, in 2012, showed that macrophages infected with T. cruzi and activated with the burst inducer phorbol 12-myristate 13-acetate (PMA) have stimulated the parasite load [36]. In conclusion, the generation and the regulation of the ROS level can help these parasites thrive in an oxidative environment [8, 35–37].

2.2.2 Murine models of Chagas disease and ROS

After the infective metacyclic forms invade host cells, macrophages, or cardiac cells, for example, they are transformed into the replicative intracellular amastigote form [6]. In response to infection, Chagas hearts present increased mitochondrial ROS [38, 39] because during *T. cruzi* infection an inefficient electron transport for ATP synthesis occurs in mitochondria [39]. Also, when deficient superoxide dismutase (SOD2 or MnSOD) mice are infected with *Trypanosoma cruzi,* the loss of the mitochondrial function increases the oxidative damage of the myocardium in Chagas cardiomyopathy and shows the importance of ROS-level regulations [40]. Moreover, ROS mobilizes intracellular iron which is essential as a cellular factor for amastigote division [30, 36]. ROS, including mitochondrial ROS, contribute to oxidative damage that persists during the chronic stage of infection and is involved in the functional impairment of the heart [40–42]. Some studies show that cardiac parasite load may vary after treatment with antioxidants but depend on the animal model and the strain used [42–44]. In fact, Gupta and collaborators [45] demonstrated that *T. cruzi* infection increases ROS production in cardiomyocytes and this effect is augmented by the pro-inflammatory cytokines. The authors argue that the ROS production by cardiomyocytes is not a defense response against *T. cruzi*. Instead, the infection promotes a mitochondrial dysfunction, including ROS production. Thus, ROS also participates in the successful infection in mammals.

3. Conclusion

Several groups have carried out research on the influence of the oxidative environment on the growth and differentiation of *Trypanosoma cruzi* in both vertebrate and invertebrate hosts. As we have learned in this chapter, the epimastigote, the non-infective and proliferative form, has its growth stimulated in the presence of oxidative compounds. Conversely, in the presence of antioxidants, or in a reductive environment, its proliferation becomes compromised. The regulated ROS levels also influence, in an orchestrated way, the differentiation of epimastigotes into metacyclic trypomastigotes (the infective form). The reductive environment increases differentiation, while ROS dramatically decreases its transformation into the infective forms. These same metacyclic forms that are formed in the rectum of the vector insect invade the vertebrate host by subverting the logic of the phagocytes that, by activation of NADPH oxidase, exacerbate the concentration of ROS in the intention to kill the pathogens. In fact, the trypomastigote forms of *T. cruzi* resist ROS and establish themselves in the cells of the vertebrate host differing into amastigotes that in cardiomyocytes coexist with increased levels of ROS when compared to uninfected hearts. However, these levels of ROS cannot decrease or increase indiscriminately.

Thus, we have followed the journey of the parasite *Trypanosoma cruzi*, both in the invertebrate and in the vertebrate hosts, that occurs under adverse redox conditions, as if in an orchestra of ROS and antioxidants, and furthermore we can observe that its journey through the intestine of the insect, along the mammalian bloodstream, and its entry and lodging in mammalian cells are finely and elegantly ruled by a redox baton.

Conflict of interest

There is no conflict of interest.

Funding

This work was supported by grants from the Conselho Nacional de Desenvolvimento Cientifico e Tecnologico (CNPq), Fundação Carlos Chagas Filho de Amparo a Pesquisa do Estado do Rio de Janeiro (FAPERJ), Coordenação de Aperfeiçoamento de Pessoal de Nível Superior (CAPES), and Instituto Nacional de Ciência e Tecnologia-Entomologia Molecular (INCT-EM).

Author details

Marcia Cristina Paes[1,2]

1 Departamento de Bioquímica, Laboratório de Interação de Tripanosomatídeos e Vetores, Instituto de Biologia, Universidade do Estado do Rio de Janeiro (UERJ), Brazil

2 Instituto Nacional de Ciência e Tecnologia, Entomologia Molecular (INCT-EM), Brazil

*Address all correspondence to: marcia.paes.uerj@gmail.com

References

[1] Chagas C. Nova tripanossomíase humana. Estudos sobre a morfologia e o ciclo evolutivo do *Schizotrypanum cruzi* n. gen., n. sp., agente etiológico de nova entidade mórbida do homem. Memórias do Instituto Oswaldo Cruz. 1909;**1**:159-218

[2] Kropf SP. Doença de Chagas, doença do Brasil: Ciência, Saúde e Nação (1909-1962). Rio de Janeiro: Editora Fiocruz; 2009

[3] Machado FS, Jelicks LA, Kirchhoff LV, Shirani J, Nagajyothi F, Mukherjee S, et al. Chagas heart disease: Report on recent developments. Cardiology in Review. 2012;**20**(2):53-65

[4] Chatelain E. Chagas disease drug discovery: Toward a new era. Journal of Biomolecular Screening. 2015;**20**:22-35

[5] Martins-Melo FR, Ramos AN Jr, Alencar CH, Heukelbach J. Mortality due to Chagas disease in Brazil from 1979 to 2009: Trends and regional differences. Journal of Infection in Developing Countries. 2012;**6**(11):817-824

[6] Rassi A, Rassi A, Marin-Neto JA. Chagas disease. The Lancet. 2010;**375**:1388-1402

[7] Graça-Souza AV, Maya-Monteiro C, Paiva-Silva G, Braz GRC, Paes MC, Sorgine MHF, et al. Adaptations against heme toxicity in blood-feeding arthropods. Insect Biochemistry and Molecular Biology. 2006;**36**:322-335

[8] Goes GR, Rocha PS, Diniz ARS, Aguiar PHN, Machado CR, Vieira LQ. *Trypanosoma cruzi* needs a signal provided by reactive oxygen species to infect macrophages. PLoS Neglected Tropical Diseases. 2016;**10**(4):e0004555

[9] Piacenza L, Zago MP, Peluffo G, Alvarez MN, Basombrio MA, Radi R. Enzymes of the antioxidant network as novel determiners of *Trypanosoma cruzi* virulence. International Journal of Parasitology. 2009;**39**:1455-1464

[10] Nogueira NP, Souza CF, Saraiva FM, Sultano PE, Dalmau SR, Bruno RE, et al. Heme-induced ROS in *Trypanosoma cruzi* activates CaMKII-like that triggers epimastigote proliferation. One helpful effect of ROS. PLoS One. 2011;**6**(10):e25935

[11] Boveris A, Chance B. The mitochondrial generation of hydrogen peroxide. General properties and effect of hyperbaric oxygen. The Biochemical Journal. 1973;**134**:707-716

[12] Droge W. Free radicals in the physiological control of cell function. Physiological Reviews. 2002;**82**:47-95

[13] Jones D. Redefining oxidative stress. Antioxidants & Redox Signaling. 2006;**8**:1865-1879

[14] Halliwell B, Gutteridge JMC. Free Radicals in Biology and Medicine. 4th ed. Oxford: Clarendon Press; 2007

[15] Sies H, Jones DP. Oxidative stress. In: Encyclopedia of Stress. 2nd ed. Vol. 3. Amsterdam: Elsevier; 2007. pp. 45-48

[16] Jones D. Radical-free biology of oxidative stress. American Journal of Physiology. Cell Physiology. 2008;**295**:C849-C868

[17] Sies H, Berndt C, Jones DP. Oxidative stress. Annual Review of Biochemistry. 2017;**86**:715-748

[18] Garcia ES, Ratcliffe NA, Whitten MM, Gonzalez MS, Azambuja P. Exploring the role of insect host factors in the dynamics of *Trypanosoma cruzi-Rhodnius prolixus* interactions. Journal of Insect Physiology. 2007;**53**:11-21

[19] Schmitt TH, Frezzatti WA, Schreier S. Hemin-induced lipid membrane disorder and increased permeability: A molecular model for the mechanism of cell lysis. Archives of Biochemistry and Biophysics. 1993;**307**:96-103

[20] Ryter SW, Tyrrell RM. The heme synthesis and degradation pathways: Role in oxidant sensitivity. Heme oxygenase has both pro- and antioxidant properties. Free Radical Biology & Medicine. 2000;**28**:289-309

[21] Finzi JK, Chiavegatto CWM, Corat KF, et al. *Trypanosoma cruzi* response to the oxidative stress generated by hydrogen peroxide. Molecular and Biochemical Parasitology. 2004;**133**:37-43

[22] Lara FA, Sant'Anna C, Lemos D, Laranja GAT, Coelho MGP, et al. Heme requirement and intracellular trafficking in *Trypanosoma cruzi* epimastigotes. Biochemical and Biophysical Research Communications. 2007;**355**:16-22

[23] Souza CF, Carneiro AB, Silveira AB, Laranja GAT, Silva-Neto MAC, et al. Heme-induced *Trypanosoma cruzi* proliferation is mediated by CaM kinase II. Biochemical and Biophysical Research Communications. 2009;**390**:541-546. DOI: 10.1016/j.bbrc.2009.09.135

[24] Nogueira NP, Saraiva FM, Sultano PE, Cunha PR, Laranja GA, Justo GA, et al. Proliferation and differentiation of *Trypanosoma cruzi* inside its vector have a new trigger: Redox status. PLoS One. 2015;**10**:e0116712

[25] Wigglesworth VB. The physiology of excretion in a blood-sucking insect. *Rhodnius prolixus.* III. The mechanism of uric acid excretion. The Journal of Experimental Biology. 1931;**8**:443-451

[26] Stiebler R, Timm BL, Oliveira PL, Hearne GR, Egan TJ, et al. On the physico-chemical and physiological requirements of hemozoin formation promoted by perimicrovillar membranes in *Rhodnius prolixus* midgut. Insect Biochemistry and Molecular Biology. 2010;**40**:284-292. DOI: 10.1016/j.ibmb.2009.12.013

[27] Ferreira CM, Stiebler R, Saraiva FM, Lechuga GC, Walter-Nuno AB, Bourguignon SC, et al. Heme crystallization in a Chagas disease vector acts as a redox-protective mechanism to allow insect reproduction and parasite infection. PLoS Neglected Tropical Diseases. 2018;**12**(7):e0006661. DOI: 10.1371/journal.pntd.0006661

[28] Nogueira NP, Saraiva FMS, Oliveira MP, et al. Heme modulates *Trypanosoma cruzi* bioenergetics inducing mitochondrial ROS production. Free Radical Biology & Medicine. 2017;**108**:183-191

[29] Augusto LS, Moretti NS, Ramos TCP, de Jesus TCL, Zhang M, Castilho BA, et al. A membrane-bound eIF2 alpha kinase located in endosomes is regulated by heme and controls differentiation and ROS levels in *Trypanosoma cruzi*. PLoS Pathogens. 2015;**11**:1-27

[30] Paiva CN, Bozza MT. Are reactive oxygen species always detrimental to pathogens? Antioxidants & Redox Signaling. 2014;**20**:1000-1037

[31] Kierszenbaum F, Knecht E, Budzko DB, Pizzimenti MC. Phagocytosis: A defense mechanism against infection with *Trypanosoma cruzi*. Journal of Immunology. 1974;**112**:1839-1844

[32] Atwood JA, Weatherly DB, Minning TA, Bundy B, Cavola C, Opperdoes FR, et al. The *Trypanosoma cruzi* proteome. Science. 2005;**309**:473-476

[33] Freire ACG, Alves CL, Goes GR, Resende BC, Moretti NS, Nunes VS, et al. Catalase expression impairs

oxidative stress-mediated signaling in *Trypanosoma cruzi*. Parasitology. 2017;**144**(11):1498-1510. DOI: 10.1017/S0031182017001044

[34] Melo RC, Fabrino DL, D'Avila H, Teixeira HC, Ferreira AP. Production of hydrogen peroxide by peripheral blood monocytes and specific macrophages during experimental infection with *Trypanosoma cruzi in vivo*. Cell Biology International. 2003;**27**:853-861

[35] Alvarez MN, Peluffo G, Piacenza L, Radi R. Intraphagosomal peroxynitrite as a macrophage-derived cytotoxin against internalized *Trypanosoma cruzi*: Consequences for oxidative killing and role of microbial peroxiredoxins in infectivity. The Journal of Biological Chemistry. 2011;**286**:6627-6640. DOI: 10.1074/jbc.M110.167247

[36] Paiva CN et al. Oxidative stress fuels *Trypanosoma cruzi* infection in mice. The Journal of Clinical Investigation. 2012;**122**(7):2531-2542

[37] Andrews NW. Oxidative stress and intracellular infections: More iron to the fire. Journal of Clinical Investigation. 2012;**122**(7):2352-2354. DOI: 10.1172/JCI64239

[38] Wen JJ, Garg NJ. Manganese superoxide dismutase deficiency exacerbates the mitochondrial ROS production and oxidative damage in Chagas disease. PLoS Neglected Tropical Diseases. 2018;**12**(7):e0006687. DOI: 10.1371/journal.pntd.0006687

[39] Wen J-J, Garg NJ. Mitochondrial complex III defects contribute to inefficient respiration and ATP synthesis in the myocardium of *Trypanosoma cruzi*-infected mice. Antioxidants & Redox Signaling. 2010;**12**:27, 10.1089/ARS.2008.2418-37

[40] Wen JJ, Garg NJ. Mitochondrial generation of reactive oxygen species is enhanced at the Q(o) site of the complex III in the myocardium of *Trypanosoma cruzi*-infected mice: Beneficial effects of an antioxidant. Journal of Bioenergetics and Biomembranes. 2008;**40**:587-598. DOI: 10.1007/s10863-008-9184-4

[41] Machado-Silva A, Cerqueira PG, Grazielle-Silva V, Gadelha FR, Peloso EF, Teixeira SMR, et al. How *Trypanosoma cruzi* deals with oxidative stress: Antioxidant defence and DNA repair pathways. Mutation Research. 2016;**767**:8-22

[42] Paiva CN, Medei E, Bozza MT. ROS and *Trypanosoma cruzi*: Fuel to infection, poison to the heart. PLoS Pathogens. 2018;**14**(4):e1006928. DOI: 10.1371/journal.ppat.1006928

[43] Dias PP, Capila RF, do Couto NF, Estrada D, Gadelha FR, Radi R, et al. Cardiomyocyte oxidants production may signal to *T. cruzi* intracellular development. PLoS Neglected Tropical Diseases. 2017;**11**(8):e0005852. DOI: 10.1371/journal.pntd.0005852

[44] Dhiman M, Garg NJ. P47phox−/− mice are compromised in expansion and activation of CD8+ T cells and susceptible to *Trypanosoma cruzi* infection. PLoS Pathogens. 2014;**10**(12):e1004516. DOI: 10.1371/journal.ppat.1004516

[45] Gupta S, Bhatia V, Wen JJ, Wu W, Huang MH, Garg NJ. *Trypanosoma cruzi* infection disturbs mitochondrial membrane potential and ROS production rate in cardiomyocytes. Free Radical Biology & Medicine. 2009;**47**(10):1414-1421. DOI: 10.1016/j.freeradbiomed.2009.08.008

Section 3

Parasite Host Cell Interaction

Chapter 9

How Does the Main Infective Stage of *T. cruzi* Enter and Avoid Degradation in Host Cells? A Description of the Pathways and Organelles Involved on These Processes

Emile Barrias, Lissa C. Reignault and Wanderley de Souza

Abstract

Trypanosoma cruzi, the etiological agent of Chagas disease, is an intracellular parasite that targets specific proteins of the host cell resulting in the generation of a unique parasitophorous vacuole (PV). As an intracellular parasite, *T. cruzi* interacts with cells from the mammalian host. Here we review aspects related with the binding of the main infective developmental stage (trypomastigote) to the host cell and its recognition by surface-exposed ligands/receptors. This process involves numerous signaling pathways and culminates in the entry of the parasite and modifications in both cells. The invasion of trypomastigotes occurs through multiple endocytic process, assembly of the PV, interaction of this vacuole with the endolysosomal system, lysis of the PV membrane, and multiplication of amastigotes within the cell in direct contact with host cell organelles.

Keywords: *Trypanosoma cruzi*, mammalian cells, host cell interaction, endocytic pathways, signaling, parasitophorous vacuole

1. Introduction

Trypanosoma cruzi, the etiological agent of Chagas disease, causes an anthropozoonosis discovered and characterized by Dr. Carlos Chagas in 1909 [1] and recognized by the World Health Organization (WHO) as one of the three most neglected tropical diseases of the world [2–4]. Nowadays, up to 8 million people are estimated to be infected with Chagas disease only in the Americas. Patients who do not receive treatment can develop severe cardiac debility and gastrointestinal organ dysfunction and may die, and 25 million are at risk of contracting the disease [3, 4]. Due to population migration and specific modes of transmission, Chagas disease is spreading beyond its natural geographical boundaries and becoming a global problem [5]. Although the protozoan has three major developmental stages, only two are capable of infecting mammals (trypomastigotes and amastigotes), and the trypomastigote

 IntechOpen

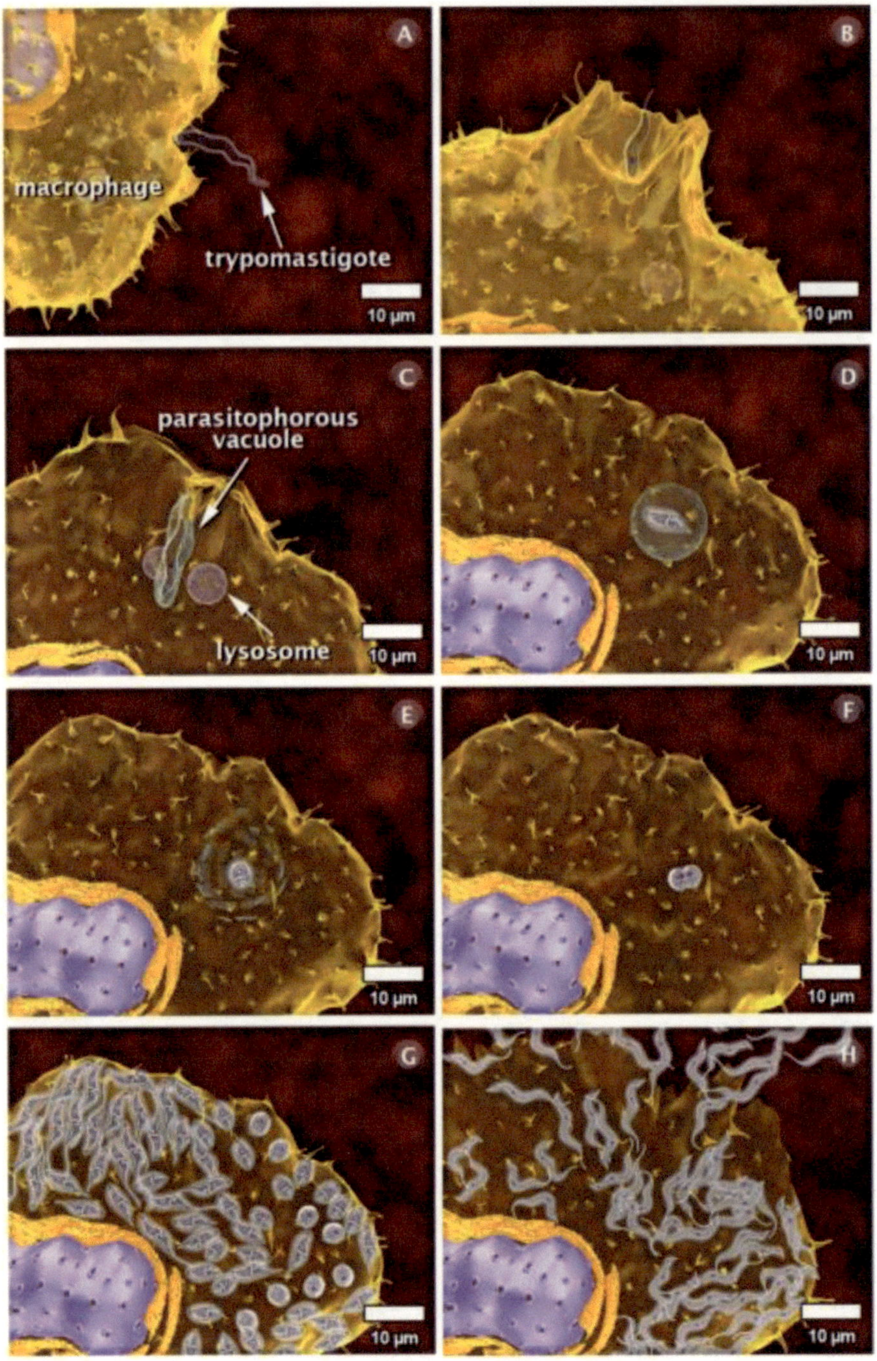

Figure 1.
Intracellular cycle of T. cruzi. (A) Attachment of the trypomastigote. (B) Internalization and recruitment and fusion of host cell lysosomes. (C) Parasitophorous vacuole is formed and lysosomal content is released. (D) Differentiation of trypomastigote to amastigotes. (E) Parasitophorous vacuole membrane disintegration. (F) Amastigote division into cytoplasm. (G) Differentiation of amastigotes into trypomastigotes. (H) Liberation of parasites into extracellular [12].

stage is the main *T. cruzi* vector [6]. Recent data indicate the existence of an infective epimastigote-like stage observed in axenic cultures as well as in the invertebrate host [7]. It is important to point out that before publication of this article, it was assumed that replicating epimastigotes present in the insect gut are not infective to mammalian host. During the vector infection (caused by a hematophagous insect of the family Reduviidae), metacyclic trypomastigotes [8], which penetrate the vertebrate host (several mammals, including man), are released along with their excreta coming in contact with conjunctiva areas or through small lesions in the own site of the bite (favored by the itch caused after the insect's bite). In turn, metacyclic trypomastigotes are able to invade virtually all cell types in the vertebrate host, especially muscle cells, fibroblasts, and macrophages [6]. At this moment, the intracellular cycle of *T. cruzi* begins, where the firing of several signaling cascades culminates with the closure of the parasitophorous vacuole (PV) where the parasite is found [9, 10]. After the PV closure, the process of differentiation of the parasite from the trypomastigote stage to the amastigote stage begins. At the same time, fragmentation of the PV membrane takes place most probably due to the increased concentration of the Tc-Tox perforin-like protein produced by the parasite [11]. After the destruction of the vacuole, the parasite, in the process of differentiation, will be found in the cytoplasm of the host cell where it will initiate its multiplication and subsequent differentiation for trypomastigotes culminating in the rupture of the host cell (**Figure 1**) [13]. The whole process of formation of the parasitophorous vacuole until its rupture counts on the participation of several organelles of the host cell. Among these, the best characterized is the participation of host cell endosomes and lysosomes. It is the fusion of these organelles with the PV membrane that probably allows the increase or expansion of the PV. In addition, this process is also responsible for the generation of an acidic environment within the PV, which probably will potentiate the action of Tc-Tox and PV membrane fragmentation [13]. Wilkowsky and colleagues [14] have shown that early and late endosomes were critical for vacuole formation. In addition, other organelles responsible for the production of proteins and energy (endoplasmic reticulum (ER), Golgi complex (GC), and mitochondria) have also been observed during the initial infection process [11]. In this chapter we will discuss the available data on the process of parasite-host cell recognition, triggering of the internalization process, and biogenesis of the PV. A better understanding of all the processes may identify new potential targets to block parasite invasion and may constitute alternative ways to treat Chagas disease.

2. *T. cruzi* trypomastigote-host cell recognition

The trypomastigote stage of *T. cruzi* has the ability to invade both professional phagocytic cells and nonprofessional phagocytic mammalian cells either in vitro or in vivo [15–17]. The kinetics of the *T. cruzi*'s intracellular cycle varies according to the strain, including time duration (Y strain, e.g., has an intracellular cycle that is completed between 5 and 7 days), since the initial interaction with trypomastigotes, triggering a signaling cascade that culminates with the formation and closure of the PV where the parasite will be located for some time [18].

In order for the interaction process and consequent internalization of the parasite to occur, there must be a recognition between molecules present on the surface of both cells. These processes are complex and involve several adhesion molecules (**Figure 2**), signaling events, and proteolytic activities [10, 19]. Some of these molecules have been identified as participants in the adhesion and invasion processes, such as gp35/50 [20], gp82 (whose expression varies according to the *T. cruzi* strain analyzed) [21], and gp90 [22]. Both gp82 and gp90 are the main metacyclic stage-specific surface molecules and are extremely important to oral infection [21, 23]. However, while glycoprotein 90 is known as a negative modulator of metacyclic trypomastigote invasion [24, 25], gp82 binds to an unidentified receptor in host cell

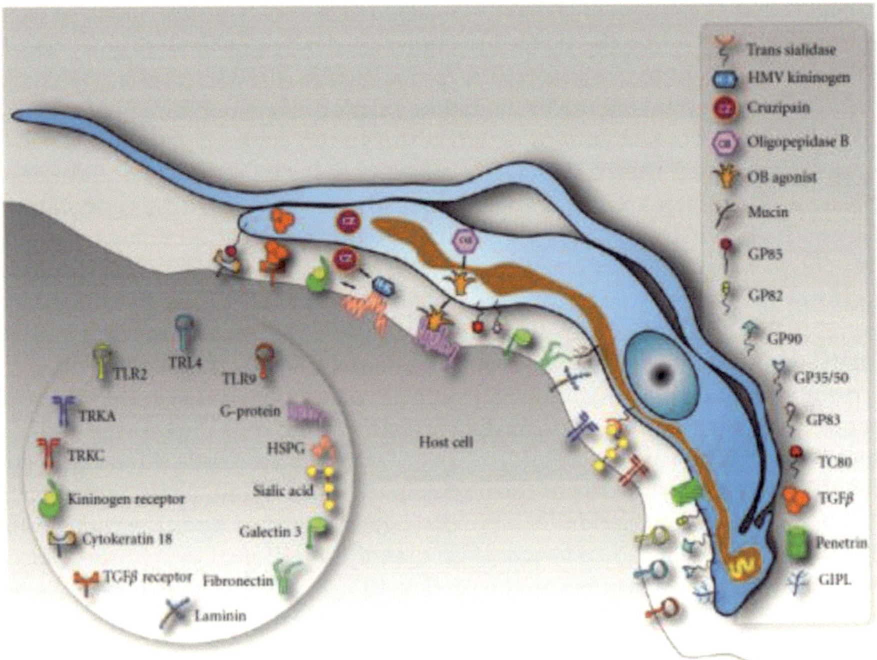

Figure 2.
Schematic model of molecules involved in T. cruzi-host cell interaction [10].

surface mediating a calcium-dependent signaling pathway that will be discussed below [20, 26]. Although the receptors presented in host cell surface that recognize gp82 and gp90 are not known, some data reinforce they are different since the target cell's interaction motif of both glycoproteins is distinct [26]. It is important to point out that these molecules are not present in tissue culture-derived trypomastigotes. At this stage, the molecules described as present in the membrane are gp85 (recognizes extracellular matrix), Ssp3, shed acute-phase antigen (SAPA) (trans-sialidase (TS)), oligopeptidases (serine proteases), and penetrin, which bind to heparin, heparan sulfate, and collagen [10].

Expression of a family of GPI-anchored glycoproteins, termed trans-sialidases/neuraminidases, present mainly on the surface of trypomastigotes is capable of modifying the exposure of surface glycoconjugates both in host cells and in the parasite itself. *T. cruzi* trans-sialidases (TS) are "shed" from the parasite membrane to the external medium through microvesicles. Microvesicles have a complex lipid bilayer and are responsible for carrying several molecules derived from the parasite, such as lipids, proteins, and nucleic acids [27–29]. Trypomastigote TS protein structure is formed by two major regions: an N-terminal catalytic region and a C-terminal region. C-terminal presents 12 amino acids repeated in tandem, named SAPA. TS can be active or inactive. Inactive TS are lectin-like proteins and are capable to bind SAcs and β-galactose [30]. Active TS are modified sialidases which, in addition to cleaving sialic acid, may transfer from sialoglycoconjugates of the host cell to β-galactose of *T. cruzi* glycoconjugates [31]. The TS gene family is complex, consisting of fourteen members divided into two groups: one translates TS present in stages present in mammalian hosts (trypomastigotes), and another translocates TS present in specific vector stages (epimastigotes). Although biochemical studies do not demonstrate the presence of TS at the amastigote stage, immunocytochemical assays have shown that TS can indeed be expressed in amastigotes [32–34]. Pereira et al. [35] observed that about 20–30% of cultured trypomastigotes had trans-sialidases/neuraminidases on their surface and these parasites were more invasive than the population that did not express the enzyme. Expression of trans-sialidases in *T. cruzi* is directly related to infectivity/virulence of the parasite since proteome studies indicate that different strains exhibit significantly different amounts of TS and TS-containing microvesicles and that host cells incubated with larger quantities containing microvesicles of TS before infection with trypomastigotes will generate a greater infection [36]. Results regarding virulence had already been related to TS expression through the analysis of virulent and non-virulent strains by transcriptome. In this work it can be observed that the expression of TS-coding genes during the differentiation process from intracellular amastigote to trypomastigote (end of intracellular cycle) is much higher in the virulent than in the non-virulent strain. Associated with this the transfection of avirulent strain with gene allowing the constitutive expression of TS also makes the release of trypomastigotes faster [37]. The TS family also includes members that have no enzymatic activity but which may also be involved in recognition between the parasite and the host cell, such as gp85 [38]. Todeschini and colleagues [39] demonstrated that inactive enzymes of the TS family are sialic acid-binding proteins and terminal β-galactopyranose (βGalp) residues. In relation to gp85/trans-sialidase, San Francisco and colleagues [40] demonstrate that this protein plays a fundamental importance in invasion since its depletion causes a decrease in *T. cruzi* virulence. The same type of result was reported by Pascuale et al. [41] since inactive TS expression in trypomastigotes of a strain that does not express these TS (iTS null) allowed a better invasion and increase of the parasitic load in mice demonstrating that the inactive form may act alternatively or complementing the active TS in pathogenesis.

Sialic acid from the host cells has a crucial importance in intracellular cycle of *T. cruzi,* a parasite that does not have the ability to synthesize sialic acids. This molecule plays an important role in protecting *T. cruzi* from lysis by serum factors and also acting in interaction with host cell. Mucins and TS function as substrate and enzyme, respectively, and sialylated mucins are localized in microdomain regions of trypomastigotes [42]. During the interaction process involving macrophages, the presence of sialic acid on the surface of trypomastigotes hinders the invasion process, since the removal of these residues through the use of neuraminidase or their blockage through the use of periodic acid or lectins from *Limax flavus* or *Limulus polyphemus* increases adhesion and internalization rates. It is possible that the presence of desialylated parasites in macrophages may increase due to the recognition of galactose/N-acetyl galactosamine receptors present on the surface of macrophages [15, 16]. In relation to sialic acid present on the surface of macrophages, there has been an increase in the entry of trypomastigote forms in cells that expose this residue when compared to cells that do not expose galactose [15, 43]. The presence of sialic acid on the surface of trypomastigotes does not yet have a fully known function though it is believed that sialic acid helps adhesion and penetration into non-phagocytic cells [34, 44].

It is currently discussed that any class of molecules exposed on the surface of mammalian cells has a great receptor potential for molecules exposed on the surface of *T. cruzi* [19]. Most of the receptor classes have carbohydrates in their composition, such as galactose, mannose, sialic acid residues [19, 44, 45], and lectin-type proteins such as galectins (binding to carbohydrate residues present on the surface of the parasite). Pineda et al. [46] described binding data from different human galectins (gal-1, gal-3, gal-4, gal-7, and gal-8) to different strains of *T. cruzi* belonging to the six different strains (DTUs). It has been observed that all galectins bind preferentially to the infective stages (amastigotes and trypomastigotes) and that many can promote higher rates of adhesion and infection to host cells and higher rates of infection to mice. In relation to galectin-1, it was observed that the presence of this glycoprotein in human and murine cardiomyocytes is able to prevent infection with trypomastigotes, one more data that goes against the modulating role of galectin in the process of internalization of trypomastigotes [47].

Galectin-3, a protein abundant in the cytoplasm of epithelial cells and macrophages, has also been described as a participant in the immune response and infection processes (in addition to the recognition process between the host and *T. cruzi*) [48, 49]. In addition, it has also been demonstrated that galectin-3 accumulates in both the parasite and phagosome entry regions and in tubules and vesicles that would derive from the endosomal system, thus suggesting that this protein is also an excellent marker of the lysis process of the PV containing this parasite [50]. The absence of galectin-3, in addition to increasing intracellular replication in vitro, is able to increase parasitemia in vivo by decreasing the secretion of pro-inflammatory cytokines and increasing cardiac fibrosis [51], which confirms the important role of this glycoprotein in the pathogenesis of Chagas disease.

In addition to all the molecules already mentioned, they also act as endothelin-1 receptors and bradykinin receptors. Both are used by trypomastigotes in the invasion of cardiovascular cells, being very important in the pathogenesis of Chagas disease [52]. Cytokeratin 18, fibronectin, laminin, and integrin are also recognized by Tc-85, forming a bridge between the parasite and the host cell [19, 53, 54]. Besides, a novel family of *T. cruzi* surface membrane proteins (designated as TcSMP) was detected in parasite surface and plays some role on host cell invasion by *T. cruzi* [54].

3. Mechanisms of internalization of *Trypanosoma cruzi* trypomastigotes in host cells

The mechanisms that lead to the internalization of trypomastigotes appear to be different when one considers the cell type where the internalization will occur. Morphological evidence shows that the parasite invades the host cell by an endocytic process that culminates in the formation of a PV. In cells of the immune system such as macrophages, which are specialized in phagocytosis events, we observe that the process of internalization occurs in two distinct ways. Dvorak and Schumunis [55] initially suggested that trypomastigotes forced the membrane of the host cell in an event where there is energy expenditure by the two cells involved. Nogueira and Cohn [56] observed the formation of projections on the surface of the host cell, which leads to a classical phagocytic process.

In cells considered as nonprofessional phagocytic, such as epithelial cells and fibroblasts, there appears to be a process of internalization where the parasite is the agent of penetration [33]. This process was confirmed by Martins et al. [57], where it was shown that metacyclic trypomastigotes (G and CL strains) require ATP to invade nonprofessional phagocytic cells. In these cell types, two different strategies are known to be involved in the invasion process and formation of the PV: one dependent on lysosomes and another one independent of lysosomes. The lysosomal-dependent pathway, first described by Tardieux et al. [58], was well characterized in nonprofessional phagocytic cell lines. Signals triggered by the recognition between *T. cruzi* and the host cell lead to the recruitment of lysosomes to the parasite's entry site, which would actively participate in the invasion process [33]. In addition, lysosome exocytosis would depend (1) on the performance of microtubules [8] and (2) on the regulation of host cell cytoplasmic Ca^{2+} levels [15]. This process of lysosomal-dependent invasion would occur in about 20% of the parasites [59]. The fusion of the lysosomes with the plasma membrane would thus end up donating the membrane for the formation of the PV in a calcium-dependent process. In addition, a similar Ca^{2+}-dependent lysosomal exocytosis mechanism was observed during injury and repair of the plasma membrane [60]. Tam et al. [61] demonstrated that the mechanism of injury-dependent endocytosis is directly related to the secretion of a specific lysosomal enzyme and acid sphingomyelinase (ASM). ASM would cleave sphingomyelin, an abundant sphingolipid in the outer leaflet of the plasma membrane [62], forming ceramide, which is internalized by the cell [63–65]. Trypomastigotes are able to take advantage of this mechanism by inducing the formation of these vesicles rich in ceramides, thus facilitating the invasion process [66]. Over the years, this entry mechanism involving plasma membrane repair pathways has been extensively studied by several groups. The participation of lysosomal proteins known as Lamp1 and Lamp2 has been shown to be essential, since the knockout of both proteins decreases the entry of trypomastigotes. This decrease in entry is not accompanied by inhibition of lysosome exocytosis to the repair region as well as phenotypic modification of the host cell or generated PV. It is believed that the decrease in parasite entry is due to problems in caveolin-mediated endocytosis and in calcium efflux [67]. Considering that there are two distinct pools of lysosomes (cortical and internal) in mammalian cells, Hissa and Andrade used cardiomyocytes and observed, through the use of specific inhibitors, that trypomastigotes enter into this cell type mainly recruiting lysosomes from the more internal cell layer (perinuclear) [52]. Some molecules known as SAPs are secreted by microvesicles which are recognized by host cells and promote an efflux of perinuclear lysosomes. These molecules probably act together with gp82, activating Ca^{2+} pathway and promoting *T. cruzi* internalization [68]. Another molecule capable to recruit lysosomes to entry site is oligopeptidase also involved in Ca^{2+} efflux [69]. Using non-phagocytic cells it is described that the entry of metacyclic trypomastigotes is predominantly due to recruitment of lysosomes, whereas entry of cultured trypomastigotes (from the same CL strain) does not involve the participation of lysosomes [70]. Recently, Rodrigues et al. [26] observed that infection of host cells by metacyclic trypomastigotes is associated with lysosome spreading and presence of gp90 (metacyclic trypomastigote surface glycoprotein). Strains expressing low amount of surface gp90 are able to recruit more lysosomes to the site of infection giving rise to a more successful infection.

In all other mechanisms used by the parasites, there would be no recruitment and exocytosis of lysosomes to the entry site, which are classified as lysosomal-independent mechanisms. In these pathways, there is an invagination of the plasma membrane of the host cell with the consequent formation of a PV without the initial presence of lysosomal markers. In 2002, Wilkowsky et al. [14] demonstrated the existence of this lysosomal-independent pathway for the invasion of *T. cruzi* into two nonprofessional phagocytic cell lines. Using Hela and CHO cells transfected with Rab5, Rab7, or dinamine-GFP, the presence of these endosomal markers in the newly formed vacuole was observed, indicating that some vacuoles fuse first with early and late endosomes and not with lysosomes. Later these events were quantified, and 50% of the parasitophorous vacuoles formed used the plasma membrane of the host cell but were enriched with PI3-kinase action products and negative for endosomal markers; 20% of the other trypomastigote-containing vacuoles were positive for EEA-1 (Rab5 effector and marker of initial endosomes) and Rab5, and approximately 20% of the vacuoles were positive for LAMP-1 (lysosomal marker). Vacuoles from these pathways are matured by the gradual fusion of early endosomes as well as of lysosomes, which allows the complete formation of the PV [16, 19, 71].
Both models of interaction are illustrated in **Figure 3**. In addition to the endolysosomal system, the cytoskeleton and autophagic processes were also related to

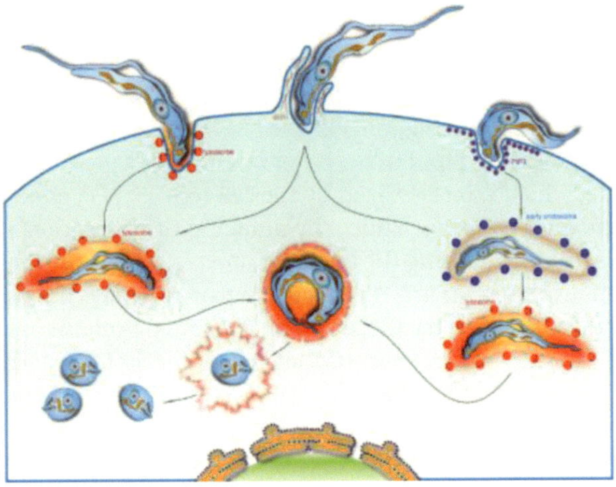

Figure 3.
Schematic model of T. cruzi internalization and parasitophorous vacuole formation using different entry pathways [10].

the complex formation process of *T. cruzi* parasitophorous vacuoles, but there is still much controversy about these involvements. The first time that a compound known to interfere with actin polymerization inhibitor was used with the aim of studying the entry of *T cruzi* was reported by Nogueira and Cohn [56]. Subsequently, Meirelles et al. [72] using the same compound demonstrated that the treatment of chicken macrophages with cytochalasin prevented the entry of the parasite without, however, preventing adhesion. This work was the first to split the entry's phase of *T. cruzi* (adhesion and internalization). On the other hand, using another cytochalasin (cytochalasin D—an actin filament polymerization inhibitor) as a tool, it was observed that the infection rate of the parasite was not significantly altered [33]. Subsequently the Tardieux group [58] demonstrated that invasion of *T. cruzi* into nonprofessional phagocytic cells is significantly enhanced by the depolymerization of the host cell actin cytoskeleton. The rapid reorganization of actin occurs as a response to the trypomastigote stage, suggesting that the direct reorganization of the actin cytoskeleton is a critical step for the entry process [18]. The reduction in the entry process was again observed in other studies, indicating that this divergence may be related to the observed interaction time [73].

In relation to the microtubules, their dynamics are important to facilitate *T. cruzi* invasion targeting lysosomes to entry site [57, 74]. It is also believed that the PV acts as a secondary center for the organization of microtubules, as regards the lysosomal fusion process at the parasite's entry site [74]. Besides a protein known as a CLASP1 (a microtubule plus-end tracking protein) is described as involved in the internalization of *T. cruzi* integrating actin pathway with microtubules and helping with the perinuclear localization of PV [75]. Microtubules are also responsible for carrying vesicles such as those positive for Vamp7 (essential for lysosomal fusion and retention of infection) to the entry site. This transport is dependent on the KIF 5 protein (a kinesin) [76]. More recently, Romano et al. [77] have shown that an autophagic protein, LC3, would also be present in the PV membrane, also demonstrating that the induction of autophagy in the host cell (pharmacological or physiological) interferes with the mechanism of trypomastigote invasion. Autophagy mechanism also plays a protective role against *T. cruzi* infection in mice by activation of a host immune response [78].

4. Endocytic mechanisms involved in trypomastigote infection

Many processes are already described as involved in the entry of *T. cruzi* trypomastigotes. As already discussed in this topic since 1972, Dvorak and Schumunis [55] had already described that the internalization of trypomastigotes in host cells could be by endocytic mechanisms. Endocytic processes are currently divided into different classes: clathrin-mediated endocytosis, endocytosis mediated by membrane microdomains (planar and caveolae), macropinocytosis, and phagocytosis. The first endocytic mechanism described as a participant in *T. cruzi* entry into host cells was phagocytosis. The participation of this mechanism was described by Nogueira and Cohn [56] through the treatment of several cell types, peritoneal macrophages, L929, HeLa, and embryo fibroblasts of calves with cytochalasin B (a drug that interferes with the extension of actin filaments). Afterward, Barbosa et al. [79] demonstrated by transmission electron microscopy that trypomastigotes are able to bind to cardiac muscle cells and induce the formation of extensive pseudopodia, a typical feature of phagocytic processes. Subsequently several reports showed that actin filaments are essential in other endocytic mechanisms, such as the macropinocytosis. Cytochalasins B and D are quite specific inhibitors of phagocytic activity. When the parasite is internalized via phagocytosis, there is internalization of CR3 receptors, β1 integrin, lysosomal membrane glycoproteins (Lpg), and Fc receptors (the latter appears only when trypomastigotes are opsonized). The participation of toll-2 receptors ("toll-like receptors 2"), as well as membrane components containing galactosyl, sialoconjugate, and glycoconjugate residues [80–82], is also demonstrated. In general phagocytosis is also a process that can be divided into different subtypes that are morphologically distinct. Using scanning electron microscopy, it was demonstrated that during internalization of trypomastigotes, the plasma membrane of peritoneal macrophages can cover the parasite in a juxtaposed way, with bilateral projections of plasma membrane forming a funnel-like structure that can follow the entire extent of the parasite's body, culminating in its total internalization (similar to what is described as classical phagocytosis). It has also been observed in the development of structures similar to the initial stages of trigger phagocytosis or macropinocytosis, in addition to the formation of structures described as coiled-type phagocytosis [19, 82, 83].

The participation of membrane rafts in the invasion of *T. cruzi* has also been demonstrated [84, 85]. Regions of membrane microdomains (rafts) are small, dynamic, cholesterol-rich membrane invagination regions, where sphingolipids, GM1 gangliosides, and caveolae (caveolae are a special type of membrane rafts) concentrate. These regions are known as signaling hotspots because they contain several proteins that can be deposited by triggering signaling cascades. This topic (signaling activation in *T. cruzi* entry process) will be discussed later in this chapter. More recently it has been demonstrated that *T. cruzi* could also use another endocytic mechanism in the invasion of its host cell: macropinocytosis. [86]. This endocytic pathway involves the internalization of large areas of plasma membrane along with significant amounts of extracellular fluid. It is important to point out the participation of dynamin as a key protein for the formation and consequent release of the early PV from the plasma membrane. Its inhibition using dynasore or its blockage through an overexpression of a dominant-negative mutant of dynamin inhibits the internalization of trypomastigotes, demonstrating that GTPase activity is also important [71, 86].

5. Signaling pathways involved in trypomastigote penetration

There are several external factors that can regulate different types of cellular responses. For these responses to occur, it is necessary that a conformational change of several proteins takes place, which means that they can now interact with other molecules leading to their activation, transducing the signal and amplifying it. The conformational change is usually dependent on the action of protein kinases or protein phosphatases. The case of the process of interaction between *Trypanosoma cruzi* and host cells is not different since it is also an external process that will require an internal response. From this perspective, several studies have focused on the investigation of different pathways that coordinate the invasion of *T. cruzi* and that modulate the gene expression of the host cell in response to this process [85, 86].

T. cruzi seems to exploit an infinity of cell surface receptors, secondary messengers, and transcription factors of different pathways to ensure its invasion and survival [86–89]. Among the signaling events, the best studied is that which leads to a calcium release in the host cells. This calcium release is one of the main responsible for regulating the process of invasion of *T. cruzi* [9, 18]. Three different models have already been proposed by different groups as being responsible for the activation of this signaling pathway. Among the described models, two involve peptidases such as oligopeptidase B and cruzipain, and the third one involves a membrane glycoprotein called gp82. Although all three mechanisms activate calcium firing, none of them are correlated. The first proposed model is based on the activation of the serine peptidase called oligopeptidase B where this enzyme present in the trypomastigote cytosol cleaves an inactive precursor to generate an active calcium agonist that is released by the parasite and binds to the receptor present on the surface of the host cell. This receptor is coupled to G protein which stimulates phospholipase C activity generating inositol 3,4,5-trisphosphate which binds to its receptor releasing calcium [90]. The second mechanism, proposed by Scharfstein et al. [91], is based on the secretion of cruzipain through the flagellar pocket region. This protein binds to a kininogen molecule that is cleaved into short kinin molecules, which in turn bind to the bradykinin receptor by stimulating the release of calcium from IP3. Cruzipain is also capable to regulate arginase activity increasing *T. cruzi* survival inside the cell through an increase in the production of IL-10 and TGF-β [92]. The model based on the activation of the glycoprotein gp82 is known to be bidirectional since it has been shown that a peak of calcium is generated not only in the host cell but also in the parasite itself. In the case of activation from this glycoprotein, the receptor is not yet known [90].

As previously described one of the activated pathways is TGF-β, where molecules secreted by trypomastigotes stimulate TGF-β receptors and activate the transcription of genes regulated by this molecule [93]. Activation of this pathway is involved in Chagas disease fibrosis development [94–96]. TGF-β was first described as being activated through cruzipain, but it is now known that trypomastigotes are capable of exposing phosphatidylserine to the outer layer of the plasma membrane. This exposure would be responsible for triggering the TGF signaling pathway in macrophages, based on the phosphorylated Smad2 nuclear translocation, leading to inhibition of iNOS in infected macrophages. This event would favor intracellular survival of the parasite [97]. More recently, the mechanism used to favor this intracellular survival was proposal by Calvet and colleagues [98] using cardiomyocytes as host cell model. In these cells the TGF-β receptor (TβRII) is localized in cardiomyocyte's costameres, which are also rich in vinculin and associated with cytoskeleton (known as a signaling domain). Its activation potentiates Smad2 phosphorylation. When *T. cruzi* infection is established, the cytoskeleton is disorganized, disrupting TβRII striations and decreasing Smad2 phosphorylation making cardiomyocytes less responsive to exogenous TGF-β stimulation.

Phosphorylation of protein tyrosine kinases is an important step in the regulation of a variety of eukaryotic cell signaling pathways [99]. In professional phagocytes, the entry of *T. cruzi* into macrophages is inhibited by treatment with genistein, a tyrosine kinase inhibitor [79]. It has also been shown that trypomastigotes of the Y strain stimulate tyrosine phosphorylation of a large number of proteins [100]. Pretreatment of fibroblasts with genistein does not inhibit entry of the parasite [18], suggesting different roles for protein tyrosine kinases in *T. cruzi* invasion of the Y strain in professional phagocytic cells or nonprofessional phagocytic cells. In cardiomyocytes and cardiac fibroblasts, tyrosine kinases appear to be essential to infection. Tyrosine kinase C (TrkC) is recognized by *T. cruzi* parasite-derived neurotrophic factor (PDNF) through neurotrophin receptor culminating in the entry of trypomastigotes into cardiac cells, while TrkA activation by the same ligand in the same cell types leads to a decrease in oxidative stress [101]. In this same sense of protection from the *T. cruzi* infection, other signaling pathways are also activated as, for example, the pathways of Erk11/Erk2 and Jak/STATs [102]. Other protein kinases also participate in *T. cruzi* invasion in host cells, such as protein kinase C (PKC), MAP kinases, and phosphatidylinositol 3-kinases (PI3-K) [86, 88, 97]. Recently the regions known as membrane microdomains have been described as signaling platforms. These regions are capable of recruiting a wide range of proteins involved in signal transduction processes. These proteins may include tyrosine kinase receptors and protein kinases such as PI3 kinase, protein kinase C, Src kinase (Lyn and Fyn) family proteins, FAK, bradykinin receptors, GTP (Rac, Rho, and Ras), and adapter proteins (Vav, Sos, and Shc). Some of these molecules have already had their share in the process of invasion of *T. cruzi* elucidated, while others have not. Proteins such as those from the Src kinases family (responsible for ITAM phosphorylation, which is essential for initiating the signal transduction cascade that triggers pathogen growth) and adapter proteins such as Vav, although not demonstrated as participants in this process, have been described by Vieira et al. [88] as possible phosphorylated proteins during the *T. cruzi* invasion in macrophages. Signaling pathways described as involved in macropinocytosis pathway (Pak1 and PKC pathways) are important to the intracellular development of infection [19]. More recently, Wnt signaling also has been shown to be an important pathway to immunomodulatory functions during *T. cruzi* infection, regulating the control of parasite replication. Activation of Wnt pathway is important to avoid a production of pro-inflammatory cytokines and indoleamine 2,3-dioxygenase activity by *T. cruzi*, helping to control the infection [103].

6. Organelles involved in parasitophorous vacuole formation

After the internalization process, *T. cruzi* resides obligatorily, and temporarily, inside a parasitophorous vacuole [10]. Over the years, several groups have described that the formation of the vacuole membrane is a complex process related to the participation of numerous molecules, depending also on the type of host cell involved in the process, i.e., phagocytic lines or nonprofessional phagocytic lines. As regards the formation of the PV membrane, it has been previously discussed in this text that several components of the host cell have already been described as components of the cytoskeleton [19, 73, 104, 105]. The organelle known as endoplasmic reticulum (ER) has been described in the literature as a donor membrane for phagosome formation, binding directly to the base of this phagosome in formation [106]. Recently, our group demonstrated by electron tomography (followed by 3-D reconstruction) and fluorescence videomicroscopy that the endoplasmic reticulum (ER) participates in the process of formation of the vacuole from initial moments until its complete maturation. The participation or not of ER in this process may be directly related to the input mechanism used by the parasite at the time of the invasion [104].

The Golgi complex (CG) plays a central role in eukaryotic cells, serving as an intermediary and bidirectional axis of protein and lipid trafficking in the endomembrane system [107]. In the case of *T. cruzi*, Carvalho et al. [82] described that after long infection times, there would be no change in GC distribution, remaining in the perinuclear region. Recently, transmission electron microscopy showed an intimate proximity between the CG and the parasite-containing vacuole. These data are of extreme relevance due to the function of GC in eukaryotic cells, i.e., because the membrane of the PV containing the parasite is still in the process of formation, requiring lipids and proteins from the host cell that would aid its closure as well as its expansion. As regards mitochondria, apparently there is no direct link between this organelle and the vacuole [104].

7. Disorganization of the parasitophorous vacuole and the endolysosomal system

As discussed above trypomastigotes of *T. cruzi* use different molecules and mechanisms to invade a host cell. Regardless of the chosen mechanism (participation of plasma membrane components or lysosomes at the site of invasion), the parasite will be located inside a vacuole.

Inside the PV the trypomastigotes release the enzyme trans-sialidase/neuraminidase, which is responsible for the removal of the sialic acid residues from the vacuole membrane. This removal makes it sensitive to the action of another enzyme, homologous to factor 9 of human complement, Tc-Tox. The lysis of the vacuole membrane by the action of Tc-Tox may be associated with the formation of pores in this membrane, which, together with the secretion by the trans-sialidase/neuraminidase parasite, will lead to the complete fragmentation of the PV membrane [35, 108].

Despite the attempt to determine the exact moment of the *T. cruzi* exit process from the parasitophorous vacuole, this step has not yet been well elucidated. Since 1989, however, our group has been trying to chart the paths necessary for this discovery. Initially Carvalho and De Souza [109] demonstrated by transmission electron microscopy that the lysing process of the PV membrane begins to occur in the first 2 hours of interaction of the trypomastigote stage in peritoneal macrophages, already suggesting the occurrence of fusion of lysosomes, after entry of the parasites. More recently, Reignault and colleagues [50] demonstrate that galectin-3 decorates *T. cruzi* vacuole acting as an important marker to be used also for the study of parasitic vacuole lysis of *T. cruzi*, as it also undergoes a process of disorganization with consequent exit of the parasite into the cytoplasm of the host cell. Transcriptome studies have shown an extensive remodeling of the intracellular *T. cruzi* in the first 4 hours of trypomastigote invasion (until the parasitophorous vacuole disintegration), and these modifications can be associated with a regulation of the initial step of host cell invasion [110]. Some modifications occurring in intracellular differentiation processes (trypomastigotes to amastigotes and subsequently to trypomastigotes again) may be involved in the process of destruction of PV. One of the modifications that seems to be involved with this process is the increase in the enzymatic activity of glutamine synthase during the amastigogenesis process. This enzyme is normally involved in the process of handling excess ammonia, and given the fact that the main energy source of the intracellular forms of *T. cruzi* is from amino acids, the ammonium generation is high. Marim et al. [111] observed that the activity blockage of this enzyme impair the progression of the intracellular cycle (amastigotes did not differentiate for trypomastigotes). In this case the role of the enzyme is to regulate the intracellular pH by controlling the content of intravacuolar ammonium (generated by the consumption of amino acids in this reduced space). It is this acidification process followed by the pH control that allows the enzymatic activity that culminates in the release of parasite in the cytoplasmic environment allowing the infection to continue [112].

8. Concluding remarks

Host cell invasion and parasite internalization are important steps in the evolution of parasitism by several pathogens. These processes present at least two important advantages: protection against the host immune response and access to a microenvironment rich in metabolic products. Substantial progress has been made in understanding the roles of proteins in infection and invasion by *T. cruzi*. Host cell intracellular signaling can combat the infection; but it can also favor parasite entry. Parasites hijack the host immune response, phagocytosis, ECM, and antiparasitic proteins for their own survival, replication, and immune evasion purposes. The complex networks are interconnected and require extensive study to identify intracellular rearrangements that facilitate parasite internalization. A multidisciplinary approach is necessary to a better understanding of parasite-host interaction and will be critical to better understand Chagas disease physiopathology, diagnosis, and treatment.

Acknowledgements

We thank all authors who work with this theme and for the contribution during the last years.

Conflict of interest

The authors declare "no conflict of interest."

Author details

Emile Barrias[1*], Lissa C. Reignault[2] and Wanderley de Souza[2]

1 National Institute of Metrology, Quality and Technology (INMETRO), Duque de Caxias, Rio de Janeiro, Brazil

2 Federal University of Rio de Janeiro (UFRJ), Rio de Janaeiro, Brazil

*Address all correspondence to: esbarrias@inmetro.gov.br

References

[1] Chagas C. Nova tripanozomíaze humana. Memórias do Instituto Oswaldo Cruz. 1909;**1**:11-80

[2] Hotez PJ, Molyneux DH, Fenwick A, Kumaresan J, Sachs SE, Sachs JD, et al. Control of neglected tropical diseases. The New England Journal of Medicine. 2007;**357**:1018-1027

[3] WHO. 2019. Available from: https://www.who.int/chagas/en/

[4] Tzizik DM, Borchardt RA. Chagas disease: An underrecognized diagnosis. Journal of the American Academy of PAs. 2018;**31**:30-33. DOI: 10.1097/01.JAA.0000547749.92933.6a

[5] Pérez-Molina JA, Molina I. Chagas disease. Lancet. 2018;**391**:82-94

[6] De Souza W. Basic cell biology of *Trypanosoma cruzi*. Current Pharmaceutical Design. 2002;**8**:211-231

[7] Kessler RL, Contreras VT, Marliére NP, Aparecida Guarneri A, Villamizar Silva LH, Mazzarotto G, et al. Recently differentiated epimastigotes from *Trypanosoma cruzi* are infective to the mammalian host. Molecular Microbiology. 2017;**104**:712-736

[8] Garcia ES, Castro DP, Figueiredo MB, Azambuja P. Immune homeostasis to microorganisms in the guts of triatomines (Reduviidae)—A review. Memórias do Instituto Oswaldo Cruz. 2010;**105**:605-610

[9] Rodriguez A, Samoff E, Riolt MG, Andrews NA. Host cell invasion by trypanosomes requires lysosomes and microtubule/kinesin-mediated transport. The Journal of Cell Biology. 1996;**134**:349-362

[10] de Souza W, Carvalho TMU, Barrias ES. Review on *Trypanosoma cruzi*: Host cell interaction. International Journal of Cell Biolog. 2010;**pii**:295394. DOI: 10.1155/2010/295394

[11] Andrews NW. From lysosomes into the cytosol: The intracellular pathway of *Trypanosoma cruzi*. Brazilian Journal of Medical and Biological Research. 1994;**27**:471-475

[12] Teixeira DE, Benchimol M, Crepaldi PH, de Souza W. Interactive multimedia to teach the life cycle of *Trypanosoma cruzi*, the causative agent of Chagas disease. PLoS Neglected Tropical Diseases. 2012;**6**(8):e1749

[13] De Souza W, Carvalho TMU. Active penetration of *Trypanosoma cruzi* into host cells: Historical considerations and current concepts. Frontiers in Immunology. 2013;**4**:1-3

[14] Wilkowsky SE, Barbieri MA, Stahl PD, Isola ELD. Regulation of *Trypanosoma cruzi* invasion of nonphagocytic cells by the endocytically active GTPases dynamin, Rab5, and Rab7. Biochemical and Biophysical Research Communications. 2002;**291**:516-521

[15] Araújo-Jorge TC. The biology of *Trypanosoma cruzi*—Macrophage interaction. Memórias do Instituto Oswaldo Cruz. 1989;**84**:441-462

[16] Araújo-Jorge TC, Barbosa HS, Meirelles MN. *Trypanosoma cruzi* recognition by macrophages and muscle cells: Perspectives after 15-year study. Memórias do Instituto Oswaldo Cruz. 1982;**87**(suppl 5):43-56

[17] Burleigh BA, Andrews N. The mechanisms of *Trypanosoma cruzi* invasion in mammalian cells. Annual Review of Microbiology. 1995;**49**:175-200

[18] Rodriguez A, Riolt MG, Ora A, Andrews NA. A trypanosome-soluble

factor induces PI3 formation, intracellular Ca^{2+} mobilization and microfilament rearrangement in host cells. The Journal of Cell Biology. 1995;**129**:1263-1273

[19] Barrias ES, De Carvalho TM, De Souza W. *Trypanosoma cruzi*: Entry into mammalian host cells and parasitophorous vacuole formation. Frontiers in Immunology. 2013;**1**:184-186

[20] Ruiz RC, Favoreto S, Dorta ML, Oshiro MEM, Ferreira AT, Manque PM, et al. Infectivity of *Trypanosoma cruzi* strains is associated with differential expression of surface glycoproteins with differential Ca^{2+} signaling activity. The Biochemical Journal. 1998;**330**:505-511

[21] Neira I, Silva FA, Cortez M, Yoshida N. Involvement of *Trypanosoma cruzi* metacyclic trypomastigote surface molecule gp82 in adhesion to gastric mucin and invasion of epithelial cells. Infection and Immunity. 2003;**71**:557-561

[22] Franco FR, Paranhos-Bacallà GS, Yamauchi LM, Yoshida N, Da Silveira JF. Characterization of a cDNA clone encoding the carboxy-terminal domain of a 90-kilodalton surface antigen of *Trypanosoma cruzi* metacyclic trypomastigotes. Infection and Immunity. 1993;**61**(10):4196-4201

[23] Cortez C, Martins RM, Alves RM, Silva RC, Bilches LC, Macedo S, et al. Differential infectivity by the oral route of *Trypanosoma cruzi* lineages derived from Y strain. PLoS Neglected Tropical Diseases. 2012;**6**(10):e1804

[24] Nogueira NF, Gonzalez MS, Gomes JE, De Souza W, Garcia ES, Azambuja P, et al. *Trypanosoma cruzi*: Involvement of glycoinositolphospholipids in the attachment to the luminal midgut surface of Rhodnius prolixus. Experimental Parasitology. 2007;**116**:120-128

[25] Malaga S, Yoshida N. Targeted reduction in expression of *Trypanosoma cruzi* surface metacyclic trypomastigote surface molecule gp82 in adhesion to gastric mucin. Microbiology. 2001;**4**(11):701-711

[26] Rodrigues JPF, Sant'ana GHT, Juliano MA, Yoshida N. Inhibition of host cell lysosome spreading by *Trypanosoma cruzi* metacyclic stage-specific surface molecule gp90 downregulates parasite invasion. Infection and Immunity. 2017;**85**(9):pii: e00302-pii: e00317

[27] Torrecilhas AC, Schumacher RI, Alves MJ, Colli W. Vesicles as carriers of virulence factors in parasitic protozoan diseases. Microbes and Infection. 2012;**14**(15):1465-1474

[28] Garcia-Silva MR, Cabrera-Cabrera F, Das Neves RF, Souto-Padron T, De Souza W, Cayota A. Gene expression changes induced by *Trypanosoma cruzi* shed microvesicles in mammalian host cells: Relevance of tRNA-derived halves. BioMed Research International. 2014;**2014**:305239

[29] De Souza W, Barrias ES. Exosomes in the pathogenic protozoan *Trypanosoma cruzi*. International Journal of Pathology and Clinical Research. 2017;**3**:1-9

[30] Parodi AJ, Pollevick GD, Mautner M, Buschiazzo A, Sanchez DO, Frasch AC. Identification of the gene(s) coding for the trans-sialidase of *Trypanosoma cruzi*. The EMBO Journal. 1992;**11**:1705-1710

[31] Previato JO, Andrade AF, Pessolani MC, Mendonça-Previato L. Incorporation of sialic acid into *Trypanosoma cruzi* macromolecules. A proposal for a new metabolic route. Molecular and Biochemical Parasitology. 1985;**16**(1):85-96

[32] Souto-Padrón T, Hath G, De Souza W. Immunocytochemical localization

of neuraminidase in *Trypanosoma cruzi*. Infection and Immunity. 1990;**58**:586-592

[33] Schenkman S, Jiang MS, Hart GW, Nussenzweig V. A novel cell surface trans-sialidase of *Trypanosoma cruzi* generates a stage-specific epitope required for invasion of mammalian cells. Cell. 1991;**65**:1117-1125

[34] Schenkman S, Eichinger D. *Trypanosoma cruzi* trans-sialidase and cell invasion. Parasitology Today. 1993;**9**:218-222

[35] Pereira ME, Zhang K, Gong Y, Herrera EM, Ming M. Invasive phenotype of *Trypanosoma cruzi* restricted to a population expressing trans-sialidase. Infection and Immunity. 1996;**64**:3884-3892

[36] Ribeiro KS, Vasconcellos CI, Soares RP, Mendes MT, Ellis CC, Aguilera-Flores M, et al. Proteomic analysis reveals different composition of extracellular vesicles released by two *Trypanosoma cruzi* strains associated with their distinct interaction with host cells. Journal of Extracellular Vesicles. 2018;**7**:1463779

[37] Belew AT, Junqueira C, Rodrigues-Luiz GF, Valente BM, Oliveira AER, Polidoro RB, et al. Comparative transcriptome profiling of virulent and non-virulent *Trypanosoma cruzi* underlines the role of surface proteins during infection. PLoS Pathogens. 2017;**13**(12):e1006767

[38] Alves MJ, Abuin G, Kuwajima VY, Colli W. Partial inhibition of trypomastigote entry into cultured mammalian cells by monoclonal antibodies against a surface glycoprotein of *Trypanosoma cruzi*. Molecular and Biochemical Parasitology. 1986;**21**:75-82

[39] Todeschini AR, Dias WB, Girard MF, Wieruszeski J, Mendonça-Previato L, Previato JO. Enzymatically inactive trans-sialidase from *T. cruzi* infection. The Journal of Biological Chemistry. 2004;**275**:32182-32186

[40] San Francisco J, Barría I, Gutiérrez B, Neira I, Muñoz C, Sagua H, et al. Decreased cruzipain and gp85/trans-sialidase family protein expression contributes to loss of *Trypanosoma cruzi* trypomastigote virulence. Microbes and Infection. 2017;**19**:55-61

[41] Pascuale CA, Burgos JM, Postan M, Lantos AB, Bertelli A, Campetella O, et al. Inactive trans-sialidase expression in iTS-null *Trypanosoma cruzi* generates virulent trypomastigotes. Frontiers in Cellular and Infection Microbiology. 2017;**4**(7):430

[42] Lantos AB, Carlevaro G, Araoz B, Ruiz Diaz P, Camara Mde L, Buscaglia CA, et al. Sialic acid glycobiology unveils *Trypanosoma cruzi* trypomastigote membrane physiology. PLoS Pathogens. 2016;**12**(4):e1005559

[43] Araujo-Jorge TC, De Souza W. Effect of carbohydrates, periodate and enzymes in the process of endocytosis of *Trypanosoma cruzi*. Acta Tropica. 1984;**41**:17-28

[44] Alves MJ, Colli W. *Trypanosoma cruzi*: Adhesion to the host cell and intracellular survival. IUBMB Life. 2007;**59**:274-279

[45] Pinazo M, Thomas M, Bustamante J, Almeida I, Lopez M, Gascon J. Biomarkers of therapeutic responses in chronic Chagas disease: State of the art and future perspectives. Memórias do Instituto Oswaldo Cruz. 2015;**110**:422-432

[46] Pineda MA, Cuervo H, Fresno M, Soto M, Bonay P. Lack of galectin-3 prevents cardiac fibrosis and effective immune responses in a murine model of *Trypanosoma cruzi* infection. The Journal of Infectious Diseases.

2015;**212**:1160-1171. DOI: 10.1093/infdis/jiv185

[47] Benatar AF, García GA, Bua J, Cerliani JP, Postan M, Tasso LM, et al. Galectin-1 prevents infection and damage induced by *Trypanosoma cruzi* on cardiac cells. PLoS Neglected Tropical Diseases. 2015;**9**(10):e0004148

[48] Souza BSF, Silva KN, Silva DN, Rocha VPC, Paredes BD, Azevedo CM. Galectin-3 knockdown impairs survival, migration, and immunomodulatory actions of mesenchymal stromal cells in a mouse model of Chagas disease cardiomyopathy. Stem Cells International. 2017: 3282656

[49] Kleshchenko Y, Moody T, Furtak VA, Ochieng J, Lima M, Villalta F. Human galectin-3 promotes *Trypanosoma cruzi* adhesion to human coronary artery smooth muscle cells. Infection and Immunity. 2004;**72**:6717-6721

[50] Reignault LC, Barrias ES, Soares Medeiros LC, De Souza W, De Carvalho TM. Structures containing galectin-3 are recruited to the parasitophorous vacuole containing *Trypanosoma cruzi* in mouse peritoneal macrophages. Parasitology Research. 2014;**113**:2323-2333

[51] Da Silva AA, Teixeira TL, Teixeira SC, Machado FC, Dos Santos MA, Tomiosso TC, et al. Galectin-3: A friend but not a foe during *Trypanosoma cruzi* experimental infection. Frontiers in Cellular and Infection Microbiology. 2017;7:463

[52] Hissa B, Andrade LDE. *Trypanosoma cruzi* uses a specific subset of host cell lysosomes for cell invasion. Parasitology International. 2015;**64**:135-138

[53] Magdesian MH, Tonelli RR, Fessel MR, Silveira MS, Schumacher RI, Linden R, et al. A conserved domain of the gp85/trans-sialidase family activates host cell extracellular signal-regulated kinase and facilitates *Trypanosoma cruzi* infection. Experimental Cell Research. 2007;**313**:210-218

[54] Martins NO, Souza RT, Cordero EM, Maldonado DC, Cortez C, Marini MM, et al. Molecular characterization of a novel family of *Trypanosoma cruzi* surface membrane proteins (TcSMP) involved in mammalian host cell invasion. PLoS Neglected Tropical Diseases. 2015;**9**(11):e0004216

[55] Dvorak JA, Schumunis GA. *Trypanosoma cruzi*: Interaction with mouse peritoneal macrophages. Experimental Parasitology. 1972;**32**:289-300

[56] Nogueira N, Cohn Z. *Trypanosoma cruzi*: Mechanism of entry and intracellular fate in mammalian cells. The Journal of Experimental Medicine. 1976;**143**:1402-1420

[57] Martins RM, Alves RM, Macedo S, Yoshida N. Starvation and rapamycin 18 differentially regulate host cell lysosome exocytosis and invasion by *Trypanosoma cruzi* metacyclic forms. Cellular Microbiology. 2011;**13**:943-954

[58] Tardieux I, Webster P, Ravesloot J, Boron W, Lunn JA, Heuser JE, et al. Lysosome recruitment and fusion are early events required for trypanosome invasion of mammalian cells. Cell. 1992;**71**:1117-1130

[59] Burleigh BA. Host cell signaling and *Trypanosoma cruzi* invasion: Do all roads lead to lysosomes? Science. 2005;**293**:36

[60] Reddy A, Caler EV, Andrews NW. Plasma membrane repair is mediated by Ca (2+)-regulated exocytosis of lysosomes. Cell. 2001;**106**:157-169

[61] Tam C, Idone V, Devlin C, Fernandes MC, Flannery A, He X, et al. Exocytosis of acid sphingomyelinase by wounded cells promotes endocytosis and plasma membrane

repair. The Journal of Cell Biology. 2010;**189**:1027-1038

[62] Koval M, Pagano RE. Intracellular transport and metabolism of sphingomyelin. Biochimica et Biophysica Acta. 1991;**1082**:113-125

[63] Gulbins E, Kolesnick R. Raft ceramide in molecular medicine. Oncogene. 2003;**22**:7070-7077

[64] Grassmé H, Riethmüller J, Gulbins E. Biological aspects of ceramide-enriched membrane domains. Progress in Lipid Research. 2007;**46**:161-170

[65] Trajkovic K, Hsu C, Chiantia S, Rajendran L, Wenzel D, Wieland F, et al. Ceramide triggers budding of exosome vesicles into multivesicular endosomes. Science. 2008;**319**:1244-1247

[66] Fernandes MC, Andrews NW. Host cell invasion by *Trypanosoma cruzi*: A unique strategy that promotes persistence. FEMS Microbiology Reviews. 2012;**36**:734-747

[67] Couto NF, Pedersane D, Rezende L, Dias PP, Corbani TL, Bentini LC, et al. LAMP-2 absence interferes with plasma membrane repair and decreases *T. cruzi* host cell invasion. PLoS Neglected Tropical Diseases. 2017;**11**(6):e0005657. DOI: 10.1371/journal.pntd.0005657

[68] Zanforlin T, Bayer-Santos E, Cortez C, Almeida IC, Yoshida N, Da Silveira JF. Molecular characterization of *Trypanosoma cruzi* SAP proteins with host-cell lysosome exocytosis-inducing activity required for parasite invasion. PLoS One. 2013;**8**(12):e83864

[69] Motta FN, Bastos IM, Faudry E, Ebel C, Lima MM, Neves D, et al. The *Trypanosoma cruzi* virulence factor oligopeptidase B (OPBTc) assembles into an active and stable dimer. PLoS One. 2012;7(1):e30431

[70] Cortez C, Real F, Yoshida N. Lysosome biogenesis/scattering increases host cell susceptibility to invasion by *Trypanosoma cruzi* metacyclic forms and resistance to tissue culture trypomastigotes. Cellular Microbiology. 2016;**18**:748-760

[71] Woosley AM, Sunwoo L, Petersen CA, Brachmann SM, Cantley LC, Burleigh BA. Novel PI3-kinase-dependent mechanisms of trypanosome invasion and vacuole maturation. Journal of Cell Science. 2003;**116**:3611-3622

[72] Meirelles MNL, Araújo-Jorge TC, De Souza W. Interaction of *Trypanosoma cruzi* with macrophages *in vitro*: Dissociation of the attachment and internalization phases by low temperature and cytochalasin B. Zeitschrift für Parasitenkunde. 1982;**68**:7-14

[73] Rosestolato CT, Dutra JDM, De Souza W, De Carvalho TM. Participation of host cell actin filaments during interaction of trypomastigotes forms of *Trypanosoma cruzi* with host cells. Cell Structure and Function. 2002;**27**:91-98

[74] Tyler KM, Luxton GW, Applewhite DA, Murphy SC, Engman DM. Responsive microtubule dynamics promote cell invasion by *Trypanosoma cruzi*. Cellular Microbiology. 2005;7:1579-1591

[75] Zhao X, Kumar P, Shah-Simpson S, Caradonna KL, Galjart N, Teygong C, et al. Host microtubule plus-end binding protein CLASP1 influences sequential steps in the *Trypanosoma cruzi* infection process. Cellular Microbiology. 2013;**15**:571-584

[76] Cueto JA, Vanrell MC, Salassa BN. Soluble N-ethylmaleimide-sensitive factor attachment protein receptors required during *Trypanosoma cruzi* parasitophorous vacuole development. Cellular Microbiology. 2017;**19**:e12713

[77] Romano PS, Cueto JA, Cassassa AF, Vanrell MC, Gotlieb RA, Colombo MI. Molecular and cellular mechanisms involved in the *Trypanosoma cruzi* host cell interplay. IUBMB Life. 2012;**64**:387-396

[78] Casassa AF, Vanrell MC, Colombo MI, Gottlieb RA, Romano PS. Autophagy plays a protective role against *Trypanosoma cruzi* infection in mice. Virulence. 2019;**10**:151-165

[79] Barbosa HS, Meirelles MNL. Evidence of participation of cytoskeleton of heart muscle cells during the invasion of *Trypanosoma cruzi*. Cell Structure and Function. 1995;**20**:275-284

[80] Hall BF, Furtado GC, Joiner KA. Characterization of host cell-derived membrane proteins of the vacuole surrounding different intracellular forms of *Trypanosoma cruzi* in J774 cells. Evidence for phagocyte receptor sorting during the early stages of parasite entry. Journal of Immunology. 1991;**147**:4313-4321

[81] Maganto-Garcia E, Punzon C, Terhorst C, Fresno M. Rab5 activation by toll-like receptor 2 is required for *Trypanosoma cruzi* internalization and replication in macrophages. Traffic. 2008;**9**:1299-1315

[82] Carvalho TMU, De Souza W, Coimbra E. Internalization of components of the host cell plasma membrane during infection by *Trypanosoma cruzi*. Memórias do Instituto Oswaldo Cruz. 1999;**94**:143-147

[83] Rittig MG, Scroppel K, Seack KH, Sander U, N'diaye EN, Parini I, et al. Coiling phagocytosis of trypanosomatids and fungal cells. Infection and Immunity. 1998;**66**:4331-4339

[84] Barrias ES, Dutra JM, De Souza W, Carvalho TMU. Participation of macrophage membrane rafts in *Trypanosoma cruzi* invasion process. Biochemical and Biophysical Research Communications. 2007;**363**:828-834

[85] Fernandes MC, Cortez M, Geraldo Yoneyama KA, Straus AH, Yoshida N, Mortara RA. Novel strategy in *Trypanosoma cruzi* cell invasion: Implication of cholesterol and host cell microdomains. International Journal for Parasitology. 2007;**37**:1431-1441

[86] Barrias ES, Reignault LC, De Souza W, Carvalho TM. *Trypanosoma cruzi* uses macropinocytosis as an additional entry pathway into mammalian host cell. Microbes and Infection. 2012;**14**:1340-1351

[87] Chuenkova MV, Furnari FB, Cavenee WK, Pereira MA. *Trypanosoma cruzi* trans-sialidase: A potent and specific survival factor for human schwann cells by means of phosphatidylinositol 3-kinase/akt signaling. Proceedings of the National Academy of Sciences. 2001;**98**:9936-9941

[88] Vieira MC, De Carvalho TMU, De Souza W. Effect of protein kinase inhibitors on the invasion process of macrophages by *Trypanososma cruzi*. Biochemical and Biophysical Research Communications. 1994;**203**:967-971

[89] Watanabe Costa R, Da Silveira JF, Bahia D. Interactions between *Trypanosoma cruzi* secreted proteins and host cell signaling pathways. Frontiers in Microbiology. 2016;**7**:388

[90] Burleigh BA, Woolsey AM. Cell signalling and *Trypanosoma cruzi* invasion. Cellular Microbiology. 2002;**4**:701-711

[91] Scharfstein J, Schmitz V, Morandi V, Capella MM, Lima AP, Morrot A, et al. Host cell invasion by *Trypanosoma cruzi* is potentiated by activation of bradykinin B(2) receptors. The

Journal of Experimental Medicine. 2000;**192**:1289-1300

[92] Ferrão PM, D'avila-Levy CM, Araujo-Jorge TC, Degrave WM, Gonçalves AS, Garzoni LC, et al. Cruzipain activates latent TGF-β from host cells during *T. cruzi* invasion. PLoS One. 2015;**10**(5):e0124832

[93] Hall B, Pereira M. Dual role for transforming growth factor β-dependent signaling in *Trypanosoma cruzi* infection of mammalian cells. Infection and Immunity. 2000;**68**:2077-2081

[94] Araújo-Jorge TC, Waghabi MC, Bailly S, Feige JJ. The TGF-β pathway as an emerging target for Chagas disease therapy. Clinical Pharmacology and Therapeutics. 2012;**92**:613-621

[95] Araújo-Jorge TC, Waghabi MC, Hasslocher-Moreno AM, Xavier SS, Higuchi MDL, Keramidas M, et al. Implication of transforming growth factor-beta1 in Chagas disease myocardiopathy. Journal of Infectious Diseases. 2002;**186**:1823-e3018

[96] Araújo-Jorge TC, Waghabi MC, Soeiro MDNC, Keramidas M, Bailly S, Feige JJ. Pivotal role for TGF-beta in infectious heart disease: The case of *Trypanosoma cruzi* infection and consequent chagasic myocardiopathy. Cytokine & Growth Factor Reviews. 2008;**19**:405-413

[97] Damatta RA, Seabra SH, Deolindo P, Arnholdt AC, Manhães L, Goldenberg S, et al. *Trypanosoma cruzi* exposes phosphatidylserine as an evasion mechanism. FEMS Microbiology Letters. 2007;**1**:29-33

[98] Calvet CM, Silva TA, de Melo TG, de Araújo-Jorge TC, Pereira MC. TGF-β receptor type II costameric localization in cardiomyocytes and host cell TGF-β response is disrupted by *Trypanosoma cruzi* infection. Parasitology. 2016;**143**(6):704-715

[99] Storz P, Toker A. 3-Phosphoinositide dependent kinase-1 (PDK-1) in PI 3-kinase signaling. Frontiers in Bioscience. 2002;**7**:886-902

[100] Ruta S, Plasman N, Zaffran Y, Capo C, Mege JL, Vray B. *Trypanosoma cruzi*-induced tyrosine phosphorylation in murine peritoneal macrophages. Parasitology Research. 1996;**82**:481-484

[101] Aridgides D, Salvador R, Pereiraperrin M. *Trypanosoma cruzi* highjacks TrkC to enter cardiomyocytes and cardiac fibroblasts while exploiting TrkA for cardioprotection against oxidative stress. Cellular Microbiology. 2013;**15**(8):1357-1366

[102] Stahl P, Schwarz RT, Debierre-Grockiego F, Meyer T. *Trypanosoma cruzi* parasites fight for control of the JAK-STAT pathway by disarming their host. JAK-STAT. 2014;**3**(4):e1012964

[103] Volpini X, Ambrosio L, Fozzatti L, Insfran C, Stempin C, Cervi L, et al. *Trypanosoma cruzi* exploits Wnt signaling pathway to promote its intracellular replication in macrophages. Frontiers in Immunology. 2018;**9**:859

[104] Reignault LC, Alcantra C, Barrias ES, de Souza W. 3D reconstruction of Trypanosoma cruzi-macrophage interaction shows the recruitment of host cell organelles towards the parasitophorous vacuole during its biogenesis. Journal of Structural Biology. 2019;**205**(2):133-146. DOI: 10.1016/j. jsb.2018.12.010

[105] Tardieux I, Nathanson MH, Andrews N. Role in host cell invasion of *Trypanosoma cruzi*-induced cytosolic free Ca^{2+} transients. The Journal of Experimental Medicine. 1994;**179**:1017-1022

[106] Huynh KK, Kay JG, Stow JL, Grinstein S. Fusion, fission, and secretion during phagocytosis. Physiology. 2007;**22**:366-372

[107] Mcdermott MI, Mousley CJ. Lipid transfer proteins and the tuning of compartmental identity in the Golgi apparatus. Chemistry and Physics of Lipids. 2016;**200**:42-61

[108] Rubin-de-Celis SS, Uemura H, Yoshida N, Schenkman S. Expression of trypomastigote trans-sialidase in metacyclic forms of *Trypanosoma cruzi* increases parasite escape from its parasitophorous vacuole. Cellular Microbiology. 2006;**8**:1888-1898

[109] Carvalho TMU, De Souza W. Early events related with the behaviour of *Trypanosoma cruzi* within an endocytic vacuole in mouse peritoneal macrophages. Cell Structure and Function. 1989;**14**:383-392

[110] Li Y, Shah-Simpson S, Okrah K, Belew AT, Choi J, Caradonna KL, et al. Transcriptome remodeling in *Trypanosoma cruzi* and human cells during intracellular infection. PLoS Pathogens. 2016;**12**(4):e1005511

[111] Crispim M, Damasceno FS, Hernández A, Barisón MJ, Pretto Sauter I, Souza Pavani R, et al. The glutamine synthetase of *Trypanosoma cruzi* is required for its resistance to ammonium accumulation and evasion of the parasitophorous vacuole during host-cell infection. PLoS Neglected Tropical Diseases. 2018;**12**(1):e0006170

[112] Ming M, Chuenkova M, Ortega-Barria E, Pereira ME. Mediation of *Trypanosoma cruzi* invasion by sialic acid on the host cell and trans-sialidase on the trypanosome. Molecular and Biochemical Parasitology. 1993;**59**:243-252

Chapter 10

Possible Role of Gap Junction Channels and Non-Junctional Channels in the Infection Caused by *Trypanosoma cruzi*

José Luis Vega, Camilo Juyumaya, Luis Rodríguez, Juan Güiza, Camila Gutíerrez, Iván Barría and Juan C. Sáez

Abstract

Chagas disease affects low-income nations with health consequences that impact the economy of those countries. Interestingly, inhibitors of channels formed by proteins of the gap junction family, such as suramin and boldine, exhibit trypanocidal activity. Gap junction proteins are integral membrane proteins present in both vertebrates and invertebrates that participate in cellular communication. These proteins form gap junction channels, which connect the cytoplasm of neighboring cells or non-junctional channels that connect the intra- and extracellular milieu. Interestingly, *Trypanosoma cruzi* modulates the expression of proteins of the gap junction family or modify the activity of the channels formed by these proteins in host cells. Moreover, Lucifer yellow microinjected into fibroblast was incorporated into associated trypanosomes of *Trypanosoma musculi*, suggesting the possibility of direct communication via gap junction channels between them. In this chapter, we summarized the current knowledge about the possible roles of gap junction family proteins in Chagas disease.

Keywords: connexin, pannexin, innexin, hemichannels, infection

1. Introduction

Chagas disease affects low-income nations with health consequences that impact the economy of these countries [1]. Research aimed at understanding their biology and identification of potential targets for drug development is the highest priority [1]. Interestingly, inhibitors of channels formed by proteins of the gap junction family such as suramin and boldine have trypanocidal activity and some of them are currently used for treatment of parasitic diseases such Human African Trypanosomiasis [2–5]. Also, studies have shown that infections caused by *Trypanosoma cruzi* (*T. cruzi*) modulate the expression of proteins of the gap junction family or modify the activity of the channels formed by these proteins in host cells [6–11]. Moreover, previous studies have shown gap junction type structures in *Trypanosoma musculi* [12]. In this chapter, we summarized the current knowledge about the role of gap junction family proteins in Chagas disease.

2. Gap junction proteins

Gap junction proteins are present in both vertebrates and invertebrates from mesozoa to mammals [13]. The protein families include connexins (Cxs), innexins (Inxs) and pannexins (Panxs) [14, 15]. They are integral membrane proteins that participate in cellular communication playing a relevant role in several physiological processes [15]. In vertebrates, Cxs and Panxs are present, while in invertebrates, only Inxs are present [15]. In humans, 21 and 3 genes encode Cx and Panx proteins, respectively [14]. Most Cx genes contain two exons and an intron of variable length [14]. The Panx1 and Panx2 genes contain 5 exons, while the Panx3 gene contains 4 exons [16]. Moreover, Inx genes have been found in the phylum *Arthropoda* [17–20], *Nematoda* [21], *Chordata* [22], *Annelida* [23], *Platyhelminthes* [24], *Cnidaria* [25], and *Mollusca* [26]. In *Drosophila melanogaster*, the Inx genes have the potential to be differentially spliced [18], while in *C. elegans*, 15% of genes are found in operons and three pairs of the innexins are polycistronic such as inx-12 and inx-13, inx-16 and inx-17, and inx-21 and inx-22 [27]. Regarding the structure of the protein topology, hydropathy plots of several Inx, Cx and Panx proteins have predicted the presence of four hydrophobic domains with transmembrane spanning regions and the extracellular loops with several highly conserved residues [28]. In contrast, the cytoplasmic loop and the carboxy terminus vary extensively in length and amino acid composition [28].

2.1 Gap junction channels

Cx and Inx proteins form gap junction channels, which connect the cytoplasm of neighboring cells [14, 15]. Moreover, Cx, Panx and Inx proteins form non-junctional channels that connect the intra- and extracellular milieu [14]. Gap junction plaques are formed by a variable number of homo- and/or heterotypic gap junction channels with distinct biophysical characteristics [29]. Structurally, they are formed by oligomers of Cx, Inx, or Panx proteins, which co-oligomerize into the same (homomeric) or mixed (heteromeric) channels [30]. Gap junction channels are essential in several physiologic functions such as electrical conduction between cardiomyocytes [31], development and regeneration of skeletal muscle [32], endocrine gland secretion [33], and ovarian folliculogenesis [34]. Also, they are implicated in pathophysiological conditions such as hereditary deafness [35], cataract [36], and tumorigenesis [37].

Gap junction proteins can also form non-junctional channels, which play important roles as autocrine/paracrine cellular communication pathways [14]. They are permeable to ions and several metabolic and signaling molecules such as glucose, glutamate, glutathione, adenosine, NAD+ (superindice) and ATP among others [14]. It has been proposed that Panxs do not form gap junctions, however, they form plasma membrane channels (named pannexons) with some properties similar to those of the non-junctional channels formed by Inxs or Cxs also called hemichannels [15].

The pannexons are permeable to ATP when are activated with certain stimuli such a low oxygen, mechanical stress, and elevated extracellular potassium ion concentration [38]. Otherwise, the Panx1 channel is selective for chloride ions and exhibits no ATP permeability when stimulated simply by depolarization to positive potentials or the C-terminal is cleaved by proteolysis [38]. Since we found increased Panx1 channel activity and increase in ATP release in cells infected with *T. cruzi* [11], it suggested that under this condition the Panx1 channel does not undergo proteolysis and adopt the large channel configuration [11, 38]. In addition, it has been described that Panx1-based channels are regulated by mechanical stress [39].

Inx-based hemichannels are activated by increased extracellular potassium ion concentration and by membrane depolarization [40]. Furthermore, Panx1- and Inx-based channels are inhibited by low concentration of carbenoxolone (<5 μM) or high probenecid concentrations (>500 μM) [41, 42].

Moreover, Cx-based hemichannels are regulated by intracellular acidification [43], intracellular Ca^{2+} [44], intracellular Na^{+} [45], pro-inflammatory cytokines [46], positive membrane potentials [47], phosphorylation [48], and S-nitrosylation [49]. They are inhibited by lanthanum chloride, carbenoxolone (>50 μM), and Cx mimetic peptides GAP16, GAP27, and GAP19 [15].

3. Gap junction and Chagas

3.1 Connexins

In 1992, Spray's group showed for the first time that gap junctions were altered between rat neonatal cardiomyocytes infected with the Tulahuen strain of *T. cruzi* (**Table 1**) [6]. They showed that junctional conductance and intercellular transfer of Lucifer yellow was reduced between cardiomyocytes infected with *T. cruzi* (**Figure 1**) [6]. In 1998, the same group described that Tulahuen strain of *T. cruzi* reduced gap junction communication between rat astrocytes and between rat leptomeningeal cells as well [7]. Also, these authors demonstrated through immunocytochemistry studies that Cx43 reactivity was significantly reduced in whole brains from rats acutely infected with *T. cruzi* [7]. In 2008, *in vitro* studies showed that Y strain of *T. cruzi* increased the amount of Cx43 at 1 hour postinfection and reduced it at 72 hour postinfection in mouse cardiomyocytes (**Figure 1**) [8]. Moreover, they demonstrated through immunoblotting analysis that the amount of Cx43 was significantly reduced in heart atria and ventricles from mice infected with Y strain of *T. cruzi* at 11 days postinfection [8]. In 2009, Waghabi and collaborators demonstrated that the number and length of Cx43 plaques were reduced in heart biopsies of human chronic chagasic patients [9]. In 2013, we described that CL Brener strain of *T. cruzi* increased the Cx43 hemichannel activity in HeLa cells stably transfected with Cx43 (**Figure 1**) [15]. Also, we observed that the number of amastigotes was 3 times higher in

Year	Strain	Cell type	Effect	Ref.
1992	Tulahuén	Cardiomyocytes	Decrease gap junction channels activity	[6]
1998	Tulahuén	Astrocytes	Decrease gap junction channels activity	[7]
1998	Tulahuén	Leptomeningeal cells	Decrease gap junction channels activity	[7]
2008	Y	Cardiomyocytes	Increase Cx43 amount at 1 h p.i.	[8]
2008	Y	Cardiomyocytes	Decrease Cx43 amount at 72 h p.i.	[8]
2009	n.d.	Heart biopsies	Decrease Cx43 levels at 11 days p.i.	[9]
2013	CL Brener	HeLa-Cx43	Increase Cx43-hemichannel activity	[15]
2014	Brazil	White adipose tissue	Increase Cx43 amount at 30 and 90 days p.i.	[10]
2014	Brazil	Brown adipose tissue	Decrease Cx43 amount at 30 and 90 days p.i.	[10]
2018	H510	Cardiomyocytes	Increase Panx1 channel activity	[11]

n.d.: not determinated; p.i.: post-infection.

Table 1.
Summary of the studies that describe the effect of T. cruzi on gap junction protein family.

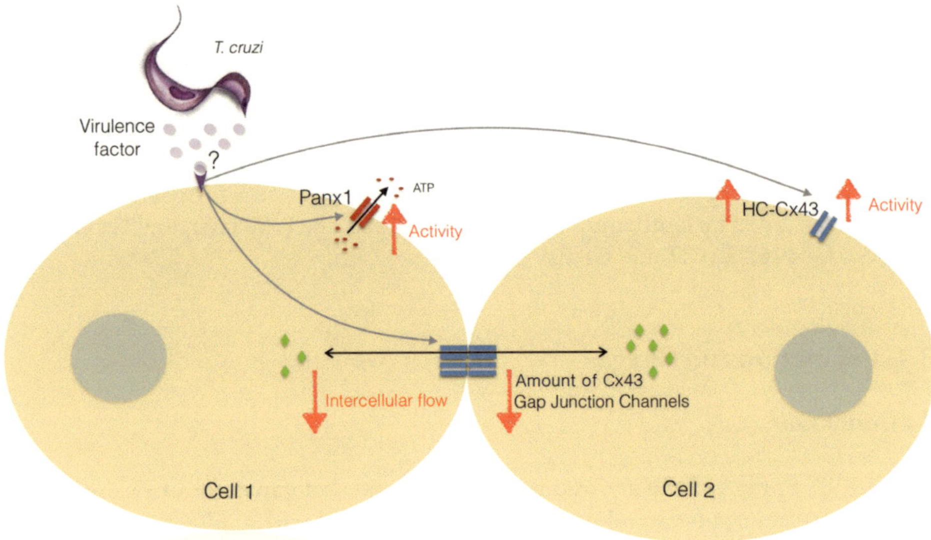

Figure 1.
Model that summarizes the main effects of the Trypanosoma cruzi on cellular communications mediated by gap junction channels. Parasites release a virulence factor that opens Panx1 channels allowing the release of ATP to the extracellular milieu [11]. Also, the parasite causes a reduction of intercellular communications mediated by gap junctions [6]. Contrarily, the parasite increases the activity of the hemichannels present in the plasma membrane of the infected cells [15].

HeLa-Cx43 as compared to HeLa parental cells at 48 hours postinfection [15]. In 2014, Burke and collaborators demonstrated that Brazil strain of *T. cruzi* reduced the amount of Cx43 in brown adipose tissue at 30 and 90 days postinfection (**Figure 1**) [10]. Moreover, *T. cruzi* infection caused an increased the amount of Cx43 protein in white adipose tissue at 30 and 90 days postinfection [10].

3.2 Pannexins

In 2018, we found that H510 strain of *T. cruzi* increased the Panx1 channel activity in rat neonatal cardiomyocytes at 1 hour postinfection [11]. Interestingly, the increased pannexon activity induced by the parasite was directly related to an elevated ATP release [11]. This is relevant because ATP has been proposed as a key molecule in *T. cruzi* host cell invasion [11, 50]. For example, blockade of $P2Y_1$ receptors with a MRS2179, a selective $P2Y_1$ antagonist, reduced *T. cruzi*-evoked Ca^{2+} transients in rat cardiomyocytes [11]. Moreover, inhibition of mitochondrial ATP production by treating parasites with rotenone plus antimycin A reduced the infectivity of the parasites [50]. Also, pre-treatment with pannexon activity-blocking drugs significantly reduced the number of intracellular parasites in cardiomyocytes infected with H510 strain of *T. cruzi* [11]. For instance, cells exposed to 100 μM ^{10}Panx1 or 400 μM probenecid showed a 114 ± 2 and 71 ± 2 parasites/500 cells, respectively, versus 5001 ± 2 parasites/500 cells in control condition [11]. Interestingly, scanning and transmission electron microscopy studies have demonstrated the presence of reminiscent gap junction at physical interactions between *Trypanosoma musculi* and mice spleen-derived adherent fibroblasts [12]. Also, Lucifer yellow microinjected into fibroblast was incorporated into associated trypanosomes, suggesting that those gap junctions were functional [12].

4. Gap junction channel blocking compounds as antiparasitic drugs

Interestingly, several drugs that block the activity of gap junction channels have been described as potent anti-parasitic agents. Among them are: probenecid, boldine and suramin, among others (**Table 2**) [2–4].

4.1 Probenecid

Human clinical trials have shown that probenecid, an inhibitor of a non-junctional channel formed by Panx1 has an antimalarial effect [2]. The authors described that 50 μM probenecid caused an increase in the sensitivity of highly resistant V1S strain of *Plasmodium falciparum* to pyrimethamine, sulfadoxine, chlorcycloguanil, and dapsone by seven-, five-, three-, and three-folds, respectively [2].

4.2 Suramin

Suramin, a general antagonist of purinergic receptors (P2Y and P2X) [51] and blocker of Panx1 channels [52], exhibits anti-parasitic properties [3]. Culture of LLC-MK2 cells treated with suramin (500 μM) during the intracellular development of *T. cruzi*, caused morphological changes in the parasites; increase in parasite width, and partial or complete detachment between flagella and cell body [3]. Interestingly, suramin is one of the 5 approved drugs for treatment of sleeping sickness [5].

4.3 Boldine

Boldine, an alkaloid obtained from Boldo tree, which blocks the activity of Cx43-formed hemichannels, and Panx1 hemichannels exhibit, an anti-parasitic activity [53, 54]. Boldine at concentrations above 500 μM reduces the epimastigotes growth of Tulahuen, LQ and DM28c strains of *T. cruzi* [4]. Since the active compound also inhibited cell respiration, it was suggested that these drugs may block the mitochondrial electron transport [4].

Agent	Chemical Formula	Class	Diseases	Target
Antiparisitic drugs				
Suramin	$C_{51}H_{40}N_6O_{23}S_6$	Acid	HAT	Cx43
Nifurtimox	$C_{10}H_{13}N_3O_5S$	Nitrofurans derivates	Chagas	n.d.
Benznidazol	$C_{12}H_{12}N_4O_3$	Nitroimidazole derivates	Chagas	n.d.
Pentamidine	$C_{19}H_{24}N_4O_2$	Amidine	HAT	n.d.
Melarsoprol	$C_{12}H_{15}AsN_6OS_2$	Arsenical	HAT	n.d.
Channel blocker drugs				
Probenecid	$C_{13}H_{19}NO_4S$	Sulfoamida	Malaria	Panx1
Carbenoxolone	$C_{34}H_{50}O_7$	Terpenes	n.d.	Panx1, Cx43, Inxs
Oleamide	$C_{18}H_{35}NO$	Amide	n.d.	Panx1
Mefloquine	$C_{17}H_{16}F_6N_2O$	Quinolines	n.d.	Panx1
Boldine	$C_{19}H_{21}NO_4$	Alkaloid	n.d.	Panx1, Cx43

HAT: Human African Trypanosomiasis; n.d.: not determined.

Table 2.
Antiparasitic drugs and channel blocker drugs.

5. Conclusions

Chagas diseases affect predominantly underprivileged areas of Latin America, but attention has been increasing lately due to the rise in people migration habits, intercontinental travels, and immune suppressed patients [1]. Unfortunately, current therapeutic options include only two compounds (nifurtimox and benznidazole) with considerable toxicity and side effects, so the new drug development is of the highest priority [1]. Hemichannels are involved in the regulation of plasma membrane permeability in ischemic insults or metabolic inhibition [55, 56]. Moreover, alterations of plasma membrane is a common phenomenon in parasite-induced infections such malaria and *T. cruzi*, among others [57, 58]. Thus, hemichannels could be key players in parasite-induced plasma membrane permeabilization. All the above data support the importance of studying the possible role of hemichannels in parasitic infections. They could be potential targets for the development of new compounds to limit parasite infections or tissue/organ damage induced by their presence.

Acknowledgements

This work was partially supported by a MINEDUC-UA project code ANT 1755 (to JLV), Fondo Nacional de Desarrollo Científico y Tecnológico (FONDECYT) 1150291 (to JCS) and ICM-Economía P09-022-F Centro Interdisciplinario de Neurociencias de Valparaíso (to JCS). JG and IB hold a CONICYT-Ph.D. fellowship, Chile.

Conflict of interest

The authors declare that the research was conducted in the absence of any commercial or financial relationships that could be construed as a potential conflict of interest.

Author details

José Luis Vega[1*], Camilo Juyumaya[1], Luis Rodríguez[1], Juan Güiza[1], Camila Gutíerrez[1], Iván Barría[2] and Juan C. Sáez[3,4]

1 Gap Junction and Parasites Laboratory (GaPaL), Antofagasta Institute, Universidad de Antofagasta, Antofagasta, Chile

2 Centro Interdisciplinario de Neurociencia de Valparaíso, Valparaíso, Chile

3 Instituto de Neurociencias, Centro Interdisciplinario de Neurociencia de Valparaíso, Universidad de Valparaíso, Chile

4 Departamento de Fisiología, Pontificia Universidad Católica de Chile, Santiago, Chile

*Address all correspondence to: joseluis.vega@uantof.cl

References

[1] Rassi A Jr, Rassi A, Marin-Neto JA. Chagas disease. Lancet. 2010;**375**:1388-1402. DOI: 10.1016/S0140-6736(10)60061-X

[2] Masseno V, Muriithi S, Nzila A. In vitro chemosensitization of *Plasmodium falciparum* to antimalarials by verapamil and probenecid. Antimicrobial Agents and Chemotherapy. 2009;**53**:3131-3134. DOI: 10.1128/AAC.01689-08

[3] Bisaggio DF, Campanati L, Pinto RC, Souto-Padrón T. Effect of suramin on trypomastigote forms of *Trypanosoma cruzi*: Changes on cell motility and on the ultrastructure of the flagellum-cell body attachment region. Acta Tropica. 2006;**98**:162-175. DOI: 10.1016/j.actatropica.2006.04.003

[4] Morello A, Lipchenca I, Cassels BK, Speisky H, Aldunate J, Repetto Y. Trypanocidal effect of boldine and related alkaloids upon several strains of *Trypanosoma cruzi*. Comparative Biochemistry and Physiology Part C: Pharmacology, Toxicology and Endocrinology. 1994;**107**:367-371. DOI: 10.1016/1367-8280(94)90063-9

[5] Singh Grewal A, Pandita D, Bhardwaj S, Lather V. Recent updates on development of drug molecules for human African trypanosomiasis. Current Topics in Medicinal Chemistry. 2016;**16**(20):2245-2265

[6] de Carvalho AC, Tanowitz HB, Wittner M, Dermietzel R, Roy C, Hertzberg EL, et al. Gap junction distribution is altered between cardiac myocytes infected with *Trypanosoma cruzi*. Circulation Research. 1992;**70**:733-742

[7] Campos de Carvalho AC, Roy C, Hertzberg EL, Tanowitz HB, Kessler JA, Weiss LM, et al. Gap junction disappearance in astrocytes and leptomeningeal cells as a consequence of protozoan infection. Brain Research. 1998;**790**:304-314. DOI: 10.1016/S0006-8993(97)01523-0

[8] Adesse D, Garzoni LR, Huang H, Tanowitz HB, de Nazareth MM, Spray DC. *Trypanosoma cruzi* induces changes in cardiac connexin43 expression. Microbes and Infection. 2008;**10**:21-28. DOI: 10.1016/j.micinf.2007.09.017

[9] Waghabi MC, Coutinho-Silva R, Feige JJ, Higuchi Mde L, Becker D, Burnstock G, et al. Gap junction reduction in cardiomyocytes following transforming growth factor-beta treatment and *Trypanosoma cruzi* infection. Memórias do Instituto Oswaldo Cruz. 2009;**104**:1083-1090. DOI: 10.1590/S0074-02762009000800004

[10] Burke S, Nagajyothi F, Thi MM, Hanani M, Scherer PE, Tanowitz HB, et al. Adipocytes in both brown and white adipose tissue of adult mice are functionally connected via gap junctions: Implications for Chagas disease. Microbes and Infection. 2014;**16**:893-901. DOI: 10.1016/j.micinf.2014.08.006

[11] Barría I, Güiza J, Cifuentes F, Zamorano P, Sáez JC, González J, et al. *Trypanosoma cruzi* infection induces pannexin-1 channel opening in cardiac myocytes. The American Journal of Tropical Medicine and Hygiene. 2018;**98**:105-112. DOI: 10.4269/ajtmh.17-0293

[12] Gugssa A, Lee CM, Gebru S, Desta D, Murray S, Baccetti B, et al. Co-culture of *Trypanosoma musculi* with spleen-derived adherent fibroblasts: Possible transfer of small molecules via connexons. Journal of Submicroscopic Cytology and Pathology. 2005;**37**:223-229

[13] Meşe G, Richard G, White TW. Gap junctions: Basic structure and function. The Journal of Investigative Dermatology. 2007;**127**:2516-2524. DOI: 10.1038/sj.jid.5700770

[14] Saez JC, Berthoud VM, Branes MC, Martinez AD, Beyer EC. Plasma membrane channels formed by connexins: Their regulation and functions. Physiological Reviews. 2003;**83**:1359-1400. DOI: 10.1152/physrev.00007.2003

[15] Vega JL, Subiabre M, Figueroa F, Schalper KA, Osorio L, González J, et al. Role of gap junctions and hemichannels in parasitic infections. BioMed Research International. 2013;**2013**:589130. DOI: 10.1155/2013/589130

[16] Baranova A, Ivanov D, Petrash N, Pestova A, Skoblov M, Kelmanson I, et al. The mammalian pannexin family is homologous to the invertebrate innexin gap junction proteins. Genomics. 2004;**83**:706-716. DOI: 10.1016/j.ygeno.2003.09.025

[17] Ganfornina MD, Sánchez D, Herrera M, Bastiani MJ. Developmental expression and molecular characterization of two gap junction channel proteins expressed during embryogenesis in the grasshopper *Schistocerca americana*. Developmental Genetics. 1999;**24**:137-150. DOI: 10.1002/(SICI)1520-6408(1999)24:1/2<137::AID-DVG13>3.0.CO;2-7

[18] Stebbings LA, Todman MG, Phillips R, Greer CE, Tam J, Phelan P, et al. Gap junctions in *Drosophila*: Developmental expression of the entire innexin gene family. Mechanisms of Development. 2002;**113**:197-205. DOI: 10.1016/S0925-4773(02)00025-4

[19] Hong SM, Kang SW, Goo TW, Kim NS, Lee JS, Kim KA, et al. Two gap junction channel (innexin) genes of the *Bombyx mori* and their expression. Journal of Insect Physiology. 2008;**54**:180-191. DOI: 10.1016/j.jinsphys.2007.09.002

[20] Calkins TL, Woods-Acevedo MA, Hildebrandt O, Piermarini PM. The molecular and immunochemical expression of innexins in the yellow fever mosquito, *Aedes aegypti*: Insights into putative life stage- and tissue-specific functions of gap junctions. Comparative Biochemistry and Physiology Part B: Biochemistry & Molecular Biology. 2015;**183**:11-21. DOI: 10.1016/j.cbpb.2014.11.013

[21] Altun Z, Chen B, Wang Z, Hall D. High resolution map of *Caenorhabditis elegans* gap junction proteins. Developmental Dynamics. 2009;**238**:1936-1950. DOI: 10.1002/dvdy.22025

[22] White T, Wang H, Mui R, Litteral J, Brink P. Cloning and functional expression of invertebrate connexins from *Halocynthia pyriformis*. FEBS Letters. 2004;**577**:42-48. DOI: 10.1016/j.febslet.2004.09.071

[23] Kandarian B, Sethi J, Wu A, Baker M, Yazdani N, Kym E, et al. The medicinal leech genome encodes 21 innexin genes: Different combinations are expressed by identified central neurons. Development Genes and Evolution. 2012;**222**:29-44. DOI: 10.1007/s00427-011-0387-z

[24] Zurabian R, Landa A, Robert L, Willms K. Immunolocalization of *Taenia solium* gap junction innexins. Parasitology. 2008;**135**:1125-1131. DOI: 10.1017/S0031182008004629

[25] Takaku Y, Hwang JS, Wolf A, Böttger A, Shimizu H, David CN, et al. Innexin gap junctions in nerve cells coordinate spontaneous contractile behavior in *Hydra* polyps. Scientific Reports. 2014;**4**:3573. DOI: 10.1038/srep03573

[26] Kelmanson IV, Shagin DA, Usman N, Matz MV, Lukyanov SA, Panchin YV. Altering electrical connections in the nervous system of the pteropod mollusc *Clione limacina* by neuronal injections of gap junction mRNA. The European Journal of Neuroscience. 2002;**16**:2475-2476. DOI: 10.1046/j.1460-9568.2002.02423.x

[27] Simonsen K, Moerman D, Naus C. Gap junctions in *C. elegans*. Frontiers in Physiology. 2014;**5**:40. DOI: 10.3389/fphys.2014.00040

[28] Phelan P. Innexins: Members of an evolutionarily conserved family of gap-junction proteins. Biochimica et Biophysica Acta - Biomembranes. 2005;**1711**:225-245. DOI: 10.1016/j.bbamem.2004.10.004

[29] Barbe MT, Monyer H, Bruzzone R. Cell-cell communication beyond connexins: The pannexin channels. Physiology (Bethesda). 2006;**21**:103-114. DOI: 10.1152/physiol.00048.2005

[30] Falk MM, Buehler LK, Kumar NM, Gilula NB. Cell-free synthesis and assembly of connexins into functional gap junction membrane channels. The EMBO Journal. 1997;**16**:2703-2716. DOI: 10.1093/emboj/16.10.2703

[31] Kanno S, Saffitz JE. The role of myocardial gap junctions in electrical conduction and arrhythmogenesis. Cardiovascular Pathology. 2001;**10**:169-177. DOI: 10.1016/S1054-8807(01)00078-3

[32] Araya R, Eckardt D, Maxeiner S, Krüger O, Theis M, Willecke K, et al. Expression of connexins during differentiation and regeneration of skeletal muscle: Functional relevance of connexin43. Journal of Cell Science. 2005;**118**:27-37. DOI: 10.1242/jcs.01553

[33] Murray SA, Davis K, Gay V. ACTH and adrenocortical gap junctions. Microscopy Research and Technique. 2003;**61**:240-246. DOI: 10.1002/jemt.10332

[34] Gershon E, Plaks V, Dekel N. Gap junctions in the ovary: Expression, localization and function. Molecular and Cellular Endocrinology. 2008;**282**:18-25. DOI: 10.1016/j.mce.2007.11.001

[35] Kelsell DP, Dunlop J, Stevens HP, Lench NJ, Liang JN, Parry G, et al. Connexin 26 mutations in hereditary non-syndromic sensorineural deafness. Nature. 1997;**387**:80-83. DOI: 10.1038/387080a0

[36] Xia CH, Chang B, Derosa AM, Cheng C, White TW, Gong X. Cataracts and microphthalmia caused by a Gja8 mutation in extracellular loop 2. PLoS One. 2012;7:e52894. DOI: 10.1371/journal.pone.0052894

[37] Naus CC, Laird DW. Implications and challenges of connexin connections to cancer. Nature Reviews Cancer. 2010;**10**:435-441. DOI: 10.1038/nrc2841

[38] Wang J, Dahl G. Pannexin1: A multifunction and multiconductance and/or permeability membrane channel. American Journal of Physiology-Cell Physiology. 2018;**315**:C290-C299. DOI: 10.1152/ajpcell.00302.2017

[39] Bao L, Locovei S, Dahl G. Pannexin membrane channels are mechanosensitive conduits for ATP. FEBS Letters. 2004;**572**:565-568. DOI: 10.1016/j.febslet.2004.07.009

[40] Dahl G, Muller KJ. Innexin and pannexin channels and their signaling. FEBS Letters. 2014;**588**:1396-1402. DOI: 10.1016/j.febslet.2014.03.007

[41] Bao L, Samuels S, Locovei S, Macagno ER, Muller KJ, Dahl G. Innexins form two types of channels. FEBS Letters. 2007;**581**:5703-5708. DOI: 10.1016/j.febslet.2007.11.030

[42] Locovei S, Bao L, Dahl G. Pannexin 1 in erythrocytes: Function without a gap. Proceedings of the National Academy of Sciences of the United States of America. 2006;**103**:7655-7659. DOI: 10.1073/pnas.0601037103

[43] Delmar M, Coombs W, Sorgen P, Duffy HS, Taffet SM. Structural bases for the chemical regulation of Connexin43 channels. Cardiovascular Research. 2004;**62**:268-275. DOI: 10.1016/j.cardiores.2003.12.030

[44] Lopez W, Gonzalez J, Liu Y, Harris AL, Contreras JE. Insights on the mechanisms of Ca(2+) regulation of connexin26 hemichannels revealed by human pathogenic mutations (D50N/Y). The Journal of General Physiology. 2013;**142**:23-35. DOI: 10.1085/jgp.201210893

[45] Li F, Sugishita K, Su Z, Ueda I, Barry WH. Activation of connexin-43 hemichannels can elevate [Ca(2+)]i and [Na(+)]i in rabbit ventricular myocytes during metabolic inhibition. Journal of Molecular and Cellular Cardiology. 2001;**33**:2145-2155. DOI: 10.1006/jmcc.2001.1477

[46] Retamal MA, Froger N, Palacios-Prado N, Ezan P, Sáez PJ, Sáez JC, et al. Cx43 hemichannels and gap junction channels in astrocytes are regulated oppositely by proinflammatory cytokines released from activated microglia. The Journal of Neuroscience. 2007;**27**:13781-13792. DOI: 10.1523/JNEUROSCI.2042-07.2007

[47] Contreras JE, Saez JC, Bukauskas FF, Bennett MV. Gating and regulation of connexin 43 (Cx43) hemichannels. Proceedings of the National Academy of Sciences of the United States of America. 2003;**100**:11388-11393. DOI: 10.1073/pnas.1434298100

[48] Pogoda K, Kameritsch P, Retamal M, Vega JL. Regulation of gap junction channels and hemichannels by phosphorylation and redox changes: A revision. BMC Cell Biology. 2016;**17**(Suppl 1):11. DOI: 10.1186/s12860-016-0099-3

[49] Retamal MA, Cortes CJ, Reuss L, Bennett MV, Saez JC. S-nitrosylation and permeation through connexin 43 hemichannels in astrocytes: Induction by oxidant stress and reversal by reducing agents. Proceedings of the National Academy of Sciences of the United States of America. 2006;**103**:4475-4480. DOI: 10.1073/pnas.0511118103

[50] Martins RM, Covarrubias C, Rojas RG, Silber AM, Yoshida N. Use of L-proline and ATP production by *Trypanosoma cruzi* metacyclic forms as requirements for host cell invasion. Infection and Immunity. 2009;**77**:3023-3032. DOI: 10.1128/IAI.00138-09

[51] Voogd TE, Vansterkenburg EL, Wilting J, Janssen LH. Recent research on the biological activity of suramin. Pharmacological Reviews. 1993;**45**:177-203

[52] Qiu F, Dahl G. A permeant regulating its permeation pore: Inhibition of pannexin 1 channels by ATP. American Journal of Physiology-Cell Physiology. 2009;**296**:C250-C255. DOI: 10.1152/ajpcell.00433.2008

[53] Hernández-Salinas R, Vielma AZ, Arismendi MN, Boric MP, Sáez JC, Velarde V. Boldine prevents renal alterations in diabetic rats. Journal Diabetes Research. 2013;**2013**:593672. DOI: 10.1155/2013/593672

[54] Yi C, Ezan P, Fernández P, Schmitt J, Sáez JC, Giaume C, et al. Inhibition of glial hemichannels by boldine treatment reduces neuronal suffering in a murine model of Alzheimer's disease. Glia. 2017;**65**:1607-1625. DOI: 10.1002/glia.23182

[55] Contreras JE, Sánchez HA, Eugenin EA, Speidel D, Theis M, Willecke K, et al. Metabolic inhibition induces opening of unapposed connexin 43 gap junction hemichannels and reduces gap junctional communication in cortical astrocytes in culture. Proceedings of the National Academy of Sciences of the United States of America. 2002;**99**:495-500. DOI: 10.1073/pnas.012589799

[56] Thompson RJ, Zhou N, MacVicar BA. Ischemia opens neuronal gap junction hemichannels. Science. 2006;**312**:924-927. DOI: 10.1126/science.1126241

[57] Ginsberg H. Alterations caused by the intraerythrocytic malaria parasite in the permeability of its host cell membrane. Comparative Biochemistry and Physiology Part A: Comparative Physiology. 1990;**95**:31-39. DOI: 10.1016/0300-9629(90)90006-E

[58] Fernandes MC, Andrews NW. Host cell invasion by *Trypanosoma cruzi*: A unique strategy that promotes persistence. FEMS Microbiology Reviews. 2012;**36**:734-747. DOI: 10.1111/j.1574-6976.2012.00333.x

Chapter 11

Trypanosoma cruzi Infection: Mechanisms of Evasion of Immune Response

Alondra Cruz Reyes and José Luis Rosales Encina

Abstract

Trypanosoma cruzi has a complex life cycle that involves a vertebrate as well as an invertebrate host. In this, last two stages are present: trypomastigotes, the flagellated and infective stage and the amastigote, which is the replicative stage. *T. cruzi* is considered one of the most successful intracellular parasites, because it cannot be eliminated by the immune system and has the capacity of invading, surviving, and replicating inside the host cells. The effects that the infection has over the immune system have been widely studied at the molecular and cellular level. However, understanding the mechanisms that the parasite uses to evade the immune system to persist in the infected individual is necessary for the effective development of drugs and/or vaccines. In this chapter, a compilation of the already described mechanisms will be carried out.

Keywords: *Trypanosoma cruzi*, immune system, *T. cruzi* immune evasion, immune modulation

1. Introduction

Chagas disease or American trypanosomiasis is a potentially life-threatening illness caused by the protozoan parasite, *Trypanosoma cruzi*. It is found mainly in 21 Latin American countries, where it is a mostly vector borne. An estimated 8 million people are infected worldwide, mostly in Latin America. Chagas disease has spread to the other continents over the last century mainly because of enhanced means of travel and global population movement to and from Latin America. It is estimated that over 10,000 people die every year from clinical manifestations of Chagas disease and more than 25 million people risk acquiring the disease [1].

Chagas disease is characterized by two stages, the initial phase is known as acute phase and lasts a few weeks and is characterized by an elevated parasitemia associated with fever, headache, nausea, that is rarely lethal and a severe hepatomegaly. The acute phase is followed by a chronic phase, which remains asymptomatic for many years. The parasites are then difficult to observe in the blood, and the other symptoms are also less severe. Most patients remain in this indeterminate stage for life (absence of any symptom), 5–10% of the people will develop anatomical and functional abnormalities at their esophagus and their colon, while 20–40% starts presenting a symptomatic chronic phase characterized by a progressive and debilitating Chagasic cardiomyopathy and sometimes mega syndromes. Arrhythmias increasing severely lead to congestive cardiac failure and death of the patients usually 10–30 years after the initial infection [2, 3].

The recognition of *T. cruzi* by the immune system relies on the activation of innate immunity and adaptative immunity. In this activation process, the recognition of pathogen-associated molecular patterns (PAMP) by Toll-like receptors of B and T cells is very important as a bridge between both types of immunity [4]. The innate and acquired immune responses are characterized by the recruitment of dendritic cells, macrophages, natural killer (NK), and B and T lymphocytes, as well as their secretion of soluble factors (cytokines and chemokines) [5]. IFN-γ plays a key role during *T. cruzi* infection, increasing the production of nitric oxide by macrophages, which inhibits the development of the intracellular from *T. cruzi* [6]. CD4+ Th1 lymphocyte that produces the cytokines interleukin-2 (IL-2) and IFN-γ can stimulate the expansion of cytotoxic CD8+ T lymphocytes, a central mechanism for systemic protection against *T. cruzi* infection [5, 7]. Recent studies have confirmed that CD8+ T cells act both indirectly by secreting IFN-γ to activate macrophages and directly through the production of perforin and their concomitant cytotoxic activity [8]. Thus, there is now a growing consensus that protective immune response against *T. cruzi* requires the activation of a Th1 immune profile, with the stimulation of CD8+ T cells, while antibodies may play a rather secondary role [9].

Although the immune response generated with *Trypanosoma cruzi* infection is potent and active, it fails to clear parasite infection, leading to the chronic phase of the Chagas disease. The parasite has developed sophisticated strategies to evade host immune responses. The *Trypanosoma cruzi* immune evasion is mainly based on altering the complement system and inhibitory effects on the mononuclear phagocyte system lifestyle [10–12], and therefore, this chapter summarizes the mechanism employed by *Trypanosoma cruzi* to escape the host's deleterious immune responses and enter into the host cells and establish persistent infection.

2. Immune evasion during the acute phase

2.1 Escape of the parasitophorous vacuole

Invading the host cell is a complex process that starts with parasite attachment to plasmatic membrane and then the compartment is formed called parasitophorous vacuole (VP), which fuses to lysosomes forming the phagolysosome (PLM). Acidification of the VP allows trypomastigotes escape from PLM to the cytoplasm, where it differentiates into amastigote forms. In the cytoplasm of infected cells, the amastigotes multiply by binary fission, they accumulate and transform to bloodform trypomastigotes, which are released in the bloodstream through the rupture of the host cell membrane to find new cells to invade [13–17].

Parasite escape from PLM occurs because the growth and development of *T. cruzi* cannot be maintained within it. Therefore, the trypomastigotes exit to the cytoplasm, and after a short period of time, they differentiate into amastigotes, and the rupture of this structure is activated by acidic pH generated by the fusion of the lysosomes [18].

To survive the acid medium into the phagolysosome, the trypomastigotes activate trans-sialidases/neuraminidase, and these enzymes transfer sialic acid from the sialyl-glycoconjugates in the membrane of the PLM to the β-galactoses of acceptor molecules (GPI-mucins) on the surface of the parasite [19, 20]. More specifically, the membrane of the phagolysosome is covered on the inner side with two important proteins known as lysosomes-associated membrane proteins (LAMP1 Y LAMP2), and these proteins are highly sialylated [21, 22]. The terminal sialylation of these proteins contributes to maintaining parasitophorous vacuole membrane integrity, whereby the removal of sialic acid residues sensitizes the membrane to the action of a transmembrane protein forming pore secreted by the parasite (Tc-TOX) [23, 24]. When the phagolysosome is broken, the trypomastigotes are released into the cytoplasm.

2.2 Evasion of the oxidative response

Previously, it was mentioned that *T. cruzi* not only invades phagocytic cells via classic phagocytosis but also can actively invade mammalian nonprofessional phagocytic cells by induced phagocytosis [25]. The main host cells targeted by *T. cruzi* are resident macrophages at the site of infection and dendritic cells, both of which play a main role in the response of the immune system as they are specialized antigen presenting cells [11].

Macrophages have two roles; on the one hand, they are important effector cells for the control and killing of the intracellular form of the parasite by oxidative and nonoxidative mechanism. On the other hand, macrophages may also serve as long-term host cells that facilitate the replication and survival of the pathogens [26]. Macrophage membrane-associated NADPH oxidase is activated during phagocytosis of trypomastigotes, resulting in a stable superoxide radical anion ($O_2^{\cdot -}$) which can be transformed to H_2O_2 by superoxide dismutase (SOD) [6, 27, 28].

IFN-γ and TNF synthesized during *T. cruzi* acute infection stimulate infected macrophages to produce high amounts of nitric oxide ($^{\cdot}NO$) derived by the enzymatic activity of inducible nitric oxide synthase (iNOS). The nitric oxide generated reacts with $O_2^{\cdot -}$ and produces peroxynitrite ($ONOO^-$), a potent oxidant and cytotoxic effector molecule against *T. cruzi* [29–33]. *Trypanosoma cruzi* has five antioxidant enzymes such as peroxidases and four iron superoxide dismutases (FeSODs) that are located at different subcellular compartments of the parasite, which reduce equivalents from NADPH and defend the parasite against host oxidative stress of the host (**Tables 1** and **2**) [27, 34].

2.3 Complement evasion

Trypomastigotes after being released in the bloodstream become a targeted of the preexisting soluble factors that potentially recognize and destroy them by different effectors [11]. Serum components, such as the complement system, consist of 35 or more soluble protein and cell surface receptors/regulators that interact with pathogen structures and activate a cascade of proteases that eliminate invading parasites, and this system comprises the first line of defense [16, 43–45].

Peroxidase	Subcellular location	Function
Tryparedoxin peroxidase (TcCPX)	Cytosol	Detoxify $ONOO^-$, H_2O_2 and small-chain organic hydroperoxides [34–37]
Tryparedoxin peroxidase (TcMPX)	Mitochondria	
Glutathione peroxidase I (TcGPXI)	Cytosol glycosome	Hydroperoxidases [38, 39]
Glutathione peroxidase II (TcGPXII)	Endoplasmic reticulum	Lipid hydroperoxidases [34, 38, 39]
Ascorbate-dependent heme peroxidase (TcAPX)	Endoplasmic reticulum	Resistance H_2O_2 [34, 40]

Table 1.
Peroxidases of Trypanosoma cruzi.

Iron superoxide dismutase (FeSODs)	Subcellular location	Specificity
TcSODB1	Cytosol	Detoxify $O_2^{\cdot -}$ and hence the formation of $ONOO^-$ [41, 42]
TcSODB1-2	Glycosome	
TcSODA	Mitochondria	
TcSODB	Mitochondria	

Table 2.
Iron superoxide dismutase of Trypanosoma cruzi.

Briefly, the action of the classical pathway is mediated by the antibodies binding to pathogen antigens, then the antibodies interact with the complement C1 protein, and it cleaves C2 and C4 to generate C2a and C4b, which join to the pathogen surface and form the C3 convertase C4b2a [16, 46]. Lectin complement pathway is the first to recognize *T. cruzi*; this is active by the binding of ficolins to the carbohydrates on the parasite surface and also mannan-binding lectins (MBLs) to the mannan on the surface of the parasite; then cysteine proteases bound to these molecules and cleave C2 and C4 and; this event also generates the C3 convertase C4b2a [47–49]. Finally, the alternative pathway is activated by the spontaneously hydrolyzed C3 or C3b originating from the other complement pathways; C3b interacts with factor B, and the latter in then cut into Bb by factor D, forming the C3 convertase C3bBb [50]. The three pathways differ in the initial steps of activation, but all three converge to produce a C3 convertase and then a C5 convertase, allowing the formation of the membrane attack complex (MAC) which is responsible for membrane lysis and subsequent pathogen elimination [16].

Trypanosoma cruzi trypomastigotes are normally resistant to the lytic effects of complement from vertebrate hosts susceptible to infection. This resistance of the trypomastigotes to direct lysis by the complement system facilitates parasite survival and infectivity but during the course of chronic infections; however, the vertebrate hosts produce antibodies that render the trypomastigotes sensitive to lysis, primarily via the alternative complement cascade and amplified by the classical pathway, and this resistance is a developmentally regulated phenomenon because the parasite is susceptible to complement lysis when it is in epimastigote form [51].

Complement activity present in normal human serum has been reported to lyse circulating forms of *T. cruzi* following activation by specific host antibodies bound to the surface of the parasites [52]. Incubation of trypomastigotes with human complement does not lead to lysis when the trypomastigotes do not have immunoglobulins on their surfaces. However, if such trypomastigotes are preincubates with sera obtained from chronically infected hosts, IgG immunoglobulins bind to their surface, and the parasites become sensitive to lysis by fresh human sera as a source of C5, thus lysis required antibodies and complement in mammals [53, 54], and on the other hand, birds, amphibians, and reptiles are naturally refractory to *T. cruzi* infection, due to complement-mediated lysis and macrophage-induced killing of the parasites [52, 55, 56]. Sera of the birds is capable of lysing infective forms of the parasite which as mentioned above are resistant to lysis by human serum, and this normal lytic activity of chicken serum seems to be independent of antibody, because it has been observed that the lytic activity of serum in vitro was not impaired in chickens that had been immunosuppressed by four different procedures, and the lysis occurs even in the absence of antibodies (e.g., bursectomized chickens are refractory to *T. cruzi* infections) in which chicken complement is probably activated through the alternative pathway [52, 57–59]. The details of the mechanisms of resistance of birds to *T. cruzi* infection need further investigation, since it could improve our understanding of the host-parasite interactions and favor the establishment of the innate mechanism of resistance and/or susceptibility of vertebrate hosts to *T. cruzi*.

2.3.1 Molecules of Trypanosoma cruzi that interfere with complement pathways

T. cruzi relies a variety of molecules that block or disrupt different steps of the complement pathways to evade complement lysis [44, 60, 61]. Below are some molecules of *T. cruzi* that it uses to get rid the action of the complement pathways.

a. Calreticulin (TcCRT): This protein is expressed in trypomastigotes, and it is mainly located in the endoplasmic reticulum (RE), but it has also been found in the Golgi, reservosomes, flagellar pocket, cell surface, cytosol, nucleus, and kinetoplast. After the infection, calreticulin moves from RE to the emerging area of the flagellum on the plasma membrane surface [62–64]. Calreticulin is a 45 kDa calcium-binding protein that binds to host mannose-binding lectin collagenous tails, preventing their interaction with parasite mannan and also interacts directly with L-ficolin preventing C4 conversion to C4b. TcCRT also interacts with C1q collagen-like domain, and therefore TcCRT inhibits both the classical and lectin pathway activation [65–67].

b. *Trypanosoma cruzi* complement regulatory protein (TcCRP) Gp160: The protein TcCRP, also called Gp160, which is GPI-anchored on the surface of trypomastigotes, can bind to C3b and C4b, thus inhibiting the formation of the classical and alternative complement C3 convertase [10, 68, 69].

c. *Trypanosoma cruzi* complement C2 receptor inhibition trispanning (TcCRIT): TcCRIT is a 32 kDa transmembrane protein mainly expressed in trypomastigotes. TcCRIT has amino acid sequence homology with the C4 beta-chain, the binding site of C2. Thus, it blocks C2 cleavage by C1s or MASP-2 into C2a and prevents C3 convertase formation, thus regulating the activation of the lectin and classical complement pathway [70–72].

d. Trypomastigote decay-accelerating factor (T-DAF): T-DAF is an 87–93 kDa glycoprotein expressed on the surface of metacyclic, bloodstream, and tissue-culture-derived trypomastigotes of *T. cruzi*. T-DAF mimics the activity of the complement regulatory protein DAF and regulates the activation of the alternative, classical, and probably the lectins pathways of the complement. In summary, T-DAF activity either accelerates the dissociation or assembly efficiency of C3 convertases [73, 74].

e. Glycoprotein 58/68: Gp58/68 is a glycoprotein expressed on the surface or released by trypomastigotes. This protein is part of a fibronectin/collagen receptor of *T. cruzi*, and for this reason, it plays an important role in the interaction of *T. cruzi* with mammalian cells. Gp58/68 allows the parasite to evade alternative pathway complement activation because it is able to inhibit the formation of cell-bound C3 convertase (decay-accelerating activity) by preventing the initial association of FB to surface fixed C3b [44, 75, 76].

On the other hand, it has been reported that *T. cruzi* metacyclic trypomastigote induces the release of membrane-derived vesicles (microvesicles) from host cells, such as lymphocytes, monocytes, and macrophages, in a Ca^{2+} dependent process. These vesicles inhibit the classical and lectin pathways of the complement by binding to C3 convertase C4b2a on the *T. cruzi* surface and inhibit its catalytic activity [44, 77].

3. Immunoregulatory effects by *Trypanosoma cruzi*

In the early stages of *T. cruzi* invasion, the host's resistance to the infection is mediated by innate immunity, which acts as the first immunological barrier. Macrophages, NK cells, and dendritic cells are the cells from the innate immune system and produce cytokines (IL-12, TNF-α, and IFN-γ) and effector molecules that control parasite replication. Dendritic cells are the connecting cells between the innate and acquired immunity, generating cytokines (IL-12) necessary for the differentiation and clonal expansion of T helper 1 (Th1) CD4+ as well as CD8+ T cells and plasma B cells. CD4+ Th1 or CD8+ T cells synthesize IFN-γ, which activates effector mechanisms in macrophages to eliminate both amastigotes and phagocytosed trypomastigotes, while cytotoxic activity by B cells lyse the extracellular trypomastigote form or facilitate the phagocytosis of parasites opsonized with IgG [3, 78, 79]. However, *T. cruzi* uses different mechanisms to modulate this action of the host immune response and then the parasite spreads to many tissues during the acute phase to finally reach the chronic stage of the Chagas disease [16]. In the acute phase, *T. cruzi* infection can induce immunosuppression [80], involving the decrease and clonal deletion of the T cells, together with strong polyclonal B cell stimulation, which ultimately restricts the development of antigen-specific lymphocytes [78, 81].

In the early steps of a primary *T. cruzi* infection, there is no activation of various host defense mechanisms, leading to silent entry [82]. There are three events that could contribute to this silent entry: the relatively slow kinetics of *T. cruzi* intracellular cycle, the parasite escape from the phagolysosome (which was previously explained), and the immunoregulatory response mediated by toll-like receptor (TLR) activation in dendritic cells [83]. A TLR-dependent activation of dendritic cells is required to induce their maturation and migration to regional lymph nodes and to activate naïve T cells [84].

Toll-like receptors (TLRs), members of a family of pattern recognition receptors, are responsible for the recognition of pathogens because they can distinguish between host molecules and molecules of the pathogens referred to as pathogen-associated molecular patterns (PAMPs). These receptors are expressed on the cell surface or in the lumen of intracellular vesicles (endosomes or lysosomes) of the antigen presenting cells (macrophages and dendritic cells), cells of the adaptive immunity (T y B lymphocytes), and nonimmune cells (epithelial and endothelial cells and fibroblasts) [84–88]. Twelve TLRs family members have been identified in mice and 10 in humans. The TLRs 1–9 are conserved in both species, whereas TLRs 11–13 is only expressed in mice and TLR10 in humans [87, 89]. TLR4 and TLR9 are homodimers, whereas TLR2 and TLR6 form heterodimers [90, 91]. Activation of the TLRs induces production of proinflammatory cytokines (MyD88-dependent signaling cascade, except TLR3) and chemokines, and therefore, these receptors are important because that activate innate immunity and molding the subsequent acquired immune responses [3, 92–94].

Infection with *T. cruzi* activates the synthesis of several inflammatory genes by different TLR pathways [87]. The parasite has molecules that can activate TLRs such as the surface molecules, glycoinositolphospholipids (GIPLs), and anchor glycoproteins and polysaccharides that contain identical GPI structures [95]. Glycosylphosphatidylinositol anchors derived from mucin-like glycoproteins (GPI-mucin) are TLR2/6 activators in dendritic cells and macrophages, and this leads to the production of the TNF-α, IL-12, and NO. On the other hand, GIPLs activate TLR4 of macrophages to increase IL-2 production. The IL-2 increase activates NK cells and B cells to secrete immunoglobulins during infection [3, 96]. *T. cruzi* genomic DNA (CpG DNA motifs) active TLR9 and TLR7 is involved in parasite RNA recognition, and this activation results in the production of IL-12 [97, 98]. CpG DNA motifs are concentrated in genomic regions encoding large gene families of surface proteins, such as mucins, trans-sialidases, and mucin-associated surface proteins (MASPs) [98]. These genes are involved in parasite immune evasion mechanisms [4].

TLR7 and TLR9 are expressed in the membrane of phagolysosome, and they are active when the parasite is lysed, but as mentioned above, parasite escapes from the lysosome, thus the activation of these receptors is reduced. On the other hand, TLR2 appeared to act as immunoregulator in the early stage of infection [99].

T. cruzi in addition to evade its destruction also has the capacity to modulate the pattern of secreted cytokines, and examples of this are *T. cruzi* membrane GPI-anchored mucin (AgC10), which can bind to the macrophage surface and induce the secretion of IL-1B but not of IL-12 or TNF-α, which are essential in a protective response. Inhibition of TNF-α and IL-12 by *T. cruzi* could be involved in the evasion of the immune response by this parasite [12, 100]. The parasite also promotes the production of IL-10 and TGF-β in infected macrophages, which inhibit the induction and effects of IL-12. A cysteine protease (cruzain) prevents macrophage activation by blocking the NF-κB P65 pathway and downregulated the expression of the proinflammatory cytokine, IL-12. Therefore, the infection of the macrophages favors the secretion of anti-inflammatory cytokines such as IL-10 and TGF-β that affect the development of protective immune response and favors the spread of infection [101, 102].

Trypanosoma cruzi induces the production of both Th1 and Th2 cytokines in infected individuals, and high expression levels have been reported for Th1 cytokines IFN-γ and IL-2, as well as for Th2 cytokines IL-4 and IL-10 [103]. The ability of *T. cruzi* to infect a host and to survive and develop and cause Chagas′ disease depends on a complex balance Th1 and Th2 cytokine production, as the display antagonistic effects, the former being protective for the host and the latter for the parasite [11].

In addition to the TLR's roles in the modulation of innate immunity, TLRs participate in the induction of the adaptive immune response [89]. This happens mainly by their action on antigen-presenting cells (APCs), either by promoting cross-presentation for CD8+ T-cell activation or by increasing the levels of costimulatory molecules and by stimulating the secretion of lineage-specific cytokines such as IL-12, IL-16, IL-1β, IL-18, and IL-23 by APCs and thus promoting Th1 and Th17 differentiation [86].

More specifically, the activation of dendritic cells by *T. cruzi* PAMPs (DNA and RNA) will lead to their maturation and production of IL-12, favoring the differentiation of Th cell precursors toward the Th1 phenotype and IFN-γ production [98, 104].

4. Polyclonal activation of B cells

During early stages of the *T. cruzi* infection, B cells are fundamental to trigger T-cell functions related to the Th1 pathway that favors the control of parasite growth [105]. B cells are known to play a key role in the humoral adaptive immune response by producing and secreting antibodies [106]. The majority B cells activated by *T. cruzi* infection are not specific for parasite antigens [107]. The infection with *T. cruzi* induces polyclonal B cell activation and hypergammaglobulinemia based on parasite-derived B cell mitogen. These antibodies are parasite-specific and bind to the trypomastigote surface interfering with the binding of IgG inhibitory antibodies and consequently preventing the elimination of the parasite [12, 108].

4.1 Factors causing antibody deficiency

4.1.1 Antigenic variability

T. cruzi expresses a variety of antigens on its surface such as mucins, trans-sialidases, and MASPs, all of them are encoded by highly polymorphic multigene families. This high variety of molecules with coordinate expression delays the activation of specific B cells and also consequently delays the production and maturation of high-affinity antibodies with neutralizing capacity and the priming of effective T-CD8+ cells, and this mechanism is known as a smoke screen [16, 19, 109, 110]. Another important point to mention is that the presence of a broad range of antigenic motifs may also be a mechanism to drive the antibody response away from catalytic sites of key parasite surface proteins which causes a weak antibody response [111].

4.1.2 Reduction of immature B cells in the bone marrow

Immature B cells are reduced because of an increase in the rate of apoptosis. Apoptosis is a common process caused by pathogens on host cells, and it probably happens as a consequence of the host immune response or by a direct effect of the pathogen [112]. Apoptosis of lymphocytes is necessary to preserve a healthy and balanced immune system, but if this occurs prior to pathogen elimination, it could reduce the effectiveness of the effector mechanisms [113, 114]. This effect may be more important if apoptosis happens in the early stages of lymphocyte development [115]. Thus, the effect of the infection with *T. cruzi* is the induction of a marked loss of immature B cells in the bone marrow and also compromises recently emigrated B cells in the periphery [116]. It has been shown that the apoptosis caused by *T. cruzi* infection occurs in addition to Fas/FasL pathway by the participation of CD11b+ myeloid cells that secrete a product of the cyclooxygenase pathway, and this event depletes immature B cells in the bone marrow [116, 117]. It is possible that the parasite takes advantage of the host cell apoptosis; moreover, it has been shown that cell apoptosis may have an additional negative effect because the elimination of apoptotic bodies by *T. cruzi* infected macrophages promotes parasite replication, favoring the chronic establishment of the parasite in the host [118], so the induction of the host cell apoptosis is another mechanisms to evade the immune response.

4.1.3 Polyclonal activation of non-specific B cells

Polyclonal activation of B cells that normally contributes to the generation of specific antibodies for conserved structures presents in pathogens but in the case of *T. cruzi* infection can be a mechanism by which parasite-specific antibodies are reduced and irrelevant antibody are increased [115]. Presentation of multiple-related sequences on the surface of the trypomastigotes or extracellular amastigotes could alter the B-cell response by inducing anergy in specific CD4+ lymphocytes and/or reduction of affinity of antibodies [109, 119]. Another consequence of polyclonal activation can be the generation of antibodies specific of the self-and/or non-related antigens [107]. Some parasite molecules such as glycoinositol-phospholipids (GIPLs) generate hypergammaglobulinemia [16, 108, 120]. In the acute infection, expansion of total antibodies is slow and starts with IgM and IgA, followed by IgG (IgG1, IgG2a, IgG2b, and IgG3) with specificities not related to *T. cruzi* antigens, while the humoral response in the chronic stage shows preferential IgG2a production [121–123]. In addition to the high variability of parasite surface antigens, parasite-derived B cell mitogens also cause polyclonal B cell activation and hypergammaglobulinemia which results in a delayed parasite-specific antibody response [108, 124, 125]. Glutamate dehydrogenase (TcGDH) [126], proline racemase [127], and trans-sialidase (TcTS) [128] are among the parasitic proteins identifying as polyclonal B cell mitogens [129].

5. Immunomodulation of T cell

Expression of identical antigenic variants on the surface of the majority of parasite population is the characteristic of the antigenic variation, with only a small subgroup expressing different variants [130, 131].

Trypanosoma cruzi releases antigen in the intracellular host cell environment, and they are accessible for presentation by the major histocompatibility complex class I [3]. The presentation of the antigens promotes the priming of a robust but delayes *T. cruzi*-specific CD8+ T-cell responses. This event is evident 5–6 days postinfection, after the first round of intracellular replication of the parasite [110]. The early immunosuppression of the CD8+ T-cell response may be associated with the density of parasites in the cytosol, because during the initial infection, previous to the first cycle of replication, low concentration of antigens is produced by amastigotes, and on the other hand, because a large repertoire of highly polymorphic and immunogenic surface protein that are coexpressed by the parasite [110, 132–134].

The parasite has developed strategies to interfere with antigen presentation, and this strategy is related with hyperpolarization of the presented antigen repertoire, which means that the repertory of CD8+ T-cells is dramatically restricted, and this phenomenon is known as immunodominance and prevents complete pathogen elimination by host cells by antigen-specific cytotoxic T-cell response [135, 136].

GPI-anchored surface proteins particularly trans-sialidase proteins are among the major known CD8+ T cell immunodominant targets in *T. cruzi* infection [110, 137], due to high expression in the infective forms and repetitive/antigenic content [137, 138].

Trans-sialidases are a polymorphic protein subfamily of the GPI-mucin superfamily and are expressed on the parasite surface and released into the extracellular medium. These proteins are coded by a multi-copy gene family, comprised by more than 1400 genes [19, 139, 140]. These enzymes remove and efficiently transfer sialic acid from host-derived glycoconjugates to parasite mucin-like glycoproteins, which are the most abundant surface components of infective forms [141]. TcTS has several potential immunogenic candidates that can generate an unfocused response. The trans-sialidase immune response is focused on a relatively small number of epitopes encoded by multiple genes, and this protein is a prominent target of *T. cruzi*-specific CD8+ T cells [19, 110, 142].

Immunodominance may occur by different mechanisms including the formation of stable MHC-I peptide complexes on the surface of APCs, higher amounts or higher affinity T cell precursors or competition of T cells for APCs [135]. CD8+ T cells that recognize TS-derived epitopes remain highly competent throughout the chronic infection, despite persistent antigen exposure [137, 143, 144].

The immunodominance of TS-derived epitopes predictably results in the outcompetition of other epitope-specific CD8+ T cell populations. However, the significance of the tight focusing of the CD8+ T cell response on only a few vast arrays of possible parasite-derived epitopes is not known [138]. The immune response targets the parasites expressing the common TcTS variants, whereas failing to identify parasites that express rare variants of the antigen [145]; thus the immunodominance seems to be an important way by which the parasite escapes from the effector mechanisms of the CD8+ T cells and persists until the chronic stage.

6. Conclusions

Trypanosoma cruzi is a parasite that has a complex life cycle that goes from the passage through an invertebrate host to different mammals displaying different developmental stages during its life cycle. Furthermore, *T. cruzi* includes several lineages that have distinct morphology, infectivity, virulence, and pathogenicity [146]. The disease reaches the symptomatic chronic stage in only 30% of patients who acquire the infection, whereas 70% shown no clinical symptoms [147]. The above characteristics suggest that during infection may occur a complex interaction between the host and the parasite, and although for more than 100 years, since the Chagas disease was described, the immunological mechanisms of resistance or susceptibility to the parasite have been studied, and it has not been possible to determine exactly which are those that effectively can eliminate the parasite. Recently, there is a growing consensus that the protective immune response against *T. cruzi* requires the activation of a Th1 immune profile, with the stimulation of CD8+ T cells, but as it is detailed in the present review, the mechanisms of evasion to the immune response by *T. cruzi* are diverse and quite effective. So not withstanding this large amount of information available, it still impossible to predict what will happen in an individual infected with this pathogen. Considering the importance of Chagas disease in the world in which there are an estimated to be around 8–10 million of infected people and that currently there are no vaccines which prevent the parasite infection, it is necessary to look for a better and complete understanding of the mechanisms of immune evasion of *Trypanosoma cruzi*.

Acknowledgements

ACR was recipient of a Ph.D. fellowship (278159) from CONACyT, México. This work was supported by a grant from Consejo Nacional de Ciencia y Tecnología (CONACyT), México (Grant 240882) to JLRE.

Conflict of interest

The authors declare that there is no conflict of interests regarding the publication on this chapter.

Author details

Alondra Cruz Reyes* and José Luis Rosales Encina
Departamento de Infectómica y Patogénesis Molecular, Centro de Investigación y de Estudios Avanzados del Instituto Politécnico Nacional, México City, México

*Address all correspondence to: alondra.cruz@cinvestav.mx

References

[1] WHO. Chagas Disease (American Trypanosomiasis). 2017. Available from: http://www.who.int/chagas/disease/en/

[2] Rassi A Jr, Rassi A, Marin-Neto JA. Chagas disease. Lancet. 2010;**375**(9723):1388-1402. DOI: 10.1016/s0140-6736(10)60061-x

[3] Junqueira C, Caetano B, Bartholomeu DC, Melo MB, Ropert C, Rodrigues MM, et al. The endless race between *Trypanosoma cruzi* and host immunity: Lessons for and beyond Chagas disease. Expert Reviews in Molecular Medicine. 2010;**12**:e29. DOI: 10.1017/S1462399410001560

[4] DosReis GA. Evasion of immune responses by *Trypanosoma cruzi*, the etiological agent of Chagas disease. Brazilian Journal of Medical and Biological Research. 2011;**44**(2):84-90

[5] Parodi C, Padilla AM, Basombrio MA. Protective immunity against *Trypanosoma cruzi*. Memórias do Instituto Oswaldo Cruz. 2009;**104**(Suppl 1):288-294

[6] Alvarez MN, Peluffo G, Piacenza L, Radi R. Intraphagosomal peroxynitrite as a macrophage-derived cytotoxin against internalized *Trypanosoma cruzi*: Consequences for oxidative killing and role of microbial peroxiredoxins in infectivity. The Journal of Biological Chemistry. 2011;**286**(8):6627-6640. DOI: 10.1074/jbc.M110.167247

[7] Padilla AM, Bustamante JM, Tarleton RL. CD8+ T cells in *Trypanosoma cruzi* infection. Current Opinion in Immunology. 2009;**21**(4):385-390. DOI: 10.1016/j.coi.2009.07.006

[8] de Alencar BC, Persechini PM, Haolla FA, de Oliveira G, Silverio JC, Lannes-Vieira J, et al. Perforin and gamma interferon expression are required for CD4+ and CD8+ T-cell-dependent protective immunity against a human parasite, *Trypanosoma cruzi*, elicited by heterologous plasmid DNA prime-recombinant adenovirus 5 boost vaccination. Infection and Immunity. 2009;**77**:4383-4385. DOI: 10.1128/IAI.01459-08

[9] Quijano-Hernandez I, Dumonteil E. Advances and challenges towards a vaccine against Chagas disease. Human Vaccines. 2011;**7**(11):1184-1191. DOI: 10.4161/hv.7.11.17016

[10] Norris KA, Bradt B, Cooper NR, So M. Characterization of a *Trypanosoma cruzi* C3 binding protein with functional and genetic similarities to the human complement regulatory protein, decay-accelerating factor. Journal of Immunology. 1991;**147**(7):2240-2247

[11] Geiger A, Bossard G, Sereno D, Pissarra J, Lemesre JL, Vincendeau P, et al. Escaping deleterious immune response in their hosts: Lessons from trypanosomatids. Frontiers in Immunology. 2016;**7**:212. DOI: 10.3389/fimmu.2016.00212

[12] Zambrano-Villa S, Rosales-Borjas D, Carrero JC, Ortiz-Ortiz L. How protozoan parasites evade the immune response. Trends in Parasitology. 2002;**18**(6):272-278

[13] Andrews NW. Living dangerously: How *Trypanosoma cruzi* uses lysosomes to get inside host cells, and then escapes into the cytoplasm. Biological Research. 1993;**26**(1-2):65-67

[14] Epting CL, Coates BM, Engman DM. Molecular mechanisms of host cell invasion by *Trypanosoma cruzi*. Experimental Parasitology. 2010;**126**(3):283-291. DOI: 10.1016/j.exppara.2010.06.023

[15] Nagajyothi F, Machado FS, Burleigh BA, Jelicks LA, Scherer PE,

Mukherjee S, et al. Mechanisms of *Trypanosoma cruzi* persistence in Chagas disease. Cellular Microbiology. 2012;**14**(5):634-643. DOI: 10.1111/j.1462-5822.2012.01764.x

[16] Cardoso MS, Reis-Cunha JL, Bartholomeu DC. Evasion of the immune response by *Trypanosoma cruzi* during acute infection. Frontiers in Immunology. 2015;**6**:659. DOI: 10.3389/fimmu.2015.00659

[17] Hall BF. *Trypanosoma cruzi*: Mechanisms for entry into host cells. Seminars in Cell Biology. 1993;**4**(5): 323-333

[18] Amer AO, Swanson MS. A phagosome of one's own: A microbial guide to life in the macrophage. Current Opinion in Microbiology. 2002;**5**(1):56-61

[19] Buscaglia CA, Campo VA, Frasch AC, Di Noia JM. *Trypanosoma cruzi* surface mucins: Host-dependent coat diversity. Nature Reviews. Microbiology. 2006;**4**(3):229-236. DOI: 10.1038/nrmicro1351

[20] Nardy AF, Freire-de-Lima CG, Perez AR, Morrot A. Role of *Trypanosoma cruzi* trans-sialidase on the escape from host immune surveillance. Frontiers in Microbiology. 2016;**7**:348. DOI: 10.3389/fmicb.2016.00348

[21] Hall BF, Webster P, Ma AK, Joiner KA, Andrews NW. Desialylation of lysosomal membrane glycoproteins by *Trypanosoma cruzi*: A role for the surface neuraminidase in facilitating parasite entry into the host cell cytoplasm. The Journal of Experimental Medicine. 1992;**176**(2):313-325

[22] Albertti LA, Macedo AM, Chiari E, Andrews NW, Andrade LO. Role of host lysosomal associated membrane protein (LAMP) in *Trypanosoma cruzi* invasion and intracellular development. Microbes and Infection. 2010;**12**(10):784-789. DOI: 10.1016/j.micinf.2010.05.015

[23] Andrews NW, Abrams CK, Slatin SL, Griffiths G. A *T. cruzi*-secreted protein immunologically related to the complement component C9: Evidence for membrane pore-forming activity at low pH. Cell. 1990;**61**(7):1277-1287

[24] Ley V, Robbins ES, Nussenzweig V, Andrews NW. The exit of *Trypanosoma cruzi* from the phagosome is inhibited by raising the pH of acidic compartments. The Journal of Experimental Medicine. 1990;**171**(2):401-413

[25] Mauel J. Intracellular survival of protozoan parasites with special reference to *Leishmania* spp., toxoplasma gondii and *Trypanosoma cruzi*. Advances in Parasitology. 1996;**38**:1-51

[26] Bogdan C, Rollinghoff M. How do protozoan parasites survive inside macrophages? Parasitology Today. 1999;**15**(1):22-28

[27] Piacenza L, Peluffo G, Alvarez MN, Martinez A, Radi R. *Trypanosoma cruzi* antioxidant enzymes as virulence factors in Chagas disease. Antioxidants & Redox Signaling. 2013;**19**(7):723-734. DOI: 10.1089/ars.2012.4618

[28] Fridovich I. Superoxide radical and superoxide dismutases. Annual Review of Biochemistry. 1995;**64**:97-112. DOI: 10.1146/annurev.bi.64.070195.000525

[29] Alvarez MN, Piacenza L, Irigoin F, Peluffo G, Radi R. Macrophage-derived peroxynitrite diffusion and toxicity to *Trypanosoma cruzi*. Archives of Biochemistry and Biophysics. 2004;**432**(2):222-232. DOI: 10.1016/j.abb.2004.09.015

[30] Denicola A, Rubbo H, Rodriguez D, Radi R. Peroxynitrite-mediated cytotoxicity to *Trypanosoma cruzi*. Archives of Biochemistry and Biophysics. 1993;**304**(1):279-286. DOI: 10.1006/abbi.1993.1350

[31] Munoz-Fernandez MA, Fernandez MA, Fresno M. Activation of human macrophages for the killing of intracellular *Trypanosoma cruzi* by TNF-alpha and IFN-gamma through a nitric oxide-dependent mechanism. Immunology Letters. 1992;**33**(1):35-40

[32] Gutierrez FR, Mineo TW, Pavanelli WR, Guedes PM, Silva JS. The effects of nitric oxide on the immune system during *Trypanosoma cruzi* infection. Memórias do Instituto Oswaldo Cruz. 2009;**104**(Suppl 1):236-245

[33] Radi R. Peroxynitrite, a stealthy biological oxidant. The Journal of Biological Chemistry. 2013;**288**(37):26464-26472. DOI: 10.1074/jbc.R113.472936

[34] Piacenza L, Peluffo G, Alvarez MN, Kelly JM, Wilkinson SR, Radi R. Peroxiredoxins play a major role in protecting *Trypanosoma cruzi* against macrophage- and endogenously-derived peroxynitrite. The Biochemical Journal. 2008;**410**(2):359-368. DOI: 10.1042/BJ20071138

[35] Wilkinson SR, Temperton NJ, Mondragon A, Kelly JM. Distinct mitochondrial and cytosolic enzymes mediate trypanothione-dependent peroxide metabolism in *Trypanosoma cruzi*. The Journal of Biological Chemistry. 2000;**275**(11):8220-8225

[36] Pineyro MD, Pizarro JC, Lema F, Pritsch O, Cayota A, Bentley GA, et al. Crystal structure of the tryparedoxin peroxidase from the human parasite *Trypanosoma cruzi*. Journal of Structural Biology. 2005;**150**(1):11-22. DOI: 10.1016/j.jsb.2004.12.005

[37] Trujillo M, Budde H, Pineyro MD, Stehr M, Robello C, Flohe L, et al. *Trypanosoma brucei* and *Trypanosoma cruzi* tryparedoxin peroxidases catalytically detoxify peroxynitrite via oxidation of fast reacting thiols. The Journal of Biological Chemistry. 2004;**279**(33):34175-34182. DOI: 10.1074/jbc.M404317200

[38] Wilkinson SR, Meyer DJ, Taylor MC, Bromley EV, Miles MA, Kelly JM. The *Trypanosoma cruzi* enzyme TcGPXI is a glycosomal peroxidase and can be linked to trypanothione reduction by glutathione or tryparedoxin. The Journal of Biological Chemistry. 2002;**277**(19):17062-17071. DOI: 10.1074/jbc.M111126200

[39] Wilkinson SR, Taylor MC, Touitha S, Mauricio IL, Meyer DJ, Kelly JM. TcGPXII, a glutathione-dependent *Trypanosoma cruzi* peroxidase with substrate specificity restricted to fatty acid and phospholipid hydroperoxides, is localized to the endoplasmic reticulum. The Biochemical Journal. 2002;**364**(Pt 3):787-794. DOI: 10.1042/BJ20020038

[40] Wilkinson SR, Obado SO, Mauricio IL, Kelly JM. *Trypanosoma cruzi* expresses a plant-like ascorbate-dependent hemoperoxidase localized to the endoplasmic reticulum. Proceedings of the National Academy of Sciences of the United States of America. 2002;**99**(21):13453-13458. DOI: 10.1073/pnas.202422899

[41] Piacenza L, Irigoin F, Alvarez MN, Peluffo G, Taylor MC, Kelly JM, et al. Mitochondrial superoxide radicals mediate programmed cell death in *Trypanosoma cruzi*: Cytoprotective action of mitochondrial iron superoxide dismutase overexpression. The Biochemical Journal. 2007;**403**(2):323-334. DOI: 10.1042/BJ20061281

[42] Mateo H, Marin C, Perez-Cordon G, Sanchez-Moreno M. Purification and biochemical characterization of four iron superoxide dismutases in *Trypanosoma cruzi*. Memórias do Instituto Oswaldo Cruz. 2008;**103**(3):271-276

[43] Mendes-Sousa AF, Nascimento AA, Queiroz DC, Vale VF, Fujiwara

RT, Araujo RN, et al. Different host complement systems and their interactions with saliva from *Lutzomyia longipalpis* (Diptera, Psychodidae) and *Leishmania infantum* promastigotes. PLoS One. 2013;**8**(11):e79787. DOI: 10.1371/journal.pone.0079787

[44] Lidani KCF, Bavia L, Ambrosio AR, de Messias-Reason IJ. The complement system: A prey of *Trypanosoma cruzi*. Frontiers in Microbiology. 2017;**8**:607. DOI: 10.3389/fmicb.2017.00607

[45] Ricklin D, Hajishengallis G, Yang K, Lambris JD. Complement: A key system for immune surveillance and homeostasis. Nature Immunology. 2010;**11**(9):785-797. DOI: 10.1038/ni.1923

[46] Lorincz Z, Gal P, Dobo J, Cseh S, Szilagyi K, Ambrus G, et al. The cleavage of two C1s subunits by a single active C1r reveals substantial flexibility of the C1s-C1r-C1r-C1s tetramer in the C1 complex. Journal of Immunology. 2000;**165**(4):2048-2051

[47] Cestari Idos S, Krarup A, Sim RB, Inal JM, Ramirez MI. Role of early lectin pathway activation in the complement-mediated killing of *Trypanosoma cruzi*. Molecular Immunology. 2009;**47**(2-3):426-437. DOI: 10.1016/j.molimm.2009.08.030

[48] Cestari I, Ramirez MI. Inefficient complement system clearance of *Trypanosoma cruzi* metacyclic trypomastigotes enables resistant strains to invade eukaryotic cells. PLoS One. 2010;**5**(3):e9721. DOI: 10.1371/journal.pone.0009721

[49] Beltrame MH, Catarino SJ, Goeldner I, Boldt AB, de Messias-Reason IJ. The lectin pathway of complement and rheumatic heart disease. Frontiers in Pediatrics. 2014;**2**:148. DOI: 10.3389/fped.2014.00148

[50] Osorio L, Rios I, Gutierrez B, Gonzalez J. Virulence factors of *Trypanosoma cruzi*: Who is who? Microbes and Infection. 2012;**14**(15):1390-1402. DOI: 10.1016/j.micinf.2012.09.003

[51] Krautz GM, Kissinger JC, Krettli AU. The targets of the lytic antibody response against *Trypanosoma cruzi*. Parasitology Today. 2000;**16**(1):31-34

[52] Kierszenbaum F, Gottlieb CA, Budzko DB. Antibody-independent, natural resistance of birds to *Trypanosoma cruzi* infection. The Journal of Parasitology. 1981;**67**(5):656-660

[53] Krettli AU, Weisz-Carrington P, Nussenzweig RS. Membrane-bound antibodies to bloodstream *Trypanosoma cruzi* in mice: Strain differences in susceptibility to complement-mediated lysis. Clinical and Experimental Immunology. 1979;**37**(3):416-423

[54] Budzko DB, Pizzimenti MC, Kierszenbaum F. Effects of complement depletion in experimental Chagas disease: Immune lysis of virulent blood forms of *Trypanosoma cruzi*. Infection and Immunity. 1975;**11**(1):86-91

[55] Meirelles MN, De Souza W. Killing of *Trypanosoma cruzi* and *Leishmania mexicana*, and survival of *Toxoplasma gondii*, in chicken macrophages in vitro. Journal of Submicroscopic Cytology. 1985;**17**(3):327-334

[56] Teixeira AR, Nascimento RJ, Sturm NR. Evolution and pathology in Chagas disease—A review. Memórias do Instituto Oswaldo Cruz. 2006;**101**(5):463-491

[57] Nery-Guimaraes F, Venancio I, Grynberg N. Refractoriness of hens towards *Trypanosoma (Schizotrypanum) cruzi*. III. Dissociation of phenomena of the refractory state and lysis of the epimastigotes by fowl serum. Memórias do Instituto Oswaldo Cruz. 1974;**72**(1-2):131-136

[58] Minter-Goedbloed E, Croon JJ. The susceptibility of chickens to *Trypanosoma (Schizotrypanum) cruzi*. Transactions of the Royal Society of Tropical Medicine and Hygiene. 1981;**75**(3):350-353

[59] Teixeira AR, Hecht MM, Guimaro MC, Sousa AO, Nitz N. Pathogenesis of Chagas' disease: Parasite persistence and autoimmunity. Clinical Microbiology Reviews. 2011;**24**(3):592-630. DOI: 10.1128/CMR.00063-10

[60] Cestari I, Evans-Osses I, Schlapbach LJ, de Messias-Reason I, Ramirez MI. Mechanisms of complement lectin pathway activation and resistance by trypanosomatid parasites. Molecular Immunology. 2013;**53**(4):328-334. DOI: 10.1016/j.molimm.2012.08.015

[61] Ramirez-Toloza G, Ferreira A. *Trypanosoma cruzi* evades the complement system as an efficient strategy to survive in the mammalian host: The specific roles of host/parasite molecules and *Trypanosoma cruzi* calreticulin. Frontiers in Microbiology. 2017;**8**:1667. DOI: 10.3389/fmicb.2017.01667

[62] Ferreira V, Molina MC, Valck C, Rojas A, Aguilar L, Ramirez G, et al. Role of calreticulin from parasites in its interaction with vertebrate hosts. Molecular Immunology. 2004;**40**(17):1279-1291

[63] Ferreira V, Valck C, Sanchez G, Gingras A, Tzima S, Molina MC, et al. The classical activation pathway of the human complement system is specifically inhibited by calreticulin from *Trypanosoma cruzi*. Journal of Immunology. 2004;**172**(5):3042-3050

[64] Souto-Padron T, Labriola CA, de Souza W. Immunocytochemical localisation of calreticulin in *Trypanosoma cruzi*. Histochemistry and Cell Biology. 2004;**122**(6):563-569. DOI: 10.1007/s00418-004-0724-7

[65] Sosoniuk E, Vallejos G, Kenawy H, Gaboriaud C, Thielens N, Fujita T, et al. *Trypanosoma cruzi* calreticulin inhibits the complement lectin pathway activation by direct interaction with L-Ficolin. Molecular Immunology. 2014;**60**(1):80-85. DOI: 10.1016/j.molimm.2014.03.014

[66] Gonzalez A, Valck C, Sanchez G, Hartel S, Mansilla J, Ramirez G, et al. *Trypanosoma cruzi* calreticulin topographical variations in parasites infecting murine macrophages. The American Journal of Tropical Medicine and Hygiene. 2015;**92**(5):887-897. DOI: 10.4269/ajtmh.14-0497

[67] Ramirez G, Valck C, Molina MC, Ribeiro CH, Lopez N, Sanchez G, et al. *Trypanosoma cruzi* calreticulin: A novel virulence factor that binds complement C1 on the parasite surface and promotes infectivity. Immunobiology. 2011;**216**(1-2):265-273. DOI: 10.1016/j.imbio.2010.04.001

[68] Norris KA, Schrimpf JE. Biochemical analysis of the membrane and soluble forms of the complement regulatory protein of *Trypanosoma cruzi*. Infection and Immunity. 1994;**62**(1):236-243

[69] Norris KA. Stable transfection of *Trypanosoma cruzi* epimastigotes with the trypomastigote-specific complement regulatory protein cDNA confers complement resistance. Infection and Immunity. 1998;**66**(6):2460-2465

[70] Inal JM. Complement C2 receptor inhibitor trispanning: From man to schistosome. Springer Seminars in Immunopathology. 2005;**27**(3):320-331. DOI: 10.1007/s00281-005-0009-9

[71] Inal JM, Sim RB. A Schistosoma protein, Sh-TOR, is a novel inhibitor of complement which binds human C2. FEBS Letters. 2000;**470**(2):131-134

[72] Inal JM, Schifferli JA. Complement C2 receptor inhibitor trispanning and the beta-chain of C4 share a binding site for complement C2. Journal of Immunology. 2002;**168**(10):5213-5221

[73] Tambourgi DV, Kipnis TL, da Silva WD, Joiner KA, Sher A, Heath S, et al. A partial cDNA clone of trypomastigote decay-accelerating factor (T-DAF), a developmentally regulated complement inhibitor of *Trypanosoma cruzi*, has genetic and functional similarities to the human complement inhibitor DAF. Infection and Immunity. 1993;**61**(9):3656-3663

[74] Joiner KA, daSilva WD, Rimoldi MT, Hammer CH, Sher A, Kipnis TL. Biochemical characterization of a factor produced by trypomastigotes of *Trypanosoma cruzi* that accelerates the decay of complement C3 convertases. The Journal of Biological Chemistry. 1988;**263**(23):11327-11335

[75] Velge P, Ouaissi MA, Cornette J, Afchain D, Capron A. Identification and isolation of *Trypanosoma cruzi* trypomastigote collagen-binding proteins: Possible role in cell-parasite interaction. Parasitology. 1988;**97** (Pt 2):255-268

[76] Fischer E, Ouaissi MA, Velge P, Cornette J, Kazatchkine MD. gp 58/68, a parasite component that contributes to the escape of the trypomastigote form of *T. cruzi* from damage by the human alternative complement pathway. Immunology. 1988;**65**(2):299-303

[77] Cestari I, Ansa-Addo E, Deolindo P, Inal JM, Ramirez MI. *Trypanosoma cruzi* immune evasion mediated by host cell-derived microvesicles. Journal of Immunology. 2012;**188**(4):1942-1952. DOI: 10.4049/jimmunol.1102053

[78] Bryan MA, Guyach SE, Norris KA. Specific humoral immunity versus polyclonal B cell activation in *Trypanosoma cruzi* infection of susceptible and resistant mice. PLoS Neglected Tropical Diseases. 2010;**4**(7):e733. DOI: 10.1371/journal.pntd.0000733

[79] Zhang L, Tarleton RL. Characterization of cytokine production in murine *Trypanosoma cruzi* infection by in situ immunocytochemistry: Lack of association between susceptibility and type 2 cytokine production. European Journal of Immunology. 1996;**26**(1):102-109. DOI: 10.1002/eji.1830260116

[80] Pinazo MJ, Espinosa G, Cortes-Lletget C, Posada Ede J, Aldasoro E, Oliveira I, et al. Immunosuppression and Chagas disease: A management challenge. PLoS Neglected Tropical Diseases. 2013;7(1):e1965. DOI: 10.1371/journal.pntd.0001965

[81] Maleckar JR, Kierszenbaum F. Inhibition of mitogen-induced proliferation of mouse T and B lymphocytes by bloodstream forms of *Trypanosoma cruzi*. Journal of Immunology. 1983;**130**(2):908-911

[82] Vaena de Avalos S, Blader IJ, Fisher M, Boothroyd JC, Burleigh BA. Immediate/early response to *Trypanosoma cruzi* infection involves minimal modulation of host cell transcription. The Journal of Biological Chemistry. 2002;**277**(1):639-644. DOI: 10.1074/jbc.M109037200

[83] Padilla AM, Simpson LJ, Tarleton RL. Insufficient TLR activation contributes to the slow development of CD8+ T cell responses in *Trypanosoma cruzi* infection. Journal of Immunology. 2009;**183**(2):1245-1252. DOI: 10.4049/jimmunol.0901178

[84] Delneste Y, Beauvillain C, Jeannin P. Innate immunity: Structure and function of TLRs. Medical Science. 2007;**23**(1):67-73. DOI: 10.1051/medsci/200723167

[85] Takeuchi O, Akira S. Pattern recognition receptors and inflammation. Cell. 2010;**140**(6):805-820. DOI: 10.1016/j.cell.2010.01.022

[86] Kulkarni R, Behboudi S, Sharif S. Insights into the role of toll-like receptors in modulation of T cell responses. Cell and Tissue Research. 2011;**343**(1):141-152. DOI: 10.1007/s00441-010-1017-1

[87] Rodrigues MM, Oliveira AC, Bellio M. The immune response to *Trypanosoma cruzi*: Role of toll-like receptors and perspectives for vaccine development. Journal of Parasitology Research. 2012;**2012**:507874. DOI: 10.1155/2012/507874

[88] Gazzinelli RT, Denkers EY. Protozoan encounters with toll-like receptor signalling pathways: Implications for host parasitism. Nature Reviews. Immunology. 2006;**6**(12): 895-906. DOI: 10.1038/nri1978

[89] Kawai T, Akira S. The role of pattern-recognition receptors in innate immunity: Update on toll-like receptors. Nature Immunology. 2010;**11**(5): 373-384. DOI: 10.1038/ni.1863

[90] Park BS, Song DH, Kim HM, Choi BS, Lee H, Lee JO. The structural basis of lipopolysaccharide recognition by the TLR4-MD-2 complex. Nature. 2009;**458**(7242):1191-1195. DOI: 10.1038/nature07830

[91] Jin MS, Kim SE, Heo JY, Lee ME, Kim HM, Paik SG, et al. Crystal structure of the TLR1-TLR2 heterodimer induced by binding of a tri-acylated lipopeptide. Cell. 2007;**130**(6):1071-1082. DOI: 10.1016/j.cell.2007.09.008

[92] Gazzinelli RT, Ropert C, Campos MA. Role of the toll/interleukin-1 receptor signaling pathway in host resistance and pathogenesis during infection with protozoan parasites. Immunological Reviews. 2004;**201**:9-25. DOI: 10.1111/j.0105-2896.2004.00174.x

[93] Iwasaki A, Medzhitov R. Regulation of adaptive immunity by the innate immune system. Science. 2010;**327**(5963):291-295. DOI: 10.1126/science.1183021

[94] Kumar H, Kawai T, Akira S. Pathogen recognition by the innate immune system. International Reviews of Immunology. 2011;**30**(1):16-34. DOI: 10.3109/08830185.2010.529976

[95] Oliveira AC, Peixoto JR, de Arruda LB, Campos MA, Gazzinelli RT, Golenbock DT, et al. Expression of functional TLR4 confers proinflammatory responsiveness to *Trypanosoma cruzi* glycoinositolphospholipids and higher resistance to infection with *T. cruzi*. Journal of Immunology. 2004;**173**(9):5688-5696

[96] Dos-Santos AL, Carvalho-Kelly LF, Dick CF, Meyer-Fernandes JR. Innate immunomodulation to trypanosomatid parasite infections. Experimental Parasitology. 2016;**167**:67-75. DOI: 10.1016/j.exppara.2016.05.005

[97] Caetano BC, Carmo BB, Melo MB, Cerny A, dos Santos SL, Bartholomeu DC, et al. Requirement of UNC93B1 reveals a critical role for TLR7 in host resistance to primary infection with *Trypanosoma cruzi*. Journal of Immunology. 2011;**187**(4):1903-1911. DOI: 10.4049/jimmunol.1003911

[98] Bartholomeu DC, Ropert C, Melo MB, Parroche P, Junqueira CF, Teixeira SM, et al. Recruitment and endo-lysosomal activation of TLR9 in dendritic cells infected with *Trypanosoma cruzi*. Journal of Immunology. 2008;**181**(2):1333-1344

[99] Gravina HD, Antonelli L, Gazzinelli RT, Ropert C. Differential use of TLR2 and TLR9 in the regulation of

immune responses during the infection with *Trypanosoma cruzi*. PLoS One. 2013;**8**(5):e63100. DOI: 10.1371/journal.pone.0063100

[100] de Diego J, Punzon C, Duarte M, Fresno M. Alteration of macrophage function by a Trypanosoma cruzi membrane mucin. Journal of Immunology. 1997;**159**(10):4983-4989

[101] Silva JS, Twardzik DR, Reed SG. Regulation of Trypanosoma cruzi infections in vitro and in vivo by transforming growth factor beta (TGF-beta). The Journal of Experimental Medicine. 1991;**174**(3):539-545

[102] Doyle PS, Zhou YM, Hsieh I, Greenbaum DC, McKerrow JH, Engel JC. The Trypanosoma cruzi protease cruzain mediates immune evasion. PLoS Pathogens. 2011;**7**(9):e1002139. DOI: 10.1371/journal.ppat.1002139

[103] Samudio M, Montenegro-James S, Cabral M, Martinez J, Rojas de Arias A, James MA. Cytokine responses in Trypanosoma cruzi-infected children in Paraguay. The American Journal of Tropical Medicine and Hygiene. 1998;**58**(1):119-121

[104] Bafica A, Santiago HC, Goldszmid R, Ropert C, Gazzinelli RT, Sher A. Cutting edge: TLR9 and TLR2 signaling together account for MyD88-dependent control of parasitemia in Trypanosoma cruzi infection. Journal of Immunology. 2006;**177**(6):3515-3519

[105] Cardillo F, Postol E, Nihei J, Aroeira LS, Nomizo A, Mengel J. B cells modulate T cells so as to favour T helper type 1 and CD8+ T-cell responses in the acute phase of Trypanosoma cruzi infection. Immunology. 2007;**122**(4):584-595. DOI: 10.1111/j.1365-2567. 2007.02677.x

[106] Wortel CM, Heidt S. Regulatory B cells: Phenotype, function and role in transplantation. Transplant Immunology. 2017;**41**:1-9. DOI: 10.1016/j.trim.2017.02.004

[107] Minoprio P, Burlen O, Pereira P, Guilbert B, Andrade L, Hontebeyrie-Joskowicz M, et al. Most B cells in acute Trypanosoma cruzi infection lack parasite specificity. Scandinavian Journal of Immunology. 1988;**28**(5):553-561

[108] Bermejo DA, Amezcua Vesely MC, Khan M, Acosta Rodriguez EV, Montes CL, Merino MC, et al. Trypanosoma cruzi infection induces a massive extrafollicular and follicular splenic B-cell response which is a high source of non-parasite-specific antibodies. Immunology. 2011;**132**(1):123-133. DOI: 10.1111/j.1365-2567.2010.03347.x

[109] Pitcovsky TA, Buscaglia CA, Mucci J, Campetella O. A functional network of intramolecular cross-reacting epitopes delays the elicitation of neutralizing antibodies to Trypanosoma cruzi trans-sialidase. The Journal of Infectious Diseases. 2002;**186**(3):397-404. DOI: 10.1086/341463

[110] Tarleton RL. CD8+ T cells in Trypanosoma cruzi infection. Seminars in Immunopathology. 2015;**37**(3):233-238. DOI: 10.1007/s00281-015-0481-9

[111] Pitcovsky TA, Mucci J, Alvarez P, Leguizamon MS, Burrone O, Alzari PM, et al. Epitope mapping of trans-sialidase from Trypanosoma cruzi reveals the presence of several cross-reactive determinants. Infection and Immunity. 2001;**69**(3):1869-1875. DOI: 10.1128/IAI.69.3.1869-1875.2001

[112] Hilleman MR. Strategies and mechanisms for host and pathogen survival in acute and persistent viral infections. Proceedings of the National Academy of Sciences of the United States of America. 2004;**101**(Suppl 2):14560-14566. DOI: 10.1073/pnas.0404758101

[113] Barcinski MA, DosReis GA. Apoptosis in parasites and parasite-induced apoptosis in the host immune system: A new approach to parasitic diseases. Brazilian Journal of Medical and Biological Research. 1999;**32**(4):395-401

[114] Lenardo M, Chan KM, Hornung F, McFarland H, Siegel R, Wang J, et al. Mature T lymphocyte apoptosis--immune regulation in a dynamic and unpredictable antigenic environment. Annual Review of Immunology. 1999;**17**:221-253. DOI: 10.1146/annurev.immunol.17.1.221

[115] Acosta Rodriguez EV, Zuniga EI, Montes CL, Merino MC, Bermejo DA, Amezcua Vesely MC, et al. Trypanosoma cruzi infection beats the B-cell compartment favouring parasite establishment: Can we strike first? Scandinavian Journal of Immunology. 2007;**66**(2-3):137-142. DOI: 10.1111/j.1365-3083.2007.01968.x

[116] Zuniga E, Acosta-Rodriguez E, Merino MC, Montes C, Gruppi A. Depletion of immature B cells during Trypanosoma cruzi infection: Involvement of myeloid cells and the cyclooxygenase pathway. European Journal of Immunology. 2005;**35**(6):1849-1858. DOI: 10.1002/eji.200526005

[117] Zuniga E, Motran CC, Montes CL, Yagita H, Gruppi A. Trypanosoma cruzi infection selectively renders parasite-specific IgG+ B lymphocytes susceptible to Fas/Fas ligand-mediated fratricide. Journal of Immunology. 2002;**168**(8):3965-3973

[118] Freire-de-Lima CG, Nascimento DO, Soares MB, Bozza PT, Castro-Faria-Neto HC, de Mello FG, et al. Uptake of apoptotic cells drives the growth of a pathogenic trypanosome in macrophages. Nature. 2000;**403**(6766):199-203. DOI: 10.1038/35003208

[119] Millar AE, Wleklinski-Lee M, Kahn SJ. The surface protein superfamily of Trypanosoma cruzi stimulates a polarized Th1 response that becomes anergic. Journal of Immunology. 1999;**162**(10):6092-6099

[120] Previato JO, Wait R, Jones C, DosReis GA, Todeschini AR, Heise N, et al. Glycoinositolphospholipid from Trypanosoma cruzi: Structure, biosynthesis and immunobiology. Advances in Parasitology. 2004;**56**:1-41

[121] Grauert MR, Houdayer M, Hontebeyrie-Joskowciz M. Trypanosoma cruzi infection enhances polyreactive antibody response in an acute case of human Chagas' disease. Clinical and Experimental Immunology. 1993;**93**(1):85-92

[122] Ouaissi A, Da Silva AC, Guevara AG, Borges M, Guilvard E. Trypanosoma cruzi-induced host immune system dysfunction: A rationale for parasite immunosuppressive factor(s) encoding gene targeting. Journal of Biomedicine & Biotechnology. 2001;**1**(1):11-17. DOI: 10.1155/S1110724301000055

[123] Spinella S, Liegeard P, Hontebeyrie-Joskowicz M. Trypanosoma cruzi: Predominance of IgG2a in nonspecific humoral response during experimental Chagas' disease. Experimental Parasitology. 1992;**74**(1):46-56

[124] Minoprio P. Parasite polyclonal activators: New targets for vaccination approaches? International Journal for Parasitology. 2001;**31**(5-6):588-591

[125] Reina-San-Martin B, Cosson A, Minoprio P. Lymphocyte polyclonal activation: A pitfall for vaccine design against infectious agents. Parasitology Today. 2000;**16**(2):62-67

[126] Montes CL, Acosta-Rodriguez EV, Mucci J, Zuniga EI, Campetella O, Gruppi AA. Trypanosoma cruzi

antigen signals CD11b+ cells to secrete cytokines that promote polyclonal B cell proliferation and differentiation into antibody-secreting cells. European Journal of Immunology. 2006;**36**(6):1474-1485. DOI: 10.1002/eji.200535537

[127] Reina-San-Martin B, Degrave W, Rougeot C, Cosson A, Chamond N, Cordeiro-Da-Silva A, et al. A B-cell mitogen from a pathogenic trypanosome is a eukaryotic proline racemase. Nature Medicine. 2000;**6**(8):890-897. DOI: 10.1038/78651

[128] Gao W, Wortis HH, Pereira MA. The Trypanosoma cruzi trans-sialidase is a T cell-independent B cell mitogen and an inducer of non-specific Ig secretion. International Immunology. 2002;**14**(3):299-308

[129] Acevedo GR, Girard MC, Gomez KA. The unsolved jigsaw puzzle of the immune response in Chagas disease. Frontiers in Immunology. 2018;**9**:1929. DOI: 10.3389/fimmu.2018.01929

[130] Borst P. Antigenic variation and allelic exclusion. Cell. 2002;**109**(1):5-8

[131] Pays E, Vanhamme L, Perez-Morga D. Antigenic variation in Trypanosoma brucei: Facts, challenges and mysteries. Current Opinion in Microbiology. 2004;**7**(4):369-374. DOI: 10.1016/j.mib.2004.05.001

[132] Tarleton RL. Immune system recognition of Trypanosoma cruzi. Current Opinion in Immunology. 2007;**19**(4):430-434. DOI: 10.1016/j.coi.2007.06.003

[133] Bartholomeu DC, Cerqueira GC, Leao AC, daRocha WD, Pais FS, Macedo C, et al. Genomic organization and expression profile of the mucin-associated surface protein (masp) family of the human pathogen Trypanosoma cruzi. Nucleic Acids Research. 2009;**37**(10):3407-3417. DOI: 10.1093/nar/gkp172

[134] dos Santos SL, Freitas LM, Lobo FP, Rodrigues-Luiz GF, Mendes TA, Oliveira AC, et al. The MASP family of Trypanosoma cruzi: Changes in gene expression and antigenic profile during the acute phase of experimental infection. PLoS Neglected Tropical Diseases. 2012;**6**(8):e1779. DOI: 10.1371/journal.pntd.0001779

[135] Tzelepis F, de Alencar BC, Penido ML, Claser C, Machado AV, Bruna-Romero O, et al. Infection with Trypanosoma cruzi restricts the repertoire of parasite-specific CD8+ T cells leading to immunodominance. Journal of Immunology. 2008;**180**(3):1737-1748

[136] Rodrigues MM, Alencar BC, Claser C, Tzelepis F. Immunodominance: A new hypothesis to explain parasite escape and host/parasite equilibrium leading to the chronic phase of Chagas' disease? Brazilian Journal of Medical and Biological Research. 2009;**42**(3):220-223

[137] Martin DL, Weatherly DB, Laucella SA, Cabinian MA, Crim MT, Sullivan S, et al. CD8+ T-cell responses to Trypanosoma cruzi are highly focused on strain-variant trans-sialidase epitopes. PLoS Pathogens. 2006;**2**(8):e77. DOI: 10.1371/journal.ppat.0020077

[138] Rosenberg CS, Martin DL, Tarleton RL. CD8+ T cells specific for immunodominant trans-sialidase epitopes contribute to control of Trypanosoma cruzi infection but are not required for resistance. Journal of Immunology. 2010;**185**(1):560-568. DOI: 10.4049/jimmunol.1000432

[139] Schenkman S, Eichinger D, Pereira ME, Nussenzweig V. Structural and functional properties of Trypanosoma trans-sialidase. Annual Review of Microbiology. 1994;**48**:499-523. DOI: 10.1146/annurev.mi.48.100194.002435

[140] Freitas LM, dos Santos SL, Rodrigues-Luiz GF, Mendes TA, Rodrigues TS, Gazzinelli RT, et al. Genomic analyses, gene expression and antigenic profile of the trans-sialidase superfamily of Trypanosoma cruzi reveal an undetected level of complexity. PLoS One. 2011;**6**(10):e25914. DOI: 10.1371/journal.pone.0025914

[141] Pereira-Chioccola VL, Acosta-Serrano A, Correia de Almeida I, Ferguson MA, Souto-Padron T, Rodrigues MM, et al. Mucin-like molecules form a negatively charged coat that protects Trypanosoma cruzi trypomastigotes from killing by human anti-alpha-galactosyl antibodies. Journal of Cell Science. 2000;**113**(Pt 7): 1299-1307

[142] Martin D, Tarleton R. Generation, specificity, and function of CD8+ T cells in Trypanosoma cruzi infection. Immunological Reviews. 2004;**201**:304-317. DOI: 10.1111/j.0105-2896.2004.00183.x

[143] Martin DL, Tarleton RL. Antigen-specific T cells maintain an effector memory phenotype during persistent Trypanosoma cruzi infection. Journal of Immunology. 2005;**174**(3):1594-1601

[144] Bixby LM, Tarleton RL. Stable CD8+ T cell memory during persistent Trypanosoma cruzi infection. Journal of Immunology. 2008;**181**(4):2644-2650

[145] Borst P, Ulbert S. Control of VSG gene expression sites. Molecular and Biochemical Parasitology. 2001;**114**(1):17-27

[146] de Freitas JM, Augusto-Pinto L, Pimenta JR, Bastos-Rodrigues L, Goncalves VF, Teixeira SM, et al. Ancestral genomes, sex, and the population structure of Trypanosoma cruzi. PLoS Pathogens. 2006;**2**(3):e24. DOI: 10.1371/journal.ppat.0020024

[147] Perlowagora-Szumlewicz A, Muller CA, Moreira CJ. Studies in search of a suitable experimental insect model for xenodiagnosis of hosts with Chagas' disease. 4—The reflection of parasite stock in the responsiveness of different vector species to chronic infection with different Trypanosoma cruzi stocks. Revista de Saúde Pública. 1990;**24**(3):165-177

Section 4

Chemotherapy

Chapter 12

Fish Oil and Inflammation: A Perspective on the Challenges of Evaluating Efficacy in *Trypanosoma cruzi* Infection

Maria Isabel Lovo-Martins, Marli Cardoso Martins-Pinge and Phileno Pinge-Filho

Abstract

Parasitic diseases constitute a big problem of ill health in both the tropics and subtropics as well as in more temperate climates and have been targeted by the Centers for Disease Control and Prevention (CDC) as priorities for public health in the USA. Parasitic infections can be caused by three types of organisms: protozoa, helminths and ectoparasites. They subsist on the host's nutrients at the host's expense. Effectively combating infections caused by parasites is essential for the survival of the organism. In this effort, cells and molecules of the immune system are susceptible to the modulating influence of fatty acids. The primary purpose of this chapter is to present a critical review of the multiple effects of fish-oil on *Trypanosoma infection.*

Keywords: *Trypanosoma cruzi*, Chagas disease, *n*-3 PUFAs, dietary fish oil, disease control

1. Introduction

Chagas disease, also known as American trypanosomiasis, is caused by infection with the hemoflagellate protozoa *Trypanosoma cruzi*. This disease was first described in 1909, by the epidemiologist Carlos Justiniano Chagas. In this pioneering work, not only the etiological agent *Trypanosoma cruzi* was described, but also its evolutionary forms, life cycle, epidemiology and clinical manifestations of the disease was fully reported [1]. More than a century after this discovery, it is estimated that Chagas disease still affects around 6–7 million people worldwide, especially in Latin America, with more than 10,000 deaths annually [2]. Currently, Chagas Disease is considered as a neglected tropical disease by the World Health Organization [3].

The progression of Chagas disease is characterized by the occurrence of three phases: acute, indeterminate and chronic. Survival during the acute phase of infection requires an inflammatory response involving cells of innate immunity, such as macrophages, dendritic cells and natural killers whereas in the chronic phase the T-lymphocyte-mediated immunity maintains parasite replication under control [4]. However, evidence suggests that the exacerbated inflammatory response of the host is one of the most determinant factors in the progression of Chagas disease, along with the virulence and tropism of the strain [5, 6].

During *T. cruzi* infection—as in other infections—the immune system acts to protect the host from infectious agents and the nutrient status is an important factor contributing to immune response [7]. Between the components from the diet, fatty acids found in oils and oily food have an important role not only in the structure of cell membranes, energy source or as hormones precursors [8], but acts directly as modulators of the immune response [9]. Specifically, the consumption of fatty acids from the family of omega-3 polyunsaturated fatty acids (*n*-3 PUFAs), found in large amounts in fish oil, has been associated with anti-inflammatory and immunomodulatory effects [10]. Taking this into account, an important issue to be raised is the effect of *n*-3 PUFAs supplementation on infectious diseases, such as Chagas disease, where an efficient—but controlled inflammation—is necessary and important for host defense [11].

Currently, daily oral supplementation with *n*-3 PUFAs is recommended by the American College of Cardiology and American Heart Association as an important adjuvant in the treatment of heart failure [12]. In the same sense, daily oral supplementation with *n*-3 PUFAs is recommended by the Brazilian Directive on Dyslipidemias and Prevention of Atherosclerosis, updated in 2017 by the Brazilian Society of Cardiology (2–4 g daily), as an important complement in the prevention of atherosclerosis and its cardioprotective benefits [13]. Despite the recognized relevance of *n*-3 PUFA supplementation in the supplementary treatment of cardiovascular diseases, and considering the important cardiac compromises that may occur during the chronic phase of Chagas disease, supplementation with *n*-3 PUFAs may in fact represent a perspective for the additional treatment of patients affected by Chagas disease. However, the immunomodulatory effects of the dietary supplementation with *n*-3 PUFAs and the relationship with the host response and resistance to *T. cruzi* infection should be carefully considered.

2. Polyunsaturated fatty acids

Polyunsaturated fatty acids (PUFAs) are a class of fatty acids with 18–22 carbons (C18–C22) containing two or more double bonds in the carbon chain. The most important PUFAs for human health and nutrition are the omega-6 (*n*-6) and omega-3 (*n*-3) families. The classification of the fatty acids between this families is made considering the position of the first double bond counting from the methyl end of the fatty acid chain [14]. Linoleic acid (LA) is considered the parent fatty acid of the *n*–6 PUFAs family, while α-linolenic acid (ALA) is considered the parent fatty acid of the *n*–3 PUFAs family. Both LA and ALA cannot be made by humans or other mammals, thus they are considered essential fatty acids and have to be supplied in the diet [15].

The *n*-6 PUFA LA (18:2n-6) could be found naturally the seeds of most plants except for cocoa, coconut, and palm. On its turn, the *n*-3 PUFA ALA (18:3n-3) is found in the seeds of flax, rape, chia, perilla, walnuts (and their vegetable oils) or even chloroplast of green leafy vegetables. In the body, both LA and ALA are metabolized to longer-chain fatty acids of 20 or 22 carbons [15, 16]. LA is metabolized to arachidonic acid (AA), a long chain *n*-6 PUFA (LC *n*-6 PUFA) whereas ALA is metabolized to eicosapentaenoic acid (EPA, 20:5n-3) and docosahexaenoic acid (DHA, 22:6n-3), a long chain *n*-3 PUFAs (LC *n*-3 PUFA). This reaction occurs through the actions of elongases enzymes, that increase the chain length; and desaturases enzymes, which add extra double bonds to the carboxyl end of the fatty acid, increasing the degree of unsaturation [17]. Humans and others mammalians could convert LA in AA; and ALA in EPA and DHA, however this process is slow.

Although there is a competition between *n*-6 PUFAs and *n*-3 PUFAs for the desaturation enzymes, which prefer the ALA instead the LA, the Western diet provides higher amounts of LA than ALA, being the AA the main unsaturated long-chain fatty acid produced [18]. Therefore, the most efficient way to increase the amount of LC *n*-3 PUFAs in the body is through directly ingestion of primary sources EPA and DHA, as the seafood. The LC *n*-3 PUFAs are found in high amounts in most seafood, especially in oily fish, in the blubber and tissues of sea mammals like seals and whales or even in supplements like fish oils, cod liver oil, krill oil, algal oils and in pharmaceutical grade preparations [10].

3. Immunomodulatory effects of long chain *n*-3 polyunsaturated fatty acids

Experimental assays have shown that diet supplementation with LC *n*-3 PUFAs results in powerful anti-inflammatory and immunomodulatory activities in a range of diseases, such as autoimmune [19], inflammatory bowel disease [20], rheumatoid arthritis [21] and even infectious diseases [22–25]. There are also clinical trials in human patients associating the use of supplements rich in LC *n*-3 PUFAs with the evolution of inflammatory diseases [26] and infections, such as sepsis [27]. Studies in patients with rheumatoid arthritis are those that present better results, with several tests showing reduction of symptoms, such as morning swelling, pain and stiffness [28].

Inflammation is a fundamental component of the body response to infections or injuries that involves the interactions among many different cell types. The leucocytes are cells from immune system found in the peripheral blood and lymph tissue that actively participate in the inflammation, being specially involved in body defense and protection [29]. Typically, inflammation is transient, exerting a protective role in the body. However, when the inflammation does not end and the acute response become chronic, this uncontrolled response leads to more injury [29]. Therefore, inflammation is the pathological mechanism behind many chronic diseases, and that is why the immunomodulatory effects of LC *n*-3 PUFAs are considered to be potentially beneficial.

Some mechanisms of the immune response modulation by LC *n*-3 PUFAs are already known, such as modification of function and composition of immune system cell membranes, change in the pattern of eicosanoids produced and in the cytokine profile, regulation of gene expression and proliferation of T lymphocytes [30]. The leukocyte membrane phospholipids from humans consuming a Western diet typically have 15–20% of AA, 0.5–1% of EPA and 2–3% of DHA. When fish oil rich in LC *n*-3 PUFAs are incorporated to the diet, increased amounts of EPA and DHA are incorporated in these phospholipids in a time and dose dependent fashion, and it is occurs at the expense of AA [31, 32]. These changes in membrane fatty acid composition subsequently modify the cell-membrane fluidity, production of eicosanoids and the formation of lipid rafts [17, 33].

4. Long chain *n-3* PUFAs and *Trypanosoma cruzi* infection

Infection with *T. cruzi* causes a strong inflammatory reaction at the inoculation site and, later, in the myocardium [34]. Approximately one-third to one-half of patients with indeterminate disease will eventually develop chronic Chagas cardiomyopathy (CCC). CCC results from the combined effects of persistent parasitism, parasite-driven tissue inflammation, micro-vascular and neurogenic dysfunction, and autoimmune responses triggered by the *T. cruzi*-infection [34, 35].

There are few studies on the effects of increased consumption of LC *n*-3 PUFAs rich foods as well as the long-term effects of LC *n*-3 PUFAs on inflammatory profile and clinical outcomes in CCC. Recently, a group of Brazilian researchers reported that patients aging >18 years, with a diagnosis of CCC, that received LC *n*-3 PUFAs capsules (1.8 g EPA and 1.2 g DHA) during an 8-week period, presented modifications in the lipid and inflammatory profile, demonstrated by a decrease in triglycerides and improvements on IL-10 concentration [36]. The same group had already supposed in 2013 that the anti-inflammatory action of LC *n*-3 PUFAs may have beneficial effects on chronic chagasic cardiomyopathy, and could be translate into a less severe progression of cardiomyopathy, with subsequent reduction in morbidity [37].

In mice, the resistance to acute infection has been shown to be dependent on interferon IFN-γ that activates macrophages to produce nitric oxide (NO) and kill the obligate intracellular amastigote form of the parasite [38–40]. In addition, TNF-α provides a second signal stimulating NO production and anti-*T. cruzi* activity in IFN-γ-activated macrophages.

TNF-α is a cytokine that appears rapidly after infections or lesions, playing a key role in fighting invasive pathogens. However, excessive TNF-α production is related to mortality and morbidity in sepsis [41], meningitis [42] and malaria [43]. Although many of the studies about TNF-α modulation by treatment with LC *n*-3 PUFAs indicate a suppressive effect on this cytokine production, there are studies reporting an increase in cytokines production by treatment with LC n-3 PUFAs. Increased TNF-α production was reported by macrophages stimulated in vitro with LPS [25]. Dietary supplementation with fish oil on experimental *Klebsiella pneumoniae* infection and in brain-infection by *Plasmodium berghei* led to increased *ex vivo* production of TNF-α and IL-1α by *in vitro* LPS-stimulated macrophages [24]. Also, the *in vivo* treatment with fish oil increased TNF-α plasma levels in mice infected with *Listeria monocytogenes* in the first 24 hours of infection, being lower in later times [44]. In addition, specifically in *T. cruzi* infected-mice that were supplemented with fish oil rich in LC *n*-3 PUFAs it was related increased TNF-α production by the spleen cells [22].

The first work on the in vivo effects of PUFAs supplementation on *T. cruzi* infection was published in 1958 by Godfrey and coworkers. They showed that oral supplementation with cod liver oil, rich in LC *n*-3 PUFA, associated with vitamins A and D, suppressed mice infections with *T. congolense* and *T. vivax*, but showed no effect on mice infected with *T. cruzi* or *T. brucei* [45]. However, when vitamin E was used together with cod liver oil, these adverse effects on host resistance on *T. congolense* and *T. vivax* infection were reversed, suggesting that diet-induced oxidative stress and vitamin E deficiency were central to this particular diet-infection relationship.

In 1995, Takeda and collaborators reported that daily oral administration of the LC *n*-3 PUFA EPA, present in great amounts in fish oil, greatly diminished host survival following an experimental infection with *T. cruzi* [46]. As in the work of 1958, they included vitamin E in EPA treatment, in order to try to avoid the oxidative stress problem. The authors reported that their EPA-treatment failed to impact tissue parasitism, but was associated with elevated capacity to produce the inflammatory cytokine TNF-α.

In contrast, more recently our research group has described that oral supplementation with fish oil on *T. cruzi* infected-mice did not change the mortality rate, but it did alter the course of parasitemia, as well as other important host responses, such the increased TNF-α production [22]. In *T. cruzi* infected-mice that were supplemented with fish oil rich in LC *n*-3 PUFAs we observed a transient, but substantial, increase in peak circulating parasitic load at the 7th day post infection. At the 12th day post infection, these fish oil-treated mice had similar levels of parasitemia in the blood compared to controls groups (mice treated with saline or corn oil). Surprisingly, besides de high peak of parasitemia, the mice that were treated with fish oil rich in LC *n*-3 PUFAs showed significantly fewer parasites in their cardiac tissue at the 12th day post infection (**Figure 1**) compared to mice treated with saline or corn oil. The oral supplementation with corn oil was used an alternative fat source rich in the *n*-6 PUFA linoleic acid but poor in LC *n*-3 PUFAs [22].

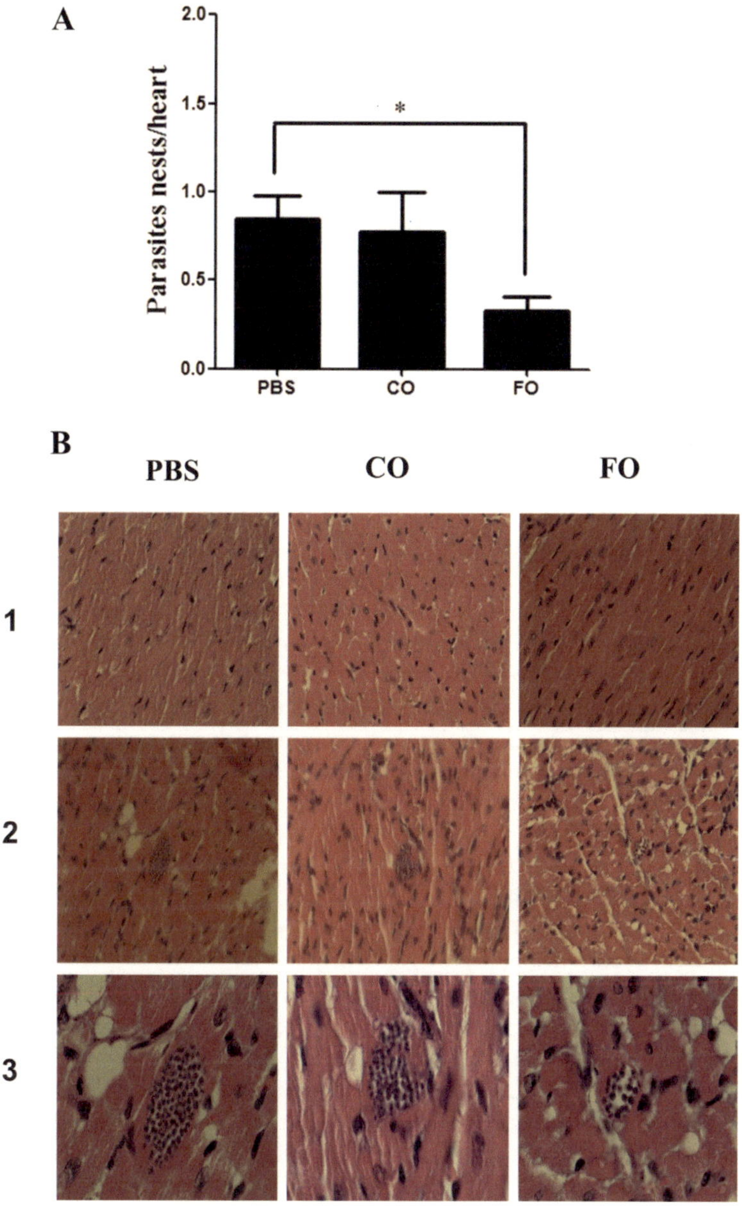

Figure 1.
Effects of FO supplementation on cardiac parasitism. From 15 days before T. cruzi infection until the 12th day post infection, C57BL/6 mice were supplemented by gavage with 0.6% (v/w) saline, corn oil, or menhaden fish oil. C57BL/6 mice were infected with 5 × 10³ blood trypomastigotes T. cruzi (Y strain). Cardiac parasitism (A): three heart sections were counted for each animal, and the results are expressed as means ± SEM of three sections from five animals per group and are representative of two independent experiments. Cardiac tissues (B) were examined by hematoxylin and eosin staining from uninfected mice (1) on day 12 after T. cruzi infection (2 and 3). Original magnifications were 400× (1 and 2) and 1000× (3). Original publication [22].

In humans, as in some experimental animals, *T. cruzi* infection is associated with anemia, thrombocytopenia, leukopenia and bone marrow hypoplasia [47–50]. Dietary supplementation with fish oil had no effect on the anemia of *T. cruzi*-infected mice [22], a finding that contrasts with the improvement in malaria-induced anemia reported by the treatment with LC *n*-3 PUFAs [46]. However, thrombocytopenia and leukopenia were less severe in *T. cruzi*-infected mice orally treated with fish oil [22].

The production of eicosanoids represents an important role in the regulation between the host's immune response and the establishment of *T. cruzi* infection (**Figure 2A**). Eicosanoids are lipid mediators of inflammation which include prostaglandins (PG), thromboxanes (TX), leukotrienes (LT) and lipoxins (LX). The initial substrate for the synthesis of eicosanoids is the lipids present in the membrane phospholipids of cells involved with inflammatory processes. In individuals with a regular western diet, AA is the most prevalent fatty acid in the inflammatory cell membrane, and it is usually the main substrate for the synthesis of eicosanoids, giving rise to series 2-series prostaglandins and thromboxanes and 4-series leukotrienes and lipoxins [30].

Acute *T. cruzi* infection in murine models is characterized by cardiac lesions associated by high levels of PGE_2 [51]. During acute *T. cruzi* infection, both PGE_2 and its $EP\text{-}_2$ receptor are involved in inflammation and cardiac inflammatory infiltrate [52], as leukotrienes, that is important for the local production of NO and cardiac parasitism control [53, 54]. Additionally, plasma PGE_2, TBX_2 e 6-oxo-$PGF_{1\alpha}$ levels are increased in murine models of acute infection [55, 56]. In addition to cardiac effects, PG are associated with immunosuppression of infected animals, with reduction in lymphocyte proliferation, on TNF-α levels, and in the microbicidal functions of macrophages [57, 58].

In vitro assays have shown that the phagocytosis of apoptotic cells by macrophages during *T. cruzi* infection potentiates the release of PGE_2 and transforming growth factor-beta (TGF-β) by these cells. These macrophages become refractory to inflammatory cytokines, consequently decreasing NO production and allowing parasite survival and growth even in an immune response environment [59]. Also, the pharmacological inhibition of the COX enzyme with aspirin in macrophages decreased the trypomastigotes internalization in these cells, with increase of interleukin-1 (IL1-β), NO and lipoxins [60]. In addition, PGE_2 elicits signaling pathways capable of instantaneously inhibiting NLRP3 inflammation activation [61]. This may be relevant in the context of *T. cruzi* infection, since the activation of the inflammatory complex NLPR3 and caspase-1 is important for parasite control during the acute phase of infection, leading to activation of trypanocidal activities in macrophages, as NO production [62].

Although PGE_2 is important for the parasite survival at the beginning of infection, the *in vivo* pharmacological blockade of COX enzymes during the acute phase of *T. cruzi* infection leads to higher levels of parasitemia, lower survival rates of experimental mice and increased cardiac parasitism [63–65]. However, COX blockade with aspirin already in the chronic phase (60 days after infection) does not cause an increase in parasitemia or mortality, but is associated with an improvement in the cardiac ejection fraction [64].

Confirming the relevance of inflammatory-lipid mediators during *T. cruzi* infection is the fact that the parasite itself synthesizes prostaglandins and thromboxanes (TX) [66]. Infective stages of *T. cruzi* have the enzyme phospholipase A-1, important for the release of fatty acids (such as AA) from the membranes and for eicosanoid synthesis, as represented in **Figure 2A**. From AA, *T. cruzi* preferentially

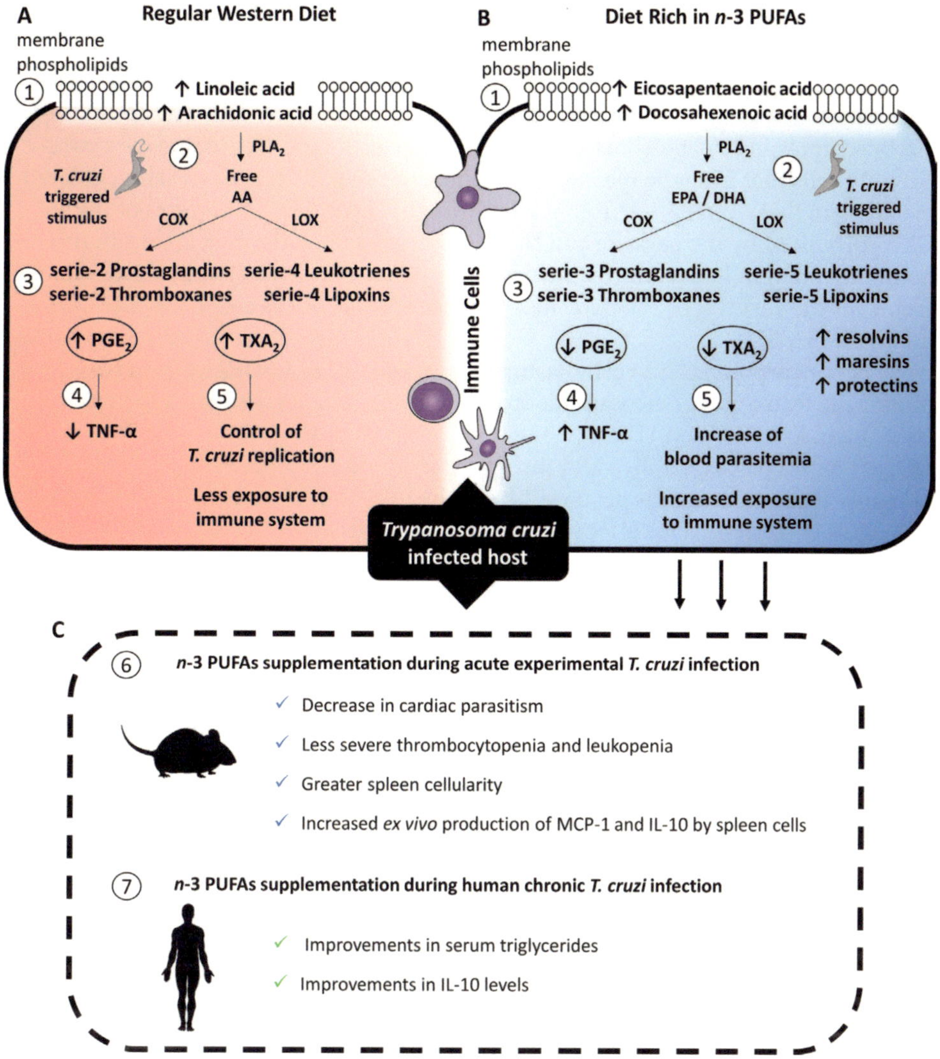

Figure 2.
Mechanism and effects of PUFAs n-3 PUFAs supplementation on Trypanosoma cruzi infection. **1**—*The traditional western diet provides to immune cells great amounts of n-6 PUFAs arachidonic acid and linoleic acid, while a diet supplementation with LC n-3 PUFAs results in eicosapentaenoic acid and docosahexaenoic acid incorporation into immune cells plasma membrane phospholipids [31, 32].* **2**—*T. cruzi infection trigger inflammatory stimulus, as the expression of cyclooxygenase (COX) and Lipoxygenase (LOX) enzymes [53, 55]. The enzyme phospholipase A^2 (PLA^2) removes fatty acids from the membrane phospholipids, mainly arachidonic acid (AA) in cells from western diet and mainly eicosapentaenoic acid (EPA) or docosahexaenoic acid (DHA) from n-3 diet.* **3**—*The enzymes COX and LOX uses the free fatty acids for eicosanoids synthesis. When AA is used as substrate 2-series prostaglandins/thromboxanes and 4-series leukotrienes/lipoxins are formed. When the substrate is EPA or DHA, a switch in the class of eicosanoids produced occurs, being produced 3-series prostaglandins/thromboxanes and 5-series leukotrienes/lipoxins [30], as well mediators that act in the resolution of inflammation: E-series resolvins generated from EPA, D-series resolvins from DHA, and protectins and maresins from DHA [72, 73].* **4**—*In the acute phase of infection, the high production of prostaglandin E^2 (PGE^2) leads to a transient immunosuppression, with decreased TNF-α production by the host [57]. The immune cells from n-3 diet produce lower amounts of PGE_2, and consequently more TNF-α [22].* **5**—*The thromboxane A^2 (TXA^2) produced during T. cruzi infection is associated with a control mechanism of parasite proliferation and less exposure to immune system [67]. The decreased TXA^2 production that occurs in cells from n-3 PUFAs diet could explain the increased parasitemia that were previously observed in vivo, leading to more T. cruzi exposition to immune system [22].* **6**—*Effects of in vivo supplementation of experimental mice during acute T. cruzi infection [22].* **7**—*Effects of in vivo supplementation of human patients with chronic Chagas cardiomyopathy [37]. Original publication [22].*

synthesizes TXA_2, in addition to small amounts of $PGE_{2\alpha}$ [67]. During acute infection, TXA_2 produced by the parasite acts on the vascular endothelium creating an inflammatory phenotype, increasing the expression of adhesion molecules and directly participating in the parasitemia control and host survival. The absence of TXA_2 receptor in infected-host cells leads to large cellular parasitism, when compared to cells that have the receptor. This indicates that *T. cruzi* has a self-regulated mechanism that controls its proliferation trough TXA_2 [67]. The effects of both TXA_2 produced by the parasite and the PGE_2 produced by the host create an immunomodulatory environment that favors the survival of the host, an indispensable factor for the survival of the parasite and maintenance of the chronic phase of the infection [63, 67].

When incorporated into cell plasma membrane, LC *n*-3 PUFAs competitively inhibit the formation of eicosanoids from AA by the enzymes cyclooxygenase (COX) and lipoxygenase (LOX), being produced, from the fatty acids EPA and DHA, 3-series prostaglandins and thromboxanes and 5-series leukotrienes and lipoxins. These eicosanoids produced from the EPA and DHA have less inflammatory activity [68, 69]. This effect occurs in part due to the reduction of AA available, but also due to a direct action of the EPA decreasing the activity and expression of the COX-2 enzyme, that is responsible for the synthesis of prostaglandins after an inflammatory stimulus [70].

We have shown that fish oil supplementation decreased the production of PGE_2 in mice uninfected (**Figure 3A**) and infected with *T. cruzi* (**Figure 3B**) [22], as also represented in **Figure 2B**. In addition, as discussed above, it is reported in the scientific literature that LC *n*-3 PUFA supplementation promotes the production of 3-series thromboxanes, rather than the production of 2-series thromboxanes, a lipid mediator described as important for the regulation and continuity of infection by *T. cruzi* [67]. The use of LC *n*-3 PUFAs to modulate the production of eicosanoids exhibit relevant differences when compared to the use of pharmacological inhibitors of COX isoforms. While pharmacological COX blockade results in a marked decrease in eicosanoid production, the LC *n*-3 PUFAs acts promoting a change in the class of eicosanoids produced, without, however, completely abolishing those produced from AA. In addition, the inflammatory pro-resolution lipid mediators resolvins, maresins and protectins produced from the LC *n*-3 PUFAs DHA and EPA could control the tissue injury resulting from the exacerbated activation of the immune response [30]. Pro-resolvins are a class of lipid mediators that act in the resolution of inflammation [71]. E-series resolvins (RvE) are generated from EPA, D-series resolvins (RvD) from DHA, and protectins and maresins from DHA. The synthesis of these pro-resolvins also involves the COX and LOX pathway [72, 73].

Recently was demonstrated that trypomastigotes and amastigotes of *T. cruzi* produce the pro-resolving lipids RvD1, RvD5, and RvE2. It has been reported that plasma RvD1 levels are elevated in *T. cruzi* infected mice and, at least in part, it is possible that this RVD1 is from the parasite itself. This mechanism suggests another way of how the parasite can modulate the environment in its favor. This modulation of the immune response by the parasite may be important and contribute to the perpetuation of the infection into the chronic phase [74].

Therefore, the elucidation of the more specific mechanisms involved with the protective effects of *n*-3 PUFAs on *T. cruzi* infection (**Figure 2C**) are important aspects to be investigated. Considering the discussion presented here, as well as all points raised in the scientific literature, it is a reasonable to consider that the immune modulation exerted by LC *n*-3 PUFAs supplementation actually favors the host and may represent a perspective for the supplementary treatment of patients affected by Chagas disease.

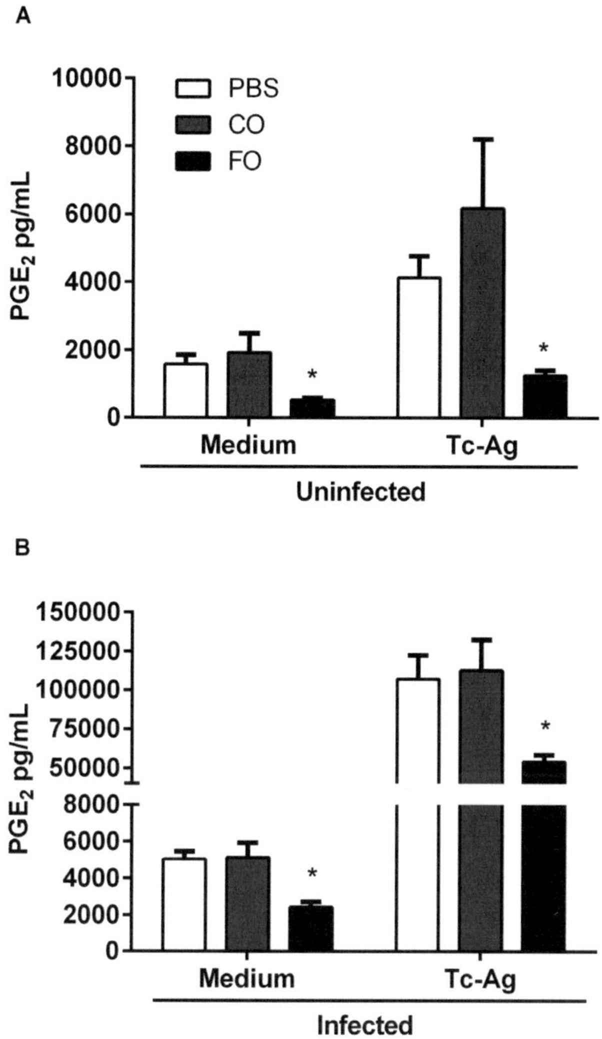

Figure 3.
Effects of fish oil supplementation on production of PGE2 by spleen cells from mice infected with T. cruzi. From 15 days before T. cruzi infection to the 7th dpi, C57BL/6 mice were supplemented by gavage with 0.6% (v/w) PBS, corn oil, or fish oil. C57BL/6 mice were infected with 5 × 103 blood trypomastigotes T. cruzi (Y strain). Splenocytes (5 × 106 cells/well) from uninfected (A) or T. cruzi-infected mice (B) were cultured with and without T. cruzi antigen (Tc-Ag). Supernatants were harvested after 8 hours, and PGE2 was quantified in supernatants by EIA. The results are expressed as means ± SEM from four animals per group and are representative of two independent experiments. Means not sharing letter are significantly different (P<0.05, 2-way ANOVA with Bonferroni post-test). Original publication [22].

5. Conclusion

The immunomodulatory effects of long-chain omega-3 polyunsaturated fatty acids (LC n-3 PUFA) are currently widely known. Dietary supplementation with LC n-3 PUFAs has been used as a complementary treatment in inflammatory diseases. However, the effects of daily supplementation with LC *n*-3 PUFAS on host resistance to infectious disease, such as the *T. cruzi* infection, are still poorly understood. Studies using a well-established mouse model of this human disease showed that fish oil supplementation improves de clinical course *T. cruzi* infection during the acute phase of infection. In fact, the potential benefits of LC *n*-3 PUFAs supplementation in humans have been the subject of recent clinical trials. The modulation of the immune response by LC *n*-3 PUFAs, mainly through the change in eicosanoids patterns produced during *T. cruzi* infection, could be one mechanism that results in improvement of the host response. However, more studies are necessary to determine whether or not oral supplementation with LC n-3 PUFA could benefit humans diagnosed with Chagas disease.

Acknowledgements

Funding: This study was financed in part by the Coordenação de Aperfeiçoamento de Pessoal de Nível Superior-Brasil (CAPES) Finance Code 001, CNPq (grant Edital Universal 14-2014, research fellowships for PP-F (CNPq 307787/2015-0), MM-P (CNPq 307544/2016-8) and MILM (PNPD-Capes 22921091) and Fundação Araucária (grant 419-2009).

Conflict of interest

The authors declare that there are no conflicts of interest.

Notes/thanks/other declarations

This study would have been impossible without the aid and support of Dr. Kevin Fritsche Department of Nutrition and Exercise Physiology, University of Missouri, Columbia, 65,211, Missouri, USA.

Author details

Maria Isabel Lovo-Martins[1], Marli Cardoso Martins-Pinge[2] and Phileno Pinge-Filho[1*]

1 Department of Pathological Sciences, Laboratory of Experimental Immunopathology, Biological Sciences Center, State University of Londrina, Londrina, Paraná, Brazil

2 Department of Physiological Sciences, Biological Sciences Center, State University of Londrina, Londrina, Paraná, Brazil

*Address all correspondence to: pingefilho@uel.br

References

[1] Chagas C. Nova tripanozomiaze humana: estudos sobre a morfolojia e o ciclo evolutivo do Schizotrypanum cruzi n. gen., n. sp., ajente etiolojico de nova entidade morbida do homem. Memórias do Instituto Oswaldo Cruz. 1909;**1**:159-218

[2] WHO. Chagas Disease (American Trypanosomiasis); 2017. Available from: http://www.who.int/mediacentre/factsheets/fs340/en/

[3] WHO. Why are Some Tropical Diseases called "neglected"? 2012. Available from: http://www.who.int/features/qa/58/en/

[4] Machado FS, Dutra WO, Esper L, Gollob KJ, Teixeira MM, Factor SM, et al. Current understanding of immunity to Trypanosoma cruzi infection and pathogenesis of Chagas disease. Seminars in Immunopathology. 2012;**34**(6):753-770. DOI: 10.1007/s00281-012-0351-7

[5] Marin-Neto JA, Cunha-Neto E, Maciel BC, Simoes MV. Pathogenesis of chronic Chagas heart disease. Circulation. 2007;**115**(9):1109-1123. DOI: 10.1161/circulationaha.106.624296

[6] Dutra WO, Menezes CA, Magalhaes LM, Gollob KJ. Immunoregulatory networks in human Chagas disease. Parasite Immunology. 2014;**36**(8): 377-387. DOI: 10.1111/pim.12107

[7] Calder PC, Kew S. The immune system: A target for functional foods? The British Journal of Nutrition. 2002;**88**(Suppl 2):S165-S177. DOI: 10.1079/bjn2002682

[8] Turchini GM, Nichols PD, Barrow C, Sinclair AJ. Jumping on the omega-3 bandwagon: Distinguishing the role of long-chain and short-chain omega-3 fatty acids. Critical Reviews in Food Science and Nutrition. 2012;**52**(9):795-803. DOI: 10.1080/10408398.2010.509553

[9] Fritsche K. Fatty acids as modulators of the immune response. Annual Review of Nutrition. 2006;**26**:45-73. DOI: 10.1146/annurev.nutr.25.050304.092610

[10] Calder PC. Very long-chain n-3 fatty acids and human health: Fact, fiction and the future. The Proceedings of the Nutrition Society. 2018;**77**(1):52-72. DOI: 10.1017/s0029665117003950

[11] Anderson M, Fritsche KL. (n-3) Fatty acids and infectious disease resistance. The Journal of Nutrition. 2002;**132**(12):3566-3576

[12] Yancy CW, Jessup M, Bozkurt B, Butler J, Casey DE Jr, Drazner MH, et al. 2013 ACCF/AHA guideline for the management of heart failure: A report of the American College of Cardiology Foundation/American Heart Association Task Force on Practice Guidelines. Journal of the American College of Cardiology 2013;**62**(16):e147–e239. DOI: 10.1016/j.jacc.2013.05.019

[13] Faludi AA, Izar MCO, Saraiva JFK, Chacra APM, Bianco HT, Afiune AN, et al. Atualização da Diretriz Brasileira de Dislipidemias e Prevenção da Aterosclerose—2017. Arquivos Brasileiros de Cardiologia. 2017;**109** (2 Supl 1):1-76. DOI: 10.5935/abc.20170121

[14] Ratnayake WM, Galli C. Fat and fatty acid terminology, methods of analysis and fat digestion and metabolism: A background review paper. Annals of Nutrition & Metabolism. 2009;**55**(1-3):8-43. DOI: 10.1159/000228994

[15] Simopoulos AP. An increase in the omega-6/omega-3 fatty acid ratio increases the risk for obesity. Nutrients. 2016;**8**(3):128. DOI: 10.3390/nu8030128

[16] Simopoulos AP. Omega-3 fatty acids in health and disease and in growth and development. The American Journal of Clinical Nutrition. 1991;**54**(3):438-463. DOI: 10.1093/ajcn/54.3.438

[17] Kaur KK, Allahbadia G, Singh M. Synthesis and functional significance of poly unsaturated fatty acids (PUFA's) in body. Acta Scientific Nutritional Health. 2018;**2**(4):8. DOI: 10.15226/jnhfs

[18] Blasbalg TL, Hibbeln JR, Ramsden CE, Majchrzak SF, Rawlings RR. Changes in consumption of omega-3 and omega-6 fatty acids in the United States during the 20th century. The American Journal of Clinical Nutrition. 2011;**93**(5):950-962. DOI: 10.3945/ajcn.110.006643

[19] Maalouly G, Ward C, Smayra V, Saliba Y, Aftimos G, Haddad F, et al. Fish oil attenuates neurologic severity of antiphospholipid syndrome in a mice experimental model. Nutritional Neuroscience. 2016;**20**:1-8. DOI: 10.1080/1028415x.2016.1206165

[20] Hokari R, Matsunaga H, Miura S. Effect of dietary fat on intestinal inflammatory diseases. Journal of Gastroenterology and Hepatology. 2013;**28**(Suppl 4):33-36. DOI: 10.1111/jgh.12252

[21] Woo SJ, Lim K, Park SY, Jung MY, Lim HS, Jeon MG, et al. Endogenous conversion of n-6 to n-3 polyunsaturated fatty acids attenuates K/BxN serum-transfer arthritis in fat-1 mice. The Journal of Nutritional Biochemistry. 2015;**26**(7):713-720. DOI: 10.1016/j.jnutbio.2015.01.011

[22] Lovo-Martins MI, Malvezi AD, da Silva RV, Zanluqui NG, VLH T, NOS C, et al. Fish oil supplementation benefits the murine host during the acute phase of a parasitic infection from Trypanosoma cruzi. Nutrition Research. 2017;**41**:73-85. DOI: 10.1016/j.nutres.2017.04.007

[23] Irons R, Anderson MJ, Zhang M, Fritsche KL. Dietary fish oil impairs primary host resistance against Listeria monocytogenes more than the immunological memory response. The Journal of Nutrition. 2003;**133**(4):1163-1169

[24] Blok WL, Vogels MT, Curfs JH, Eling WM, Buurman WA, van der Meer JW. Dietary fish-oil supplementation in experimental gram-negative infection and in cerebral malaria in mice. The Journal of Infectious Diseases. 1992;**165**(5):898-903

[25] Blok WL, Rabinovitch M, Zilberfarb V, Netea MG, Buurman WA, van der Meer JW. The influence of dietary fish-oil supplementation on cutaneous Leishmania amazonensis infection in mice. Cytokine. 2002;**19**(5):213-217

[26] Rajaei E, Mowla K, Ghorbani A, Bahadoram S, Bahadoram M, Dargahi-Malamir M. The effect of omega-3 fatty acids in patients with active rheumatoid arthritis receiving DMARDs therapy: Double-blind randomized controlled trial. Global Journal of Health Science. 2015;**8**(7):18-25. DOI: 10.5539/gjhs.v8n7p18

[27] Ferguson JF, Mulvey CK, Patel PN, Shah RY, Doveikis J, Zhang W, et al. Omega-3 PUFA supplementation and the response to evoked endotoxemia in healthy volunteers. Molecular Nutrition & Food Research. 2014;**58**(3):601-613. DOI: 10.1002/mnfr.201300368

[28] Miles EA, Calder PC. Influence of marine n-3 polyunsaturated fatty acids on immune function and a systematic review of their effects on clinical outcomes in rheumatoid arthritis. The British Journal of Nutrition. 2012;**107**(Suppl 2):S171-S184. DOI: 10.1017/S0007114512001560

[29] Fritsche KL. The science of fatty acids and inflammation. Advances in Nutrition (Bethesda, Md).

2015;6(3):293s-301s. DOI: 10.3945/an.114.006940.

[30] Calder PC. Marine omega-3 fatty acids and inflammatory processes: Effects, mechanisms and clinical relevance. Biochimica et Biophysica Acta. 2015;**1851**(4):469-484. DOI: 10.1016/j.bbalip.2014.08.010

[31] Faber J, Berkhout M, Vos AP, Sijben JW, Calder PC, Garssen J, et al. Supplementation with a fish oil-enriched, high-protein medical food leads to rapid incorporation of EPA into white blood cells and modulates immune responses within one week in healthy men and women. The Journal of Nutrition. 2011;**141**(5):964-970. DOI: 10.3945/jn.110.132985

[32] Yaqoob P, Pala HS, Cortina-Borja M, Newsholme EA, Calder PC. Encapsulated fish oil enriched in alpha-tocopherol alters plasma phospholipid and mononuclear cell fatty acid compositions but not mononuclear cell functions. European Journal of Clinical Investigation. 2000;**30**(3):260-274

[33] Calder PC, Yaqoob P, Harvey DJ, Watts A, Newsholme EA. Incorporation of fatty acids by concanavalin A-stimulated lymphocytes and the effect on fatty acid composition and membrane fluidity. Biochemical Journal. 1994;**300**(Pt 2):509-518

[34] Groom ZC, Protopapas AD, Zochios V. Tropical diseases of the myocardium: A review. International Journal of General Medicine. 2017;**10**:101-111. DOI: 10.2147/IJGM.S130828

[35] Ribeiro AL, Nunes MP, Teixeira MM, Rocha MO. Diagnosis and management of Chagas disease and cardiomyopathy. Nature Reviews. Cardiology. 2012;**9**(10):576-589. DOI: 10.1038/nrcardio.2012.109

[36] Silva PSD, Mediano MFF, Silva G, Brito PD, Cardoso CSA, Almeida CF, et al. Omega-3 supplementation on inflammatory markers in patients with chronic Chagas cardiomyopathy: A randomized clinical study. Nutrition Journal. 2017;**16**(1):36. DOI: 10.1186/s12937-017-0259-0

[37] Silva PS, Sperandio da Silva GM, de Souza AP, Cardoso CS, Fonseca CA, Brito PD, et al. Effects of omega-3 polyunsaturated fatty acid supplementation in patients with chronic chagasic cardiomyopathy: Study protocol for a randomized controlled trial. Trials. 2013;**14**:379. DOI: 10.1186/1745-6215-14-379

[38] Gazzinelli RT, Oswald IP, Hieny S, James SL, Sher A. The microbicidal activity of interferon-gamma-treated macrophages against Trypanosoma cruzi involves an L-arginine-dependent, nitrogen oxide-mediated mechanism inhibitable by interleukin-10 and transforming growth factor-beta. European Journal of Immunology. 1992;**22**(10):2501-2506. DOI: 10.1002/eji.1830221006

[39] Petray P, Rottenberg ME, Grinstein S, Orn A. Release of nitric oxide during the experimental infection with Trypanosoma cruzi. Parasite Immunology. 1994;**16**(4):193-199

[40] Vespa GN, Cunha FQ, Silva JS. Nitric oxide is involved in control of Trypanosoma cruzi-induced parasitemia and directly kills the parasite in vitro. Infection and Immunity. 1994;**62**(11):5177-5182

[41] Kaech C, Bochud PY, Calandra T. Cytokines and *Escherichia coli* Sepsis. EcoSal Plus. 2006;**2**(1):1-20. DOI: 10.1128/ecosalplus.8.8.15

[42] Armah H, Wired EK, Dodoo AK, Adjei AA, Tettey Y, Gyasi R. Cytokines and adhesion molecules expression in the brain in human cerebral malaria. International Journal of Environmental Research and Public Health. 2005;**2**(1):123-131

[43] Perlmann P, Troye-Blomberg M. Malaria blood-stage infection and its control by the immune system. Folia Biologica. 2000;**46**(6):210-218

[44] Puertollano MA, Cruz-Chamorro L, Puertollano E, Perez-Toscano MT, Alvarez de Cienfuegos G, de Pablo MA. Assessment of interleukin-12, gamma interferon, and tumor necrosis factor alpha secretion in sera from mice fed with dietary lipids during different stages of *Listeria monocytogenes* infection. Clinical and Diagnostic Laboratory Immunology. 2005;**12**(9):1098-1103. DOI: 10.1128/cdli.12.9.1098-1103.2005

[45] Godfrey DG. Influence of dietary cod liver oil upon *Trypanosoma congolense*, *T. cruzi*, *T. vivax* and *T. brucei*. Experimental Parasitology. 1958;**7**(3):255-268

[46] Takeda GK, Starobinas N, Marcondes MC, Mello EA, Russo M, Stolf AM. Oral administration of fish-oil induces high levels of seric TNF in Trypanosoma cruzi infected C57BL/6 mice. Acta Tropica. 1995;**60**(3):215-219

[47] Cardoso JE, Brener Z. Hematological changes in mice experimentally infected with Trypanosoma cruzi. Memórias do Instituto Oswaldo Cruz. 1980;**75**(3-4):97-104

[48] Repka D, Rangel HA, Atta AM, Gavino VA, Piedrabuena AE. Experimental Chagas' disease in mice infected with one LD50 of parasite. Revista Brasileira de Biologia. 1985;**45**(3):309-316

[49] Marcondes MC, Borelli P, Yoshida N, Russo M. Acute Trypanosoma cruzi infection is associated with anemia, thrombocytopenia, leukopenia, and bone marrow hypoplasia: Reversal by nifurtimox treatment. Microbes and Infection. 2000;**2**(4):347-352

[50] Malvezi AD, Cecchini R, de Souza F, Tadokoro CE, Rizzo LV, Pinge-Filho P. Involvement of nitric oxide (NO) and TNF-alpha in the oxidative stress associated with anemia in experimental Trypanosoma cruzi infection. FEMS Immunology and Medical Microbiology. 2004;**41**(1):69-77. DOI: 10.1016/j.femsim.2004.01.005

[51] Abdalla GK, Faria GE, Silva KT, Castro EC, Reis MA, Michelin MA. Trypanosoma cruzi: The role of PGE2 in immune response during the acute phase of experimental infection. Experimental Parasitology. 2008;**118**(4):514-521. DOI: 10.1016/j.exppara.2007.11.003

[52] Guerrero NA, Camacho M, Vila L, Iniguez MA, Chillon-Marinas C, Cuervo H, et al. Cyclooxygenase-2 and Prostaglandin E2 Signaling through Prostaglandin Receptor EP-2 Favor the Development of Myocarditis during Acute Trypanosoma cruzi Infection. PLoS Neglected Tropical Diseases. 2015;**9**(8):e0004025. DOI: 10.1371/journal.pntd.0004025

[53] Pavanelli WR, Gutierrez FR, Mariano FS, Prado CM, Ferreira BR, Teixeira MM, et al. 5-lipoxygenase is a key determinant of acute myocardial inflammation and mortality during Trypanosoma cruzi infection. Microbes and Infection. 2010;**12**(8-9):587-597. DOI: 10.1016/j.micinf.2010.03.016

[54] Panis C, Mazzuco TL, Costa CZ, Victorino VJ, Tatakihara VL, Yamauchi LM, et al. Trypanosoma cruzi: Effect of the absence of 5-lipoxygenase (5-LO)-derived leukotrienes on levels of cytokines, nitric oxide and iNOS expression in cardiac tissue in the acute phase of infection in mice. Experimental Parasitology. 2011;**127**(1):58-65. DOI: 10.1016/j.exppara.2010.06.030

[55] Celentano AM, Gorelik G, Solana ME, Sterin-Borda L, Borda E, Gonzalez Cappa SM. PGE2 involvement

in experimental infection with Trypanosoma cruzi subpopulations. Prostaglandins. 1995;**49**(3):141-153

[56] Cardoni RL, Antunez MI. Circulating levels of cyclooxygenase metabolites in experimental Trypanosoma cruzi infections. Mediators of Inflammation. 2004;**13**(4):235-240. DOI: 10.1080/09637480400003022

[57] Pinge-Filho P, Tadokoro CE, Abrahamsohn IA. Prostaglandins mediate suppression of lymphocyte proliferation and cytokine synthesis in acute Trypanosoma cruzi infection. Cellular Immunology. 1999;**193**(1):90-98

[58] Michelin MA, Silva JS, Cunha FQ. Inducible cyclooxygenase released prostaglandin mediates immunosuppression in acute phase of experimental Trypanosoma cruzi infection. Experimental Parasitology. 2005;**111**(2):71-79. DOI: 10.1016/j.exppara.2005.05.001

[59] Freire-de-Lima CG, Nascimento DO, Soares MB, Bozza PT, Castro-Faria-Neto HC, de Mello FG, et al. Uptake of apoptotic cells drives the growth of a pathogenic trypanosome in macrophages. Nature. 2000;**403**(6766):199-203. DOI: 10.1038/35003208

[60] Malvezi AD, Panis C, da Silva RV, de Freitas RC, Lovo-Martins MI, Tatakihara VL, et al. Inhibition of cyclooxygenase-1 and cyclooxygenase-2 impairs Trypanosoma cruzi entry into cardiac cells and promotes differential modulation of the inflammatory response. Antimicrobial Agents and Chemotherapy. 2014;**58**(10):6157-6164. DOI: 10.1128/aac.02752-14

[61] Sokolowska M, Chen LY, Liu Y, Martinez-Anton A, Qi HY, Logun C, et al. Prostaglandin E2 inhibits NLRP3 inflammasome activation through EP4 receptor and intracellular cyclic AMP in human macrophages. Journal of Immunology. 2015;**194**(11):5472-5487. DOI: 10.4049/jimmunol.1401343

[62] Goncalves VM, Matteucci KC, Buzzo CL, Miollo BH, Ferrante D, Torrecilhas AC, et al. NLRP3 controls Trypanosoma cruzi infection through a caspase-1-dependent IL-1R-independent NO production. PLoS Neglected Tropical Diseases. 2013;**7**(10):e2469. DOI: 10.1371/journal.pntd.0002469

[63] Sterin-Borda L, Gorelik G, Goren N, Cappa SG, Celentano AM, Borda E. Lymphocyte muscarinic cholinergic activity and PGE2 involvement in experimental Trypanosoma cruzi infection. Clinical Immunology and Immunopathology. 1996;**81**(2):122-128

[64] Mukherjee S, Sadekar N, Ashton AW, Huang H, Spray DC, Lisanti MP, et al. Identification of a functional prostanoid-like receptor in the protozoan parasite, Trypanosoma cruzi. Parasitology Research. 2013;**112**(4):1417-1425. DOI: 10.1007/s00436-012-3271-5

[65] Hideko Tatakihara VL, Cecchini R, Borges CL, Malvezi AD, Graca-de Souza VK, Yamada-Ogatta SF, et al. Effects of cyclooxygenase inhibitors on parasite burden, anemia and oxidative stress in murine Trypanosoma cruzi infection. FEMS Immunology and Medical Microbiology. 2008;**52**(1):47-58. DOI: 10.1111/j.1574-695X.2007.00340.x

[66] Machado FS, Mukherjee S, Weiss LM, Tanowitz HB, Ashton AW. Bioactive lipids in Trypanosoma cruzi infection. Advances in Parasitology. 2011;**76**:1-31. DOI: 10.1016/b978-0-12-385895-5.00001-3

[67] Ashton AW, Mukherjee S, Nagajyothi FN, Huang H, Braunstein VL, Desruisseaux MS, et al. Thromboxane A2 is a key regulator of pathogenesis during Trypanosoma cruzi

infection. The Journal of Experimental Medicine. 2007;**204**(4):929-940. DOI: 10.1084/jem.20062432

[68] Chapkin RS, Akoh CC, Miller CC. Influence of dietary n-3 fatty acids on macrophage glycerophospholipid molecular species and peptidoleukotriene synthesis. Journal of Lipid Research. 1991;**32**(7):1205-1213

[69] Trebble TM, Wootton SA, Miles EA, Mullee M, Arden NK, Ballinger AB, et al. Prostaglandin E2 production and T cell function after fish-oil supplementation: Response to antioxidant cosupplementation. The American Journal of Clinical Nutrition. 2003;**78**(3):376-382

[70] Bagga D, Wang L, Farias-Eisner R, Glaspy JA, Reddy ST. Differential effects of prostaglandin derived from omega-6 and omega-3 polyunsaturated fatty acids on COX-2 expression and IL-6 secretion. Proceedings of the National Academy of Sciences of the United States of America. 2003;**100**(4):1751-1756. DOI: 10.1073/pnas.0334211100

[71] Serhan CN. Pro-resolving lipid mediators are leads for resolution physiology. Nature. 2014;**510**(7503): 92-101. DOI: 10.1038/nature13479

[72] Hong S, Gronert K, Devchand PR, Moussignac RL, Serhan CN. Novel docosatrienes and 17S-resolvins generated from docosahexaenoic acid in murine brain, human blood, and glial cells. Autacoids in anti-inflammation. The Journal of Biological Chemistry. 2003;**278**(17):14677-14687. DOI: 10.1074/jbc.M300218200

[73] Mas E, Croft KD, Zahra P, Barden A, Mori TA. Resolvins D1, D2, and other mediators of self-limited resolution of inflammation in human blood following n-3 fatty acid supplementation. Clinical Chemistry. 2012;**58**(10):1476-1484. DOI: 10.1373/clinchem.2012.190199

[74] Colas RA, Ashton AW. Trypanosoma cruzi produces the specialized proresolving mediators resolvin D1, resolvin D5, and resolvin E2. Infection and Immunity. 2018;**86**(4):e00688. DOI: 10.1128/iai.00688-17

Chapter 13

Parasite, Compartments, and Molecules: Trick versus Treatment on Chagas Disease

Marcos André Vannier-Santos, Giselle V. Brunoro, Maria de Nazaré C. Soeiro, Solange L. DeCastro and Rubem F.S. Menna-Barreto

Abstract

Chagas disease, caused by the protozoan *Trypanosoma cruzi*, is endemic to Latin America, standing out as a socio-economic problem for low-income tropical populations. Such disease affects millions of people worldwide and emerges in nonendemic areas due to migration and climate changes. The current chemotherapy is restricted to two nitroderivatives (benznidazole and nifurtimox), which is unsatisfactory due to limited efficacy (particularly in chronic phase) and adverse side effects. *T. cruzi* life cycle is complex, including invertebrate and vertebrate hosts and three developmental forms (epimastigotes, trypomastigotes, and amastigotes). In this chapter, we will discuss promising cellular and molecular targets present in the vertebrate-dwelling forms of the parasite (trypomastigotes and amastigotes). Among the cellular targets, the mitochondrion is the most frequently studied; while among the molecular ones, we highlight squalene synthase, C14α-sterol demethylase, and cysteine proteases. In this scenario, proteomics becomes a valuable tool for the identification of other molecular targets, and some previously identified candidates will be also discussed. Multidisciplinary studies are needed to identify novel key molecules in *T. cruzi* in order to increase trypanocidal activity and reduce mammalian toxicity, ensuring the development of novel drugs for Chagas disease.

Keywords: *Trypanosoma cruzi*, Chagas disease, chemotherapy, drug targets, organelles, proteomics, oxidative metabolism

1. Introduction

Chagas disease, or American trypanosomiasis, was described in 1909 by the Brazilian physician Carlos Chagas, who identified the causative agent—*Trypanosoma cruzi*—the transmission vector, major reservoirs, mechanism of human infection, as well as some clinical manifestations [1]. This disease is primarily transmitted to humans by the feces of hematophagous insects, of the family Reduviidae, subfamily Triatominae. The impressive decrease in Chagas disease prevalence from 16–18 million people by 1990 to 6–8 million people by 2010 was essentially the consequence of the launching of transnational programs in Latin America focused on the elimination of domestic vectors and blood donors screening supported by Pan American

IntechOpen

Health Organization/World Health Organization (PAHO/WHO) [2]. Despite these successful control interventions, the Chagas disease prevalence was estimated to reach 1.0–2.4% of the Brazilian population [1]. This disease, classically associated with poor and rural populations, underwent an urbanization process in the 1970s and 1980s to Latin American cities and later on beyond endemic countries, creating new epidemiological, social, and political challenges [3–5].

With the success of vector and blood bank control programs, congenital [6, 7], and oral [8, 9] transmissions have become important sources of new cases of Chagas disease. Congenital infections represent an estimated 22% of new cases in Latin America [2], occurring also in nonendemic countries [10, 11]. The oral route, which is probably the most frequent mechanism among vectors and wild mammals, has recently become relevant, due to environmental changes caused by deforestation [12]. *T. cruzi* DNA was recently shown in 10% among 140 samples of açai-based products marketed in Rio de Janeiro and Pará States in Brazil [13].

Chagas disease results from the establishment of *T. cruzi* in host tissues, involving an initial acute phase followed by a chronic phase, classified as indeterminate, cardiac, and/or digestive syndromes. The acute phase is characterized by detectable parasitemia and is commonly asymptomatic [14]. Without treatment, approximately 5–10% of symptomatic patients die during this phase due to encephalomyelitis or severe cardiac failure and rarely due to cardiac arrest [15]. After 2–3 months, the infection enters the chronic phase, and without successful treatment, it is lifelong. Approximately, two-thirds of infected individuals have the indeterminate form of the chronic phase, which is asymptomatic and defined by the presence of *T. cruzi* antibodies and normal electrocardiographic and radiologic exams. The remaining infected individuals, due to an unbalanced inflammatory response and persistent low parasitism, will develop years or even decades later symptomatic chronic disease with cardiac (20–30%) and/or digestive (15–20%) disorders.

The current etiological treatment for Chagas disease is restricted to two nitroheterocyclic drugs: benznidazole (Bz/LAFEPE, Abarax®, ELEA and Bz/Chemo Research, Exeltis) and nifurtimox (Nif, LAMPIT®, Bayer) (**Figure 1**). Bz has been recently FDA-approved for use in children aged 2–12 years, being the first treatment approved in the United States for Chagas disease [16]. The results obtained with these two nitroderivatives vary according to the phase of Chagas disease, the period, and dose of treatment, as well as the age and geographical origin of the patients [17]. Both drugs have often shown successful results with high parasitological cure rates during the acute phase, but the effectiveness decreases with advance of the infection; therefore, early detection and intervention are crucial for reaching high cure rates [18]. The high incidence of collateral effects, especially for adults, leads to treatment abandonment rates reaching over 30% of the patients [19–21]. In contrast, children have a markedly higher tolerance for treatment [1, 14].

There are significant drawbacks on the use of these drugs, mostly related to the limited efficacy in the chronic phase [22], and so, new alternative therapies

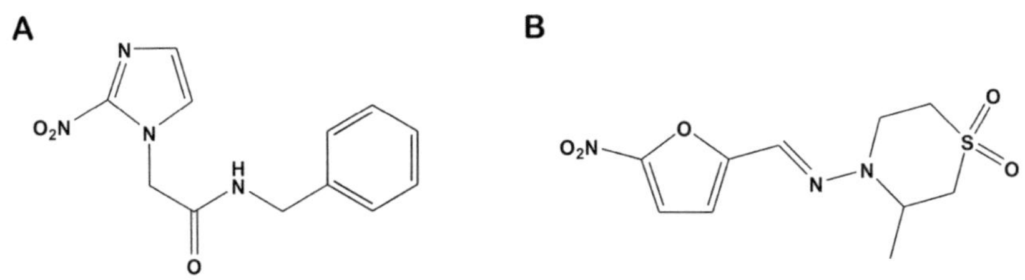

Figure 1.
Clinical drugs for Chagas disease treatment: (A) benznidazole and (B) nifurtimox.

are urgently required. In the last decades, many chemical diversity libraries from several pharmaceutical companies have been screened in the search of novel anti-*T. cruzi* candidates. In these programs, different approaches have been used including target-oriented studies, combination therapies, new formulations for drugs in use, and drug repurposing, and thus, in the present review, some of these points will be addressed.

2. *Trypanosoma cruzi* and drug targets

One important point to be addressed in the search of alternative molecular targets in *T. cruzi* is their presence in parasite forms dwelling in vertebrates. Once the parasite stages present different metabolic profiles [23, 24], the most promising targets are involved in crucial metabolic pathways, such as key enzymes related to antioxidant metabolism or sterol biosynthesis. In this section, we revised some of the most studied targets for drug intervention.

2.1 Mitochondrion, glycosomes, and oxidative metabolism

Mitochondrion plays a pivotal role in the oxidative stress, since the electron leakage from the electron transport chain (ETC particularly from complexes I, III and coenzyme Q) leads to the partial reduction of oxygen, being the main source of reactive oxygen species (ROS) in the cells [25]. During electron leakage, ROS were produced that interfere with different biological processes [26]. Such production leads to the increase in the expression of antioxidant enzymes such as superoxide dismutase (SOD), trypanothione reductase (TR), and peroxidases in response to the oxidative burst, and TR is considered one of the most promising chemotherapy targets in Chagas disease [27]. In trypanosomatids, mitochondrial metabolism is quite similar to that of other eukaryotes. Complex I (NADH: ubiquinone oxidoreductase) is expressed (almost 19 subunits were detected) but its functionality is still controversial [26]. In this way, glucose metabolism results mostly in succinate (complex II substrate) in trypanosomatids, derived from glycosomal and mitochondrial NADH-dependent fumarate reductase activities [28, 29]. Since complex I is not functional, oxidative phosphorylation is exclusively dependent of complex II in these protozoa. On the other hand, complexes III (ubiquinol:cytochrome c oxidoreductase) and IV (cytochrome c oxidase) of high eukaryotes and trypanosomatids display no differences, being complex III considered the major mitochondrial source of ROS production [30]. Many studies pointed to the susceptibility of *T. cruzi* mitochondrion to a great variety of compounds, and such mitochondrial damage (ultrastructural swelling, decreased mitochondrial membrane potential, etc.) may comprise early or late events in trypanocidal agent activity (**Figure 2**) [31].

Glycosomes are organelles crucial for the energetic and antioxidant metabolisms of the parasite, and the compartmentalization of their enzymes (including the majority of the enzymes of the glycolytic pathway) has also been reported to be directly involved in the maintenance of *T. cruzi* viability, indicating this organelle as a potential drug target [32]. Among the glycosomal oxidative, scavengers are SOD isoforms, tryparedoxin, and peroxidases [26]. Up to now, no specific inhibitors of glycosomal enzymes showed promising trypanocidal activity [33].

2.1.1 Mitochondrial ETC

T. cruzi mitochondrion is the most recurrent cellular drug target described in mechanistic studies; however, the exact molecular machinery involved in the

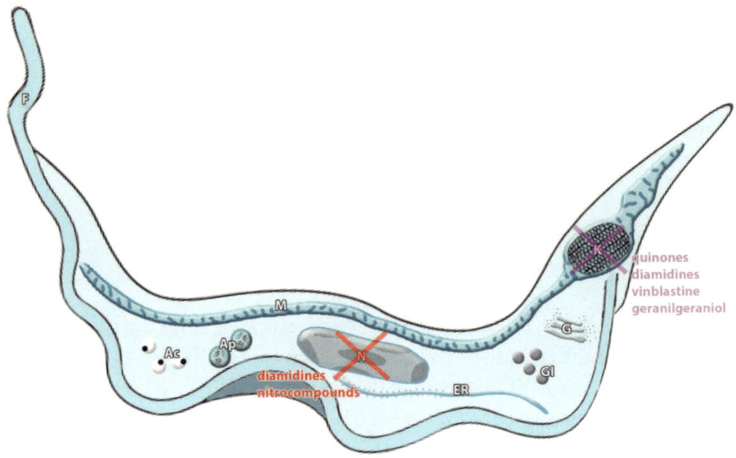

Figure 2.
Most recurrent cellular drug targets in T. cruzi mammalian stages. M: mitochondrion; K: kinetoplast; N: nucleus; F: flagellum; Gl: glycosomes; G: golgi; Ac: acidocalcisomes; ER: endoplasmic reticulum; and Ap: autophagosomes.

susceptibility of this organelle to different classes of compounds is still unclear [31]. No hypothesis about the molecular mechanism involved in the mitochondrial effect of the great majority of trypanocidal drugs was postulated hitherto, and the damage specificity in this organelle is very debatable. Some specific inhibitors of ETC complexes have already been tested on *T. cruzi*. Rotenone, a well-known complex I inhibitor, has a controversial activity in the parasite. However, rotenone at high doses inhibited the activity of *T. cruzi* NADH-dependent enzymes [34]. The existence of complex I activity was not strongly supported by such inhibition and could be caused by nonspecific binding to other electron carriers. On the other hand, depolarization of mitochondrial membrane, ROS production, and apoptotic-like phenotype was detected in parasites after the treatment with inhibitors of complexes III and IV, antimycin A, and potassium cyanide, respectively [24]. Structural and functional similarities between mammalian and trypanosomatidae complexes are suggestive of high toxicity.

2.1.2 Trypanothione reductase

The presence of some antioxidant components that are absent in mammalian cells makes this pathway a promising target of drug intervention in trypanosomatids. The unusual spermidine-glutathione adduct named trypanothione or N1,N8-bis(glutathionyl)spermidine found solely in these parasites functions as an electron donor in many pathways by neutralizing diverse reactive species through redox reactions, also providing reducing equivalents to intermediate molecules in other antioxidant pathways and in biosynthetic pathways such as DNA synthesis [35, 36]. The catalysis of NADPH-dependent reduction of trypanothione disulfide to $T(SH)_2$ is performed by trypanothione reductase (TR), enzyme that has been proposed as a molecular target, based on the specific inhibition of antioxidant defenses of the parasite [37, 38]. The central role of trypanothione makes other enzymes that influence its production also interesting drug targets such as trypanothione synthetase, ornithine decarboxylase (ODC), S-adenosylmethionine (AdoMet) decarboxylase, γ-glutamylcysteine synthetase as well as polyamine transporters [39, 40].

In the last decades, many TR inhibitors were developed, but only a few had a positive correlation between trypanocidal activity and binding to the enzyme demonstrated [41–43]. Recently, a high-throughput screening of 1.8 million compounds was performed, and specific inhibitors of *Leishmania* TR were identified. Since this enzyme is considered well-conserved among trypanosomatids, this study could represent a critical step for the identification of inhibitors also for *T. cruzi* TR [44]. Up to the moment, the inhibition of trypanothione metabolism of this parasite was poorly assessed in animal models, and no clinical trial has been reported involving this target (**Figure 3**).

2.1.3 ROS inducers

Varying the dose or the time of drug treatment, injury to the mitochondrion usually leads to ROS production [45]. Despite many compounds having induced mitochondrial alterations, generating ROS, the molecular mechanistic action was not elucidated in most studies. In this section, we will discuss only quinones and nitrocompounds, compounds with oxidative mechanisms of action well-characterized.

Quinones: Chemical properties of quinoidal carbonyls lead to the direct ROS generation [46], and trypanocidal effect of natural quinones and derivatives have been assessed [47–50]. In epimastigotes, the oxidative activity of β-lapachone was first reported almost 40 years ago [51, 52], and increase in ROS levels has also been related to the treatment of *T. cruzi* with other naphthoquinones [53, 54]. In 2009, we proposed the trypanocidal mechanism of action of naphthofuranquinones. Such quinones strongly impaired the parasite mitochondrion by the deviation of the electrons from ubiquinone, culminating in this organelle depolarization, loss of respiratory rates, inhibition of complexes I–III activities, and ROS production [54]. Increased levels of ROS were also detected in parasites after the treatment with other classes of compounds such as pyrazyl/pyridylhydrazones and thiosemicarbazones [55–57].

Nitrocompounds: These compounds are usually avoided in medicinal chemistry approaches because the presence of a nitro group creates concerns regarding toxicity issues associated with DNA damage [58]. Regarding Chagas disease, fexinidazole evaluated *in vivo*, led to high cure rates and reduced myocarditis [59]. Subsequently, a phase II clinical trial was performed in chronic chagasic patients in Bolivia using fexinidazole treatment (NCT02498782), and it was observed that parasitemia was cleared; however, after recruiting 47 participants, some safety and tolerability issues arose, and it was decided to conclude the trial without the inclusion of new participants. After a 12 month follow-up, a high efficacy rate was evidenced, without relapses [60]. Therefore, a new proof of concept study was initiated in Spain in 2017, with the results expected in 2019 (EudraCT Number 2016-004905-15).

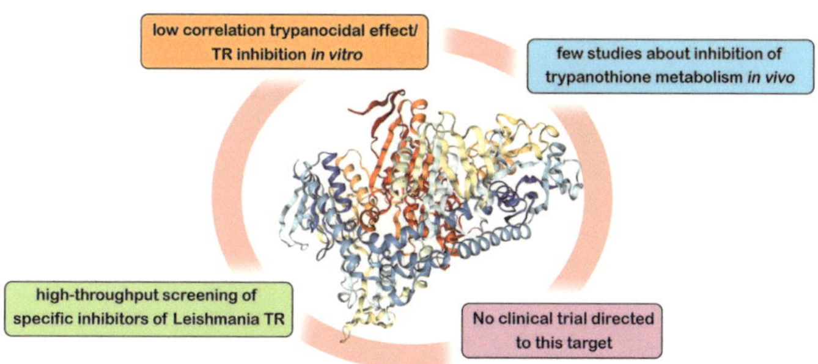

Figure 3.
Landmarks in the investigation of TR as a drug target. Despite many efforts up to now, specific inhibitors of this enzyme presented neither important trypanocidal activity in vitro nor in vivo. TR DPB ID: 1BZL.

Surprisingly, trypanocidal action of Nif and Bz is still controversial. Nif has the oxidative activity demonstrated in the early 1980s, being hydrogen peroxide and superoxide anion production detected, while no reactive species was found after the treatment with Bz [61]. Bz and Nif are considered prodrugs that require activation by nitroreductases—NTR-I, an oxygen insensitive class catalyzing the two-electron reduction of the nitro group and NTR-II, an oxygen-sensitive class catalyzing one-electron reduction [62]. The mechanistic proposal of Nif involves nitroanion radical metabolization by NTR-II, followed by reoxidation by molecular oxygen to form superoxide anion (•O_2^-), which is converted to hydrogen peroxide (H_2O_2) under catalysis by SOD (**Figure 4**) [63]. On the other hand, low molecular weight thiol reduction together with no redox cycling in trypanocidal doses supported the hypothesis that oxidative effect was not involved in the parasite killing by Nif [64]. Additionally, NTR-I activity has been related to the trypanocidal effect of Nif and Bz through a two-electron reduction in the nitro group. In an oxygen-independent way, the production of nitroso and hydroxylamine intermediates led to amine generation, using NADH as a cofactor. The cleavage of the Nif furane ring produces a highly reactive unsaturated open chain nitrile [65].

Recently, some highly potent 3-nitro-1H-1,2,4-triazole derivatives emerged as excellent substrates for NTR-I, but the enzymatic activity was not required for the trypanocidal activity [66]. Alternative enzymes have been associated with the reduction of nitro compounds in *T. cruzi*, indicative of a secondary action for these drugs, and further studies about the molecular mechanisms involved must be performed. The high trypanocidal activity together with the identification of exclusive nitroreductases in trypanosomatids supports the hypothesis of selectivity [63].

2.1.4 Polyamines

Polyamines (PA) are ubiquitous organic polycations that play a plethora of ubiquitous biological roles in most cell types, including bacteria, protozoa, and higher organisms [67], with significant metabolic differences, therefore comprising promising drug targets for protozoal diseases [68]. PA metabolism among parasitic protozoa is defective in a number of pathways as compared to mammalian cells. *T. cruzi* parasites lack ODC and so are auxotrophic for the diamine putrescine [69]. Therefore, the protozoan relies on the diamine uptake from the extracellular milieu via surface transporters or permeases [70], and so, these mechanisms comprise targets for chemotherapy agent development [71], and pentamidine was shown to inhibit polyamine transport by *T. cruzi* [72]. Putrescine uptake is required for the massive infection [73] and scape from stress conditions [74]. Spermidine is synthesized by the transfer of an aminopropyl group from decarboxylated S-adenosyl-L-methionine to putrescine and takes part in the biogenesis of T[SH]2 a pivotal adduct in oxidative stress endurance and involved in anti-*T. cruzi* drug

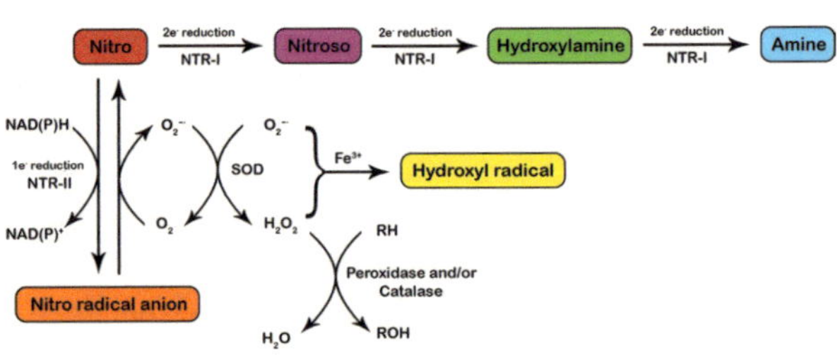

Figure 4.
Metobolization of nitrocompounds by NTR-1 and NTR-II. ROS is generated by reoxidation (one electron route). Two-electron reduction produces hydroxylamine intermediates and reactive nitroso [63].

resistance [75]. In this regard, the putrescine analog DAB (1,4-diamino-2-butanone) promotes oxidative stress in *T. cruzi* [76] and leads to *T. cruzi* mitochondrial destruction [77]. It is noteworthy that DAB not only is involved on reactive species production but also inhibit putrescine synthesis [78] and incorporation [79].

Polyamines may play multiple functions in parasite endurance under oxidative stress conditions, not only for TSH is a spermidine adduct but also because these polycations *per se* may be antioxidant, protecting *T. cruzi* from oxidative stress [80]. Polyamines are also relevant for controlling differentiation, including *T. cruzi* metacyclogenesis [81]. Thus, the enzymes involved in polyamine and TSH metabolism provide important drug targets for potential anti-*T. cruzi* therapy [40].

2.2 Biosynthesis of sterols

Sterols are essential lipid molecules, performing numerous cellular roles associated with membrane and signal functions [82]. Cholesterol is biosynthesized in humans, whereas ergosterol or other 24-alkylated sterols are biosynthesized in opportunistic fungi and parasitic protozoa and such difference is exploited in the drug development [83]. *T. cruzi* and related trypanosomatids have a strict requirement for endogenous sterols (ergosterol and analogs) for survival that cannot be replaced by cholesterol found in the host. Thus, the biosynthesis of sterols is a major target in the drug development for Chagas disease [84]. Among enzymes of the sterol metabolism, squalene synthase (SQS) and C14α-sterol demethylase (CYP51) have been intensively investigated as drug targets (**Figure 5**).

2.2.1 Squalene synthase (SQS)

This enzyme catalyzes the dimerization of two molecules of farnesyl pyrophosphate (FPP) to produce squalene. This enzyme is under study as a possible target for cholesterol-lowering agents in humans [85]. SQS is a membrane-bound enzyme

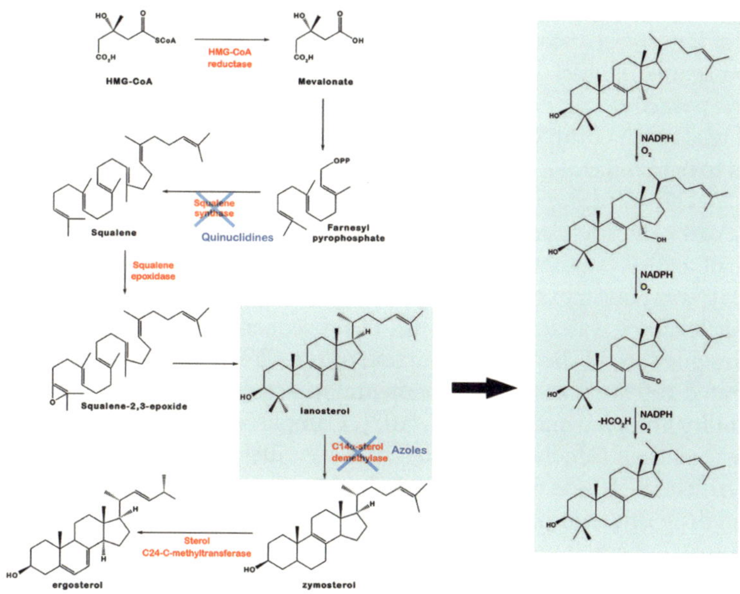

Figure 5.
Ergosterol biosynthesis is an important drug target in T. cruzi. In red, the enzymes described as molecular targets and in blue, the classes of the speciffic inhibitors. Inside gray boxes, the intermediary steps of the conversion of lanosterol into 4,4-dimethyl-5a-cholesta-8,1,4,24-3b-ol by sterol 14-α-demethylase (CYP51).

in *T. cruzi* epimastigotes, being distributed between glycosome and mitochondrial/microsomal vesicles [86]. FPP is a branching point in isoprenoid biosynthesis: conversion to squalene and sterols by SQS or synthesis of other essential isoprenoids. The quinuclidine-based inhibitors of mammalian SQS, 3-(biphenyl-4-yl)-3-hydroxyquinuclidine (BPQ-OH) ER27856, E5700, and ER-119884 were assayed against *T. cruzi*, leading to the *in vitro* inhibition of epimastigote and intracellular amastigote proliferation, depletion of endogenous squalene and sterols, and marked ultrastructural alterations [86, 87] and *in vivo* E5700 led to 100% survival and parasitemia negativation [88]. However, E5700 and ER-119884 have no selectivity toward the parasite enzyme in comparative assays with the recombinant human enzyme [89]. In 2014, the X-ray crystallographic structure of SQS from *T. cruzi* was reported, confirming the binding of the enzyme to distinct classes of inhibitors such as the quinuclidines E5700 and ER119884 and the thiocyanate WC-9 opening possibilities to the development of alternative inhibitors [90].

In a screening of compounds containing the 4-phenoxyphenoxy skeleton, 4-phenoxyphenoxyethyl thiocyanate (WC-9) was highlighted due to the high activity against the proliferative form epimastigotes (low micromolar) and intracellular amastigotes (nanomolar) and a potent inhibitor of the enzymatic activity of both glycosomal and mitochondrial isoforms of SQS [91, 92]. Since then, different series of WC-9 analogs have been developed [91, 93], including seleno-containing analogs resulting in compounds, such as 4-phenoxyphenoxyethyl selenocyanate, with EC_{50} values at low nanomolar level and selectivity index (SI) higher than 900 [94].

2.2.2 C14α-sterol demethylase (CYP51)

This enzyme catalyzes the oxidative removal of the 14α-methyl group from of catalyzing the oxidative removal of the 14α-methyl group from sterol precursors such as lanosterol or eburicol, via a repetitive three-step process that uses NADPH and oxygen to produce 4,4-dimethyl-5α-cholesta-8,14,24-trien-3β-ol [83]. CYP51s are the most conserved cytochrome P450 enzymes [84]. Series of azoles originally developed for the treatment of fungal infections targets this enzyme leading to accumulation of lanosterol and other sterol intermediates and displaying activity *in vitro* and *in vivo* against *T. cruzi* [95–97]. This line of investigation led to the selection of the triazole posaconazole [98, 99] and E1224 (fosravuconazole) [100, 101] for Phase II clinical trials with chronic patients, which, however, led to therapeutic failure as compared to benznidazole, with parasitemia relapses: NCT01162967 (Chagazol) [102], NCT01377480 (Stopchagas) [103], and NCT01489228 (E1224 trial) [104].

VNI, a carboxamide-containing β-phenyl-imidazole, identified from a Novartis collection of azoles, was active in acute and chronic mouse models using Tulahuen strain [105]; whereas in experiments with other parasite strains, no complete parasitological clearance was achieved [106]. In subsequent work, VFV, a fluoro analog of VNI, designed to fill the deepest portion of the CYP51 substrate-binding cavity demonstrated 100% efficacy in experimental infection, displaying favorable oral bioavailability and pharmacokinetics [107]. Comparison between VNI and VFV, in murine models of infection, revealed that regardless of the treatment scheme or delivery vehicle, VFV was more potent in both genders [108]. VT-1161, a 1-tetrazole-based drug undergoing phase II antifungal clinical trials, is active *in vitro* and *in vivo* against *T. cruzi*. It was structurally characterized in a complex with TcCYP51, allowing for the optimization of new tetrazole-based analogs and presents good pharmacokinetic properties and an excellent safety profile [109]. Friggeri et al. [110] synthesized imidazolyl-2-phenylethanol derivatives, and several of them were active against intracellular amastigotes and inhibited TcCYP51. In sequence, eight new derivatives were prepared and assayed against the parasite, and the most active was a piperazinyl-carbamate derivative at nanomolar range, low cytotoxicity, and good chemical and metabolic stability [111]. Recently, a series of pyrazolo[3,4-e][1,4]thiazepin analogs, novel CYP51 inhibitors, were investigated revealing *in vitro* and *in vivo* activity against *T. cruzi*, with several analogs displaying effect at low micromolar dosis and low host toxicity [112].

2.3 Cysteine proteases

Cysteine proteases are intensively used as molecular targets in trypanosomatid disease drug discovery efforts. Target-based screening, structure-based drug design, and medicinal chemistry approaches targeting cysteine proteases are strategies intensively used in the development of drugs for diseases caused by pathogenic trypanosomatids. *T. cruzi* cysteine protease named cruzipain (or cruzain) is a cathepsin-L-like protease of the papain family and is essential for the intracellular replication, differentiation, and immune evasion of the parasite [113, 114]. Three-dimensional structures of cruzain with different ligands have been reported, allowing the design and synthesis of new hit compounds [115]. Based on the interaction with its active site, enzyme inhibitors have been classified as irreversible, forming covalent bonds with cysteine sulfur, or as reversible, forming 1,2-adducts with cysteine that are generally unstable [116]. Among irreversible peptidyl inhibitors of cruzipain, we highlight diazomethyl ketones, allyl sulfones, vinyl sulfonamides, and vinyl sulfones, including K777 and its arginine variant WRR-483 [117–119]. Among nonpeptidyl inhibitors of cruzipain, we found thiosemicarbazones, thiazolylhydrazones, thiazoles, and oxadiazoles (**Figure 6**) [120].

Another group of compounds that has been studied as cruzipain reversible inhibitors are those containing a nitrile head: purine nitriles [121], nitrile analogs of odanacatib [122, 123], and nonpeptidic nitriles [124]. Salas-Sarduy et al. [125] identified two new cruzipain inhibitory scaffolds from GlaxoSmithKline HAT and Chagas chemical boxes, both containing a nitrile moiety, with major structural differences between them. Benzimidazoles and oxidiazoles have also been explored as noncovalent cruzain inhibitors, using an approach combining high throughput and virtual screenings [126, 127].

Development of cruzipain inhibitors by structure activity relationship (SAR) studies, combinatorial chemistry, HTS, and virtual screening are also employed in repositioning strategies [128]. Bromocriptine (antiparkinson and antidiabetic

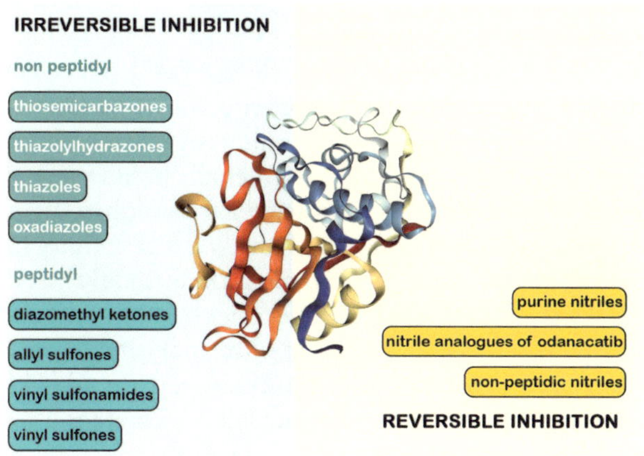

Figure 6.
Cruzipain as a molecular target in T. cruzi. Irreversible and reversible inhibition was demonstrated by different classes of compounds. Cruzipain PDB ID: 3Io6.

drug), amiodarone (antiarrhythmic drug), and levothyroxine (hypothyroidism drug) were selected in a screening campaign for cruzain inhibitors of the DrugBank database [129], clofazimine (antileprosy drug) and benidipine (antihypertensive) from the Merck Index 12th database [130, 131], and etofyllin clofibrate (antilipemic drug) and piperacillin, cefoperazone, and flucloxacillin (β-lactam antibiotics) from a collection of 3180 FDA drugs [132].

Calpains are calcium-dependent nonlysosomal cysteine peptidases highly conserved among eukaryotes, but their precise biological function is not completely clear. In mammalian cells, calpains participate in many different calcium processes including proliferation, differentiation, cytoskeletal assembly, cellular signaling, among many others; however, *T. cruzi* calpains do not present a mapped active catalytic site up to now [133]. In man, uncontrolled activity of calpains has been associated with muscular and neurological disorders such as Alzheimer, Parkinson, multiple sclerosis, and arthritis, and the terapeutic effect of specific calpain inhibitors was suggested [134, 135]. Recently, the repurposing of calpain inhibitors was also postulated for neglected tropical diseases, including Chagas disease [133]. In *T. cruzi*, only the inhibitor MDL28170 was tested, and the trypanocidal activity at low micromolar range was shown on all three parasite forms, impairing the ultrastructural architecture of Golgi and reservosomes [136, 137]. Despite the study's scarcity, calpains inhibition has been suggested as an attractive antitrypanosomatid approach even without the confirmation of their proteolytic activity in these parasites.

2.4 Nuclear and kinetoplast DNA

Both *T. cruzi* kinetoplast and nucleus may be targeted by different classes of compounds with antiparasitic activity [138, 139]. Early studies [140] revealed that hydroxystilbamidine led to the disorganization of kinetoplast DNA (kDNA). Later other compounds such as vinblastine, geranylgeraniol, diaminobenzidine, and aromatic diamidines were reported to affect kDNA arrangement causing its fragmentation [138, 141, 142]. Trypanosomatid nucleus and kinetoplast display topoisomerases that show significant structural differences from host orthologs advocating their potential as drug targets [143]. Interestingly, *T. cruzi* topoisomerase I is inactivated by ROS [144], so the oxidative stress induced by both immune response and trypanocidal agents may also affect parasite chromatin organization.

Classic aromatic diamidines have been shown to bind noncovalently and through a nonintercalative manner to the minor groove of DNA; several hypotheses regarding their mode of action were proposed. They could act by complexation with DNA and subsequently lead to a selective inhibition of DNA-dependent enzymes and/or through the direct inhibition of transcription [145]. Thus, evidences suggest that diamidines interfere in the kinetoplast function of trypanosomatids through a selective association to the unique AT-rich regions of kDNA minicircles, perhaps involving DNA-processing enzymes [146]. Medicinal chemistry studies pointed to arylimidamides (AIAs) as the most promising antimicrobial diamidines [106]. DB702, DB786, DB811, and DB889 presented anti-*T. cruzi* activity in the low-micromolar range and led to ultrastructural alterations mainly associated with the nucleus and mitochondrion [147, 148]. On the other hand, recent study with novel bis-AIAs revealed their higher potency and *in silico* analysis showed DNA as the main target, but no DNA ultrastructural alterations were found [149].

On the other hand, enzymes involved in nucleic acid metabolism could be also promising targets. Topoisomerases play a crucial role for the DNA dynamics during the transcription, replication, or even in the repair. Due to their participation in essential cellular processes, interfering with DNA topology, and consequently leading to physiological implications, topoisomerases have been described as molecular targets for cancer and also parasitic illnesses such as Chagas disease. Up to now, innumerous topoisomerase inhibitors presented antitrypanosomatidae activities such as camptothecin, doxorubicin, etoposide, suramin, among many others [150]. Recently, voacamine and an isobenzofuranone derivative induced important morphological alterations in different trypanosomatids, including *T. cruzi*. In Leishmania parasites, the most affected organelle was mitochondrion (severely swelled presenting membrane depolarization), where the derivative led to kinetoplast network disorganization [151, 152].

3. Proteomic insights for the target identification in the parasite

The evaluation of the proteomic profile in trypanosomatids is particularly interesting because these protozoa exhibit open reading frames in long polycistronic regions, and the regulation of gene expression occurs only post-transcriptionally, justifying the importance on monitoring the protein expression by proteomic approach [153]. This section will focus on proteomic analysis of parasite forms dwelling in mammalian hosts. The first large-scale analysis was performed by Atwood and colleagues in 2005 [23], identifying 1486 proteins of culture-derived trypomastigotes, and 30 trans-sialidases, enzymes that play an important role in parasite host cell invasion, were among the top-scoring proteins exclusively detected in this developmental form. *T. cruzi* surface subproteome and basic proteins analysis confirmed the high distribution of trans-sialidases in this life form [154, 155]. Specific trypomastigote surface analysis also revealed membrane-associated enzymes that are involved in biosynthetic pathway of phospholipid and glycolysis [155]. The evaluation of chromatin fraction of this stage revealed RNA-binding proteins and histones, representing 29% of chromatin protein content [156], providing new insights into gene expression and histone modifications involved in the parasite cycle regulation. Likewise, *T. cruzi* glycoproteome was assessed, and trypomastigote-specific glycoproteins were identified, including mucin family members [157, 158].

In metacyclic trypomastigotes, Atwood and co-workers [23] identified 2339 proteins, and different antioxidant enzymes were among the main proteins detected. The presence of these enzymes in this stage could be related to the parasite adaptation to the oxidative environment inside the vertebrate host circulation and particularly inside the phagocytes. The analysis of metacyclogenesis revealed increased expression of cytoskeletal proteins as well as proteins related to energetic and oxidative metabolisms, suggestive of the morphological and metabolic reorganization [159]. Plasma membrane subproteome pointed to a large repertoire of surface proteins in this parasite stage, including trans-sialidases, mucins, and GP63 protease [160]. Such glycoprotein diversity confers adaptation of the parasite to distinct environmental conditions. Morever, secretome of metacyclic trypomastigotes also demonstrated trans-sialidases and other surface molecules, playing a role in parasite invasion during acute and chronic infections [161, 162]. The blockage of this process could be an interesting strategy in novel drug development.

The first proteomic analysis of bloodstream trypomastigotes was performed by our group in 2015, identifying a total 5901 proteins [163]. In this work, a comparison among the proteomic maps of trypomastigotes (bloodstream, cultured-derived, and metacyclic forms) was also assessed, and 2202 proteins related to the parasite surface, cytoskeleton, redox metabolism, cell signaling, and energetic metabolism were exclusively detected in bloodstream forms. Overall, the proteomic profile of bloodstream form comprises an important tool to discover potential new drug targets and novel antigens for vaccines or diagnostics. The differences in the trypomastigote proteomic profiles were expected due to their environment, and huge number of stage-specific proteins in bloodstream forms, probably triggered by the exposure to the host immune system reinforces the necessity for drug validation on this developmental form. In relation to proteomic evaluation of trypanocidal action of drugs, β-lapachone-derived naphthoimidazoles induced the increase in the abundance of 27 proteins, involved in stress response, cell structure, energetic metabolism, nucleic and amino acid metabolisms, oxidative metabolism, among other pathways [164]. This large-scale study revealed an important set of proteins belonging to metabolic pathways that play pivotal functions for this parasite form, providing new insights for the understanding of the parasite biology and of potential drugable molecules for the treatment of Chagas disease.

In 2005, 1871 proteins of culture-derived amastigotes were identified, preferably involved in endoplasmic reticulum to Golgi trafficking, suggesting an intense traffic at this stage [23]. The analysis of amastigogenesis evidenced high abundance of glycolytic enzymes in amastigotes as well as the lower abundance of flagellar components, compatible with the morphology of this stage [165]. Later, the surface subproteome of vertebrate-dwelling parasite forms was characterized, displaying molecules involved in cell division, signal transduction, and lipid metabolism, crucial for the parasite intracellular self-maintenance [155].

Another interesting target is the posttranslational modification of parasite proteins. Acetylation at lysine residues exerts important role in both vertebrates and microbial cells. The NAD^+-dependent lysine deacetylases are termed sirtuins. Humans present seven different sirtuins, whereas *T. cruzi*, solely two. TcSIR2RP1 andTcSIR2RP3 are found in the cytosolic and mitochondrial compartments, respectively. Parasites overexpressing TcSIR2RP1 display enhanced metacyclogenesis and host cell infection [166]. The sirtuin antagonist salermide diminishes intracellular parasite proliferation and parasitemia in murine infection [167]. Thus, acetylation of *T. cruzi* proteins may provide useful targets for the development of antiparasitic agents [167, 168]. In addition, mammalian sirtuin targeting may be beneficial in chronic chagasic cardiomyopathy [169].

4. Conclusions

The clinical chemotherapy for Chagas disease (Nif and Bz) led to a parasitological cure in the great number of congenital, adult acute, or early chronic cases [170]. However, undesirable side effects and the resistance of some parasite strains [171], together with the limited efficacy in symptomatic chronic cases, drive the continuous search for novel chemotherapeutic agents [33]. Drug repurposing or even combinations with the current drugs could be options to minimize this problem [59, 120]. In this direction, phenotypic strategy has been considered the most valuable approach for the screening of antiparasitic compounds [172].

High throughput screening complemented by whole-cell phenotypic assays represents the more feasible option in the search for novel anti-*T. cruzi* compounds, also leading to an increment in sensitivity [173]. Preclinical *in silico* combined to *in vitro* assays represents an essential step for the recognition of *T. cruzi* drug targets, generating knowledge about the metabolism, biochemical pathways, or biological processes allowing the target validation. In this way, the identification of selective targets comprises a logical startpoint to the reduction of the side effects in the hosts [153, 163]. Ultrastructural observations can predict the potential primary cellular targets due to their earlier alterations triggered by pharmacologic stimuli, helping the prospection of molecular modes of action of antiparasitic agents [141, 174].

Large-scale proteomics represents an alternative approach for the assessment of molecular mechanisms of trypanocidal drugs. Indeed, despite its potential, such technique is poorly employed in this context. The importance of the use of clinical relevant *T. cruzi* stages in drug screening was evidenced by the remarkable differences found among the proteomic profile of parasite forms [23, 163]. The proteomic analysis of bloodstream trypomastigotes identified a huge variety of proteins from distinct biological processes, pointed to more than 2000 proteins present only in bloodstream forms (not in trypomastigotes from other sources), reinforcing the importance of the chemotherapeutic tests using recent isolated parasites from animals' blood [163].

In the current scenario, the CYP51 still represents one the most promising alternatives. Large-scale screening pointed to the high activity of CYP51 inhibitors *in vitro* (even higher than Bz), but not producing *T. cruzi* elimination [175, 176]. Unfortunately, the clinical trial with posaconazole and E1224 was also unsatisfactory [102]. In Argentina, a drug–drug interaction analysis of the combination Bz and E1224 in healthy volunteers demonstrated no improvement in tolerability or safety parameters [177], and a clinical study using this combination is planned [178]. Also, the use of new scaffolds with the design of inhibitors much more selective toward CYP51 has demonstrated its high trypanocidal activity in association or not with other licensed drugs [108, 179].

Among the parasite antioxidant defenses, the most promising drug target is TR, and the specific inhibitor development has been proposed in the last three decades [36]. However, no active inhibitors of this enzyme were described up to now [180]. The hypothesis that *T. cruzi* is more susceptible to oxidative species than the vertebrate hosts is an old misleading concept, due to the highly efficient scavengers described in the parasite [51, 181]. Another aspect to be considered is the central role of ROS production and the mitochondrion for the trypanocial action of a great variety of preclinical compounds [31]. The mitochondrial swelling is frequently observed in *T. cruzi* after the treatment with different drugs, but this phenotype is rarely associated with a specific molecular mechanism, except for DNA ligands such as aromatic amidines [106]. The molecular mechanistic proposal opens the possibility of the mitochondrial dysfunction to be as a random consequences of indirect effect triggered by the impaired homeostasis, resulting in redox imbalance [31].

Bioinformatic, proteomic, and ultrastructural analyses are pivotal tools in the identification of drug targets; however, the use of specific inhibitors must be validated before the following studies. The absence of the predicted biological activity or even the specific binding to the respective molecular target is not uncommon [178, 182]. To improve the safety, mechanisms of action characterization should be performed in parallel to the high-throughput screening of the trypanocidal activity [183]. In case of obligatory intracellular parasites as *T. cruzi*, the direct analysis in a phenotypic assay is also essential, considering permeability of the host cells besides the therapeutic window related to the mammalian host toxicity aspects [182]. The adequate choice of the experimental design is also crucial. Animal models, parasite strains, and treatment protocols for preclinical assays (*in vitro* and *in vivo*) must be standardized, in order to reach better translation to humans [184].

Acknowledgements

This research was funded by grants from the Conselho Nacional de Desenvolvimento Científico e Tecnológico (CNPq), Fundação de Amparo à Pesquisa do Rio de Janeiro (Faperj) and by Fundação Oswaldo Cruz (Fiocruz).

Conflict of interest

None to declare.

Author details

Marcos André Vannier-Santos[1], Giselle V. Brunoro[2], Maria de Nazaré C. Soeiro[3], Solange L. DeCastro[3] and Rubem F.S. Menna-Barreto[3]*

1 Oswaldo Cruz Institute, FIOCRUZ, Rio de Janeiro, Brazil

2 Centre of Excellence in New Target Discovery, Butantan Institute, São Paulo, Brazil

3 Laboratory of Cellular Biology Cellular, Oswaldo Cruz Institute, FIOCRUZ, Rio de Janeiro, Brazil

*Address all correspondence to: rubemsadok@gmail.com

References

[1] Dias JCP, Ramos AN Jr, Gontijo ED, Luquetti A, Shikanai-Yasuda MA, Coura JR, et al. 2 nd Brazilian consensus on Chagas disease, 2015. Revista da Sociedade Brasileira de Medicina Tropical. 2016;**49**:3-60. DOI: 10.1590/0037-8682-0505-2016

[2] WHOChagas disease in Latin America: An epidemiological update based on 2010 estimates. Relevé Épidémiologique Hebdomadaire. 2015;**90**:33-43

[3] Schmunis GA. Epidemiology of Chagas disease in non-endemic countries: The role of international migration. Memórias do Instituto Oswaldo Cruz. 2007;**102**(Suppl 1): 75-85. DOI: 10.1590/S0074-02762007005000093

[4] Antinori S, Galimberti L, Bianco R, Grande R, Galli M, Corbellino M. Chagas disease in Europe: A review for the internist in the globalized world. European Journal of Internal Medicine. 2017;**43**:6-15. DOI: 10.1016/j.ejim.2017.05.001

[5] Monge-Maillo B, López-Vélez R. Challenges in the management of Chagas disease in Latin-American migrants in Europe. Clinical Microbiology and Infection. 2017;**23**:290-295. DOI: 10.1016/j.cmi.2017.04.013

[6] Messenger LA, Bern C. Congenital Chagas disease: Current diagnostics, limitations and future perspectives. Current Opinion in Infectious Diseases. 2018;**31**:415, 421. DOI: 10.1097/QCO.0000000000000478

[7] Carlier Y, Torrico F, Sosa-Estani S, Russomando G, Luquetti A, Freilij H, et al. Congenital Chagas disease: Recommendations for diagnosis, treatment and control of newborns, siblings and pregnant women. PLoS Neglected Tropical Diseases. 2011;**5**:e1250. DOI: 10.1371/journal.pntd.0001250

[8] Shikanai-Yasuda MA, Carvalho NB. Oral transmission of Chagas disease. Clinical Infectious Diseases. 2012;**54**: 845-852. DOI: 10.1093/cid/cir956

[9] Silva-dos-Santos D, Barreto-de-Albuquerque J, Guerra B, Moreira OC, Berbert LR, Ramos MT, et al. Unraveling Chagas disease transmission through the oral route: Gateways to *Trypanosoma cruzi* infection and target tissues. PLoS Neglected Tropical Diseases. 2017;**11**:e0005507. DOI: 10.1371/journal.pntd.0005507

[10] Centers for Disease Control and Prevention. Congenital transmission of Chagas disease—Virginia, 2010. Morbidity and Mortality Weekly Report. 2012;**61**:477-479

[11] Rodari P, Angheben A, Gennati G, Trezzi L, Bargiggia G, Maino M, et al. Congenital Chagas disease in a non-endemic area: Results from a control programme in Bergamo province, Northern Italy. Travel Medicine and Infectious Disease. 2018;**25**:31-34. DOI: 10.1016/j.tmaid.2018.04.011

[12] Coura JR, Junqueira AC. Surveillance, health promotion and control of Chagas disease in the Amazon region—Medical attention in the Brazilian Amazon region: A proposal. Memórias do Instituto Oswaldo Cruz. 2015;**110**:825-830. DOI: 10.1590/0074-02760150153

[13] Ferreira RTB, Cabral ML, Martins RS, Araujo PF, da Silva SA, Britto C, et al. Detection and genotyping of *Trypanosoma cruzi* from açai products commercialized in Rio de Janeiro and Pará, Brazil. Parasites & Vectors. 2018;**11**. DOI: 10.1186/s13071-018-2699-6

[14] WHO, editor. Control of Chagas Disease: Second Report of the WHO Expert Committee. Geneva: WHO; 2002

[15] Prata A. Clinical and epidemiological aspects of Chagas disease. The Lancet Infectious Diseases. 2001;**1**:92-100. DOI: 10.1016/S1473-3099(01)00065-2

[16] FDA. FDA approves first U.S. treatment for Chagas disease. US Food Drug Adm; 2017. Available from: https://www.fda.gov/newsevents/newsroom/pressannouncements/ucm573942.htm [Accessed: September 20, 2018]

[17] Coura J, de Castro SL. A critical review on Chagas disease chemotherapy. Memórias do Instituto Oswaldo Cruz. 2002;**97**:3-24. DOI: 10.1590/S0074-02762002000100001

[18] Coura JR, Borges-Pereira J. Chronic phase of Chagas disease: Why should it be treated? A comprehensive review. Memórias do Instituto Oswaldo Cruz. 2011;**106**:641-645

[19] Jackson Y, Alirol E, Getaz L, Wolff H, Combescure C, Chappuis F. Tolerance and safety of Nifurtimox in patients with chronic Chagas disease. Clinical Infectious Diseases. 2010;**51**:e69-e75. DOI: 10.1086/656917

[20] Sperandio da Silva GM, Mediano MFF, Alvarenga Americano do Brasil PE, da Costa Chambela M, da Silva JA, de Sousa AS, et al. A clinical adverse drug reaction prediction model for patients with Chagas disease treated with benznidazole. Antimicrobial Agents and Chemotherapy. 2014;**58**:6371-6377. DOI: 10.1128/AAC.02842-14

[21] Pérez-Molina JA, Molina I. Chagas disease. The Lancet. 2018;**391**:82-94. DOI: 10.1016/S0140-6736(17)31612-4

[22] Rassi A Jr, Marin Neto JA, Rassi A. Chronic Chagas cardiomyopathy: A review of the main pathogenic mechanisms and the efficacy of aetiological treatment following the BENznidazole evaluation for interrupting Trypanosomiasis (BENEFIT) trial. Memórias do Instituto Oswaldo Cruz. 2017;**112**:224-235. DOI: 10.1590/0074-02760160334

[23] Atwood JA, Weatherly DB, Minning TA, Bundy B, Cavola C, Opperdoes FR, et al. The *Trypanosoma cruzi* proteome. Science. 2005;**309**:473-476. DOI: 10.1126/science.1110289

[24] Gonçalves RLS, Barreto RFSM, Polycarpo CR, Gadelha FR, Castro SL, Oliveira MF. A comparative assessment of mitochondrial function in epimastigotes and bloodstream trypomastigotes of *Trypanosoma cruzi*. Journal of Bioenergetics and Biomembranes. 2011;**43**:651-661. DOI: 10.1007/s10863-011-9398-8

[25] Venditti P, Di Stefano L, Di Meo S. Mitochondrial metabolism of reactive oxygen species. Mitochondrion. 2013;**13**:71-82. DOI: 10.1016/j.mito.2013.01.008

[26] Bombaça ACS, Menna-Barreto RFS. The oxidative metabolism in trypanosomatids: Implications for these protozoa biology and perspectives for drugs development. In: Leon L, Torres-Santos EC, editors. Differ. Asp. Chemother. Trypanos. First. New York: Nova Science Publishers; 2017. pp. 93-129

[27] Beltran-Hortelano I, Perez-Silanes S, Galiano S. Trypanothione reductase and superoxide dismutase as current drug targets for *Trypanosoma cruzi*: An overview of compounds with activity against Chagas disease. Current Medicinal Chemistry. 2017;**24**:1066-1138. DOI: 10.2174/0929867323666161227094049

[28] Boveris A, Hertig CM, Turrens JF. Fumarate reductase and other

mitochondrial activities in *Trypanosoma cruzi*. Molecular and Biochemical Parasitology. 1986;**19**:163-169

[29] Coustou V, Besteiro S, Rivière L, Biran M, Biteau N, Franconi J-M, et al. A mitochondrial NADH-dependent fumarate reductase involved in the production of succinate excreted by procyclic *Trypanosoma brucei*. The Journal of Biological Chemistry. 2005;**280**:16559-16570. DOI: 10.1074/jbc.M500343200

[30] Mehta A, Shaha C. Apoptotic death in *Leishmania donovani* promastigotes in response to respiratory chain inhibition: Complex II inhibition results in increased pentamidine cytotoxicity. The Journal of Biological Chemistry. 2004;**279**:11798-11813. DOI: 10.1074/jbc.M309341200

[31] Menna-Barreto RFS, de Castro SL. The double-edged sword in pathogenic trypanosomatids: The pivotal role of mitochondria in oxidative stress and bioenergetics. BioMed Research International. 2014;**2014**:1-14. DOI: 10.1155/2014/614014

[32] Barros-Alvarez X, Gualdrón-López M, Acosta H, Cáceres AJ, Graminha MAS, Michels PAM, et al. Glycosomal targets for anti-trypanosomatid drug discovery. Current Medicinal Chemistry. 2014;**21**:1679-1706. DOI: 10.2174/09298673113209990139

[33] Menna-Barreto RFS, d CSL. Clear shot at primary aim: Susceptibility of *Trypanosoma cruzi* organelles, structures and molecular targets to drug treatment. Current Topics in Medicinal Chemistry. 2016;**17**:1212-1234. DOI: 10.2174/1568026616666161025161858

[34] Hernandez FR, Turrens JF. Rotenone at high concentrations inhibits NADH-fumarate reductase and the mitochondrial respiratory chain of Trypanosoma brucei and T. cruzi. Mol Biochem Parasitol. 1998;**93**:135-137. DOI: 10.1016/S0166-6851(98)00015-2

[35] Fairlamb AH, Blackburn P, Ulrich P, Chait BT, Cerami A. Trypanothione: A novel bis(glutathionyl)spermidine cofactor for glutathione reductase in trypanosomatids. Science. 1985;**227**:1485-1487

[36] Leroux AE, Krauth-Siegel RL. Thiol redox biology of trypanosomatids and potential targets for chemotherapy. Molecular and Biochemical Parasitology. 2016;**206**:67-74. DOI: 10.1016/j.molbiopara.2015.11.003

[37] Krieger S, Schwarz W, Ariyanayagam MR, Fairlamb AH, Krauth-Siegel RL, Clayton C. Trypanosomes lacking trypanothione reductase are avirulent and show increased sensitivity to oxidative stress. Molecular Microbiology. 2000;**35**:542-552

[38] Krauth-Siegel RL, Bauer H, Schirmer RH. Dithiol proteins as guardians of the intracellular redox milieu in parasites: Old and new drug targets in trypanosomes and malaria-causing plasmodia. Angewandte Chemie, International Edition. 2005;**44**:690-715. DOI: 10.1002/anie.200300639

[39] Flohé L. The trypanothione system and its implications in the therapy of trypanosomatid diseases. International Journal of Medical Microbiology. 2012;**302**:216-220. DOI: 10.1016/j.ijmm.2012.07.008

[40] Maya JD, Salas CO, Aguilera-Venegas B, Diaz MV, López-Muñoz R. Key proteins in the polyamine-trypanothione pathway as drug targets against *Trypanosoma cruzi*. Current Medicinal Chemistry. 2014;**21**:1757-1771

[41] Beig M, Oellien F, Garoff L, Noack S, Krauth-Siegel RL, Selzer

PM. Trypanothione reductase: A target protein for a combined in vitro and in silico screening approach. PLoS Neglected Tropical Diseases. 2015;**9**:e0003773. DOI: 10.1371/journal.pntd.0003773

[42] Vázquez K, Paulino M, Salas CO, Zarate-Ramos JJ, Vera B, Rivera G. Trypanothione reductase: A target for the development of anti-*Trypanosoma cruzi* drugs. Mini Reviews in Medicinal Chemistry. 2017;**17**:939-946. DOI: 10.2174/1389557517666170315145410

[43] Talevi A, Carrillo C, Comini MA. The thiol-polyamine metabolism of *Trypanosoma cruzi*: Molecular targets and drug repurposing strategies. Current Medicinal Chemistry. 2018;**25**. DOI: 10.2174/0929867325666180926151059. [Epub ahead of print]

[44] Ilari A, Genovese I, Fiorillo F, Battista T, DII, Fiorillo A, et al. Toward a drug against all kinetoplastids: From LeishBox to specific and potent trypanothione reductase inhibitors. Molecular Pharmaceutics. 2018;**15**:3069-3078. DOI: 10.1021/acs.molpharmaceut.8b00185

[45] Docampo R, Gadelha FR, Moreno SN, Benaim G, Hoffmann ME, Vercesi AE. Disruption of Ca^{2+} homeostasis in *Trypanosoma cruzi* by crystal violet. The Journal of Eukaryotic Microbiology. 1993;**40**:311-316

[46] Monks TJ, Jones DC. The metabolism and toxicity of quinones, quinonimines, quinone methides, and quinone-thioethers. Current Drug Metabolism. 2002;**3**:425-438

[47] Salas CO, Faúndez M, Morello A, Maya JD, Tapia RA. Natural and synthetic naphthoquinones active against *Trypanosoma cruzi*: An initial step towards new drugs for Chagas disease. Current Medicinal Chemistry. 2011;**18**:144-161

[48] Schmidt TJ, Khalid SA, Romanha AJ, Alves TM, Biavatti MW, Brun R, et al. The potential of secondary metabolites from plants as drugs or leads against protozoan neglected diseases—Part II. Current Medicinal Chemistry. 2012;**19**:2176-2228

[49] da Silva Júnior EN, Jardim GAM, Menna-Barreto RFS, de Castro SL. Anti-*Trypanosoma cruzi* compounds: Our contribution for the evaluation and insights on the mode of action of naphthoquinones and derivatives. Journal of the Brazilian Chemical Society. 2014;**25**:1780-1798. DOI: 10.5935/0103-5053.20140180

[50] Pieretti S, Haanstra JR, Mazet M, Perozzo R, Bergamini C, Prati F, et al. Naphthoquinone derivatives exert their Antitrypanosomal activity via a multi-target MECHANISM. PLoS Neglected Tropical Diseases. 2013;**7**:e2012. DOI: 10.1371/journal.pntd.0002012

[51] Boveris A, Stoppani AOM. Hydrogen peroxide generation in *Trypanosoma cruzi*. Experientia. 1977;**33**:1306-1308. DOI: 10.1007/BF01920148

[52] Cruz FS, Docampo R, de Souza W. Effect of beta-lapachone on hydrogen peroxide production in *Trypanosoma cruzi*. Acta Tropica. 1978;**35**:35-40

[53] Lara LS, Moreira CS, Calvet CM, Lechuga GC, Souza RS, Bourguignon SC, et al. Efficacy of 2-hydroxy-3-phenylsulfanylmethyl-[1,4]-naphthoquinone derivatives against different *Trypanosoma cruzi* discrete type units: Identification of a promising hit compound. European Journal of Medicinal Chemistry. 2018;**144**:572-581. DOI: 10.1016/j.ejmech.2017.12.052

[54] Menna-Barreto RFS, Goncalves RLS, Costa EM, Silva RSF, Pinto AV, Oliveira MF, et al. The effects on *Trypanosoma cruzi* of novel synthetic

naphthoquinones are mediated by mitochondrial dysfunction. Free Radical Biology & Medicine. 2009;**47**:644-653. DOI: 10.1016/j.freeradbiomed.2009.06.004

[55] Soares ROA, Echevarria A, Bellieny MSS, Pinho RT, de Leo RMM, Seguins WS, et al. Evaluation of thiosemicarbazones and semicarbazones as potential agents anti-*Trypanosoma cruzi*. Experimental Parasitology. 2011;**129**:381-387. DOI: 10.1016/j.exppara.2011.08.019

[56] Martins SC, Lazarin-Bidóia D, Desoti VC, Falzirolli H, da Silva CC, Ueda-Nakamura T, et al. 1,3,4-Thiadiazole derivatives of R-(+)-limonene benzaldehyde-thiosemicarbazones cause death in *Trypanosoma cruzi* through oxidative stress. Microbes and Infection. 2016;**18**:787-797. DOI: 10.1016/j.micinf.2016.07.007

[57] Lapier M, Zuniga-Lopez MC, Aguilera-Venegas B, Adam R, Abarca B, Ballesteros R, et al. Evaluation of the novel antichagasic activity of [1,2,3]Triazolo[1,5-a]pyridine derivatives. Current Topics in Medicinal Chemistry. 2017;**17**:399-411

[58] Patterson S, Wyllie S. Nitro drugs for the treatment of trypanosomatid diseases: Past, present, and future prospects. Trends in Parasitology. 2014;**30**:289-298. DOI: 10.1016/j.pt.2014.04.003

[59] Bahia MT, Diniz L d F, Mosqueira VCF. Therapeutical approaches under investigation for treatment of Chagas disease. Expert Opinion on Investigational Drugs. 2014;**23**:1225-1237. DOI: 10.1517/13543784.2014.922952

[60] DNDi. Fexinidazole (Chagas). Drugs Neglected Dis Initiat DNDi. 2018. Available from: https://www.dndi.org/diseases-projects/portfolio/fexinidazole-chagas [Accessed: November 2, 2018]

[61] Docampo R, Moreno SN. Free radical metabolites in the mode of action of chemotherapeutic agents and phagocytic cells on *Trypanosoma cruzi*. Reviews of Infectious Diseases. 1984;**6**:223-238

[62] Peterson FJ, Mason RP, Hovsepian J, Holtzman JL. Oxygen-sensitive and -insensitive nitroreduction by *Escherichia coli* and rat hepatic microsomes. The Journal of Biological Chemistry. 1979;**254**:4009-4014

[63] Fairlamb AH, Patterson S. Current and future prospects of nitro-compounds as drugs for Trypanosomiasis and Leishmaniasis. Current Medicinal Chemistry. 2018;**25**. DOI: 10.2174/0929867325666180426164352. [Epub ahead of print]

[64] Boiani M, Piacenza L, Hernández P, Boiani L, Cerecetto H, González M, et al. Mode of action of nifurtimox and N-oxide-containing heterocycles against *Trypanosoma cruzi*: Is oxidative stress involved? Biochemical Pharmacology. 2010;**79**:1736-1745. DOI: 10.1016/j.bcp.2010.02.009

[65] Hall BS, Wilkinson SR. Activation of benznidazole by trypanosomal type I nitroreductases results in glyoxal formation. Antimicrobial Agents and Chemotherapy. 2012;**56**:115-123. DOI: 10.1128/AAC.05135-11

[66] Papadopoulou MV, Bloomer WD, Rosenzweig HS, O'Shea IP, Wilkinson SR, Kaiser M, et al. Discovery of potent nitrotriazole-based antitrypanosomal agents: In vitro and in vivo evaluation. Bioorganic & Medicinal Chemistry. 2015;**23**:6467-6476. DOI: 10.1016/j.bmc.2015.08.014

[67] Vannier-Santos MA, Suarez-Fontes AM. Role of polyamines in parasite cell architecture and function. Current Pharmaceutical Design. 2017;**23**:3342-3358. DOI: 10.2174/1381612823666170703163458

[68] Roberts S, Ullman B. Parasite polyamines as pharmaceutical targets. Current Pharmaceutical Design. 2017;**23**:3325-3341. DOI: 10.2174/1381612823666170601101644

[69] Algranati ID. Polyamine metabolism in *Trypanosoma cruzi*: Studies on the expression and regulation of heterologous genes involved in polyamine biosynthesis. Amino Acids. 2010;**38**:645-651. DOI: 10.1007/s00726-009-0425-6

[70] Hasne M-P, Coppens I, Soysa R, Ullman B. A high-affinity putrescine-cadaverine transporter from *Trypanosoma cruzi*. Molecular Microbiology. 2010;**76**:78-91. DOI: 10.1111/j.1365-2958.2010.07081.x

[71] Reigada C, Phanstiel O, Miranda MR, Pereira CA. Targeting polyamine transport in *Trypanosoma cruzi*. European Journal of Medicinal Chemistry. 2018;**147**:1-6. DOI: 10.1016/j.ejmech.2018.01.083

[72] Díaz MV, Miranda MR, Campos-Estrada C, Reigada C, Maya JD, Pereira CA, et al. Pentamidine exerts in vitro and in vivo anti *Trypanosoma cruzi* activity and inhibits the polyamine transport in *Trypanosoma cruzi*. Acta Tropica. 2014;**134**:1-9. DOI: 10.1016/j.actatropica.2014.02.012

[73] Hasne M-P, Soysa R, Ullman B. The *Trypanosoma cruzi* diamine transporter is essential for robust infection of mammalian cells. PLoS One. 2016;**11**:e0152715. DOI: 10.1371/journal.pone.0152715

[74] Reigada C, Sayé M, Vera EV, Balcazar D, Fraccaroli L, Carrillo C, et al. *Trypanosoma cruzi* polyamine transporter: Its role on parasite growth and survival under stress conditions. The Journal of Membrane Biology. 2016;**249**:475-481. DOI: 10.1007/s00232-016-9888-z

[75] Mesías AC, Sasoni N, Arias DG, Pérez Brandán C, Orban OCF, Kunick C, et al. Trypanothione synthetase confers growth, survival advantage and resistance to anti-protozoal drugs in *Trypanosoma cruzi*. Free Radical Biology & Medicine. 2019;**130**:23-34. DOI: 10.1016/j.freeradbiomed.2018.10.436

[76] Soares CO, Colli W, Bechara EJH, Alves MJM. 1,4-Diamino-2-butanone, a putrescine analogue, promotes redox imbalance in *Trypanosoma cruzi* and mammalian cells. Archives of Biochemistry and Biophysics. 2012;**528**:103-110. DOI: 10.1016/j.abb.2012.09.005

[77] Menezes D, Valentim C, Oliveira MF, Vannier-Santos MA. Putrescine analogue cytotoxicity against *Trypanosoma cruzi*. Parasitology Research. 2006;**98**:99-105. DOI: 10.1007/s00436-005-0010-1

[78] Reis IA, Martinez MP, Yarlett N, Johnson PJ, Silva-Filho FC, Vannier-Santos MA. Inhibition of polyamine synthesis arrests trichomonad growth and induces destruction of hydrogenosomes. Antimicrobial Agents and Chemotherapy. 1999;**43**:1919-1923

[79] Vannier-Santos MA, Menezes D, Oliveira MF, de Mello FG. The putrescine analogue 1,4-diamino-2-butanone affects polyamine synthesis, transport, ultrastructure and intracellular survival in Leishmania amazonensis. Microbiology. 2008;**154**:3104-3111. DOI: 10.1099/mic.0.2007/013896-0

[80] Hernández SM, Sánchez MS, de Tarlovsky MNS. Polyamines as a defense mechanism against lipoperoxidation in *Trypanosoma cruzi*. Acta Tropica. 2006;**98**:94-102. DOI: 10.1016/j.actatropica.2006.02.005

[81] Vanrell MC, Losinno AD, Cueto JA, Balcazar D, Fraccaroli LV, Carrillo

C, et al. The regulation of autophagy differentially affects *Trypanosoma cruzi* metacyclogenesis. PLoS Neglected Tropical Diseases. 2017;**11**:e0006049. DOI: 10.1371/journal.pntd.0006049

[82] Nes WD. Biosynthesis of cholesterol and other sterols. Chemical Reviews. 2011;**111**:6423-6451. DOI: 10.1021/cr200021m

[83] Leaver D. Synthesis and biological activity of sterol 14α-demethylase and sterol C24-methyltransferase inhibitors. Molecules. 2018;**23**:1753. DOI: 10.3390/molecules23071753

[84] Lepesheva GI, Friggeri L, Waterman MR. CYP51 as drug targets for fungi and protozoan parasites: Past, present and future. Parasitology. 2018;**145**:1820-1836. DOI: 10.1017/S0031182018000562

[85] Do R, Kiss R, Gaudet D, Engert J. Squalene synthase: A critical enzyme in the cholesterol biosynthesis pathway. Clinical Genetics. 2009;**75**:19-29. DOI: 10.1111/j.1399-0004.2008.01099.x

[86] Urbina JA, Concepcion JL, Rangel S, Visbal G, Lira R. Squalene synthase as a chemotherapeutic target in *Trypanosoma cruzi* and Leishmania mexicana. Molecular and Biochemical Parasitology. 2002;**125**:35-45

[87] Braga MV, Urbina JA, de Souza W. Effects of squalene synthase inhibitors on the growth and ultrastructure of *Trypanosoma cruzi*. International Journal of Antimicrobial Agents. 2004;**24**:72-78. DOI: 10.1016/j.ijantimicag.2003.12.009

[88] Urbina JA, Concepcion JL, Caldera A, Payares G, Sanoja C, Otomo T, et al. In vitro and in vivo activities of E5700 and ER-119884, two novel orally active squalene synthase inhibitors, against *Trypanosoma cruzi*. Antimicrobial Agents and Chemotherapy. 2004;**48**:2379-2387. DOI: 10.1128/AAC.48.7.2379-2387.2004

[89] Sealey-Cardona M, Cammerer S, Jones S, Ruiz-Perez LM, Brun R, Gilbert IH, et al. Kinetic characterization of squalene synthase from *Trypanosoma cruzi*: Selective inhibition by quinuclidine derivatives. Antimicrobial Agents and Chemotherapy. 2007;**51**:2123-2129. DOI: 10.1128/AAC.01454-06

[90] Shang N, Li Q, Ko T-P, Chan H-C, Li J, Zheng Y, et al. Squalene synthase As a target for Chagas disease therapeutics. PLoS Pathogens. 2014;**10**:e1004114. DOI: 10.1371/journal.ppat.1004114

[91] Rodriguez JB. WC-9 a Lead drug with great prospects for American Trypanosomiasis and Toxoplasmosis. Mini Reviews in Medicinal Chemistry. 2016;**16**:1195-1200

[92] Urbina JA, Concepcion JL, Montalvetti A, Rodriguez JB, Docampo R. Mechanism of action of 4-phenoxyphenoxyethyl thiocyanate (WC-9) against Trypanosoma cruzi, the causative agent of Chagas' disease. Antimicrobial Agents and Chemotherapy. 2003;**47**:2047-2050

[93] Liñares GEG, Ravaschino EL, Rodriguez JB. Progresses in the field of drug design to combat tropical protozoan parasitic diseases. Current Medicinal Chemistry. 2006;**13**:335-360

[94] Chao MN, Storey M, Li C, Rodríguez MG, Di Salvo F, Szajnman SH, et al. Selenium-containing analogues of WC-9 are extremely potent inhibitors of *Trypanosoma cruzi* proliferation. Bioorganic & Medicinal Chemistry. 2017;**25**:6435-6449. DOI: 10.1016/j.bmc.2017.10.016

[95] Urbina JA. Ergosterol biosynthesis and drug development for Chagas disease. Memórias do Instituto Oswaldo Cruz. 2009;**104**(Suppl 1):311-318

[96] Lepesheva GI, Villalta F, Waterman MR. Targeting *Trypanosoma cruzi* sterol

14α-demethylase (CYP51). Advances in Parasitology. 2011;**75**:65-87. DOI: 10.1016/B978-0-12-385863-4.00004-6

[97] Urbina JA, McKerrow JH. Drug susceptibility of genetically engineered *Trypanosoma cruzi* strains and sterile cure in animal models as a criterion for potential clinical efficacy of anti-*T. cruzi* drugs. Antimicrobial Agents and Chemotherapy. 2015;**59**:7923-7924. DOI: 10.1128/AAC.01714-15

[98] Urbina JA, Payares G, Contreras LM, Liendo A, Sanoja C, Molina J, et al. Antiproliferative effects and mechanism of action of SCH 56592 against Trypanosoma (Schizotrypanum) cruzi: In vitro and in vivo studies. Antimicrobial Agents and Chemotherapy. 1998;**42**:1771-1777

[99] Molina J, Martins-Filho O, Brener Z, Romanha AJ, Loebenberg D, Urbina JA. Activities of the triazole derivative SCH 56592 (posaconazole) against drug-resistant strains of the protozoan parasite Trypanosoma (Schizotrypanum) cruzi in immunocompetent and immunosuppressed murine hosts. Antimicrobial Agents and Chemotherapy. 2000;**44**:150-155

[100] Urbina JA, Payares G, Sanoja C, Lira R, Romanha AJ. In vitro and in vivo activities of ravuconazole on *Trypanosoma cruzi*, the causative agent of Chagas disease. International Journal of Antimicrobial Agents. 2003;**21**:27-38

[101] Diniz L de F, Caldas IS, Guedes PM da M, Crepalde G, de Lana M, Carneiro CM, et al. Effects of ravuconazole treatment on parasite load and immune response in dogs experimentally infected with *Trypanosoma cruzi*. Antimicrobial Agents and Chemotherapy. 2010;**54**:2979-2986. DOI: 10.1128/AAC.01742-09

[102] Molina I, Gómez i, Prat J, Salvador F, Treviño B, Sulleiro E, et al. Randomized trial of posaconazole and benznidazole for chronic Chagas' disease. The New England Journal of Medicine. 2014;**370**:1899-1908. DOI: 10.1056/NEJMoa1313122

[103] Morillo CA, Waskin H, Sosa-Estani S, del Carmen Bangher M, Cuneo C, Milesi R, et al. Benznidazole and Posaconazole in eliminating parasites in asymptomatic *T. cruzi* carriers. Journal of the American College of Cardiology. 2017;**69**:939-947. DOI: 10.1016/j.jacc.2016.12.023

[104] Torrico F, Gascon J, Ortiz L, Alonso-Vega C, Pinazo M-J, Schijman A, et al. Treatment of adult chronic indeterminate Chagas disease with benznidazole and three E1224 dosing regimens: A proof-of-concept, randomised, placebo-controlled trial. The Lancet Infectious Diseases. 2018;**18**:419-430. DOI: 10.1016/S1473-3099(17)30538-8

[105] Villalta F, Dobish MC, Nde PN, Kleshchenko YY, Hargrove TY, Johnson CA, et al. VNI cures acute and chronic experimental Chagas disease. The Journal of Infectious Diseases. 2013;**208**:504-511. DOI: 10.1093/infdis/jit042

[106] Soeiro M de NC, de Souza EM, da Silva CF, Batista D da GJ, Batista MM, Pavão BP, et al. *In vitro* and *in vivo* studies of the Antiparasitic activity of sterol 14α-demethylase (CYP51) inhibitor VNI against drug-resistant strains of *Trypanosoma cruzi*. Antimicrobial Agents and Chemotherapy. 2013;**57**:4151-4163. DOI: 10.1128/AAC.00070-13

[107] Lepesheva GI, Hargrove TY, Rachakonda G, Wawrzak Z, Pomel S, Cojean S, et al. VFV as a new effective CYP51 structure-derived drug candidate for Chagas disease and visceral Leishmaniasis. The Journal of Infectious Diseases. 2015;**212**:1439-1448. DOI: 10.1093/infdis/jiv228

[108] Guedes-da-Silva FH, Batista DGJ, Da Silva CF, De Araújo JS, Pavão BP, Simões-Silva MR, et al. Antitrypanosomal activity of sterol 14α-demethylase (CYP51) inhibitors VNI and VFV in the Swiss mouse models of Chagas disease induced by the *Trypanosoma cruzi* Y strain. Antimicrobial Agents and Chemotherapy. 2017;**61**:e02098-16. DOI: 10.1128/AAC.02098-16

[109] Hoekstra WJ, Hargrove TY, Wawrzak Z, da Gama Jaen Batista D, da Silva CF, Nefertiti ASG, et al. Clinical candidate VT-1161's antiparasitic effect *in vitro*, activity in a murine model of Chagas disease, and structural characterization in complex with the target enzyme CYP51 from *Trypanosoma cruzi*. Antimicrobial Agents and Chemotherapy. 2016;**60**:1058-1066. DOI: 10.1128/AAC.02287-15

[110] Friggeri L, Hargrove TY, Rachakonda G, Williams AD, Wawrzak Z, DSR, et al. Structural basis for rational design of inhibitors targeting *Trypanosoma cruzi* sterol 14α-demethylase: Two regions of the enzyme molecule potentiate its inhibition. Journal of Medicinal Chemistry. 2014;**57**:6704-6717. DOI: 10.1021/jm500739f

[111] De Vita D, Moraca F, Zamperini C, Pandolfi F, Di Santo R, Matheeussen A, et al. In vitro screening of 2-(1H-imidazol-1-yl)-1-phenylethanol derivatives as antiprotozoal agents and docking studies on *Trypanosoma cruzi* CYP51. European Journal of Medicinal Chemistry. 2016;**113**:28-33. DOI: 10.1016/j.ejmech.2016.02.028

[112] Ferreira de Almeida Fiuza L, Peres RB, Simões-Silva MR, da Silva PB, Batista D d GJ, da Silva CF, et al. Identification of Pyrazolo[3,4-e][1,4] thiazepin based CYP51 inhibitors as potential Chagas disease therapeutic alternative: In vitro and in vivo evaluation, binding mode prediction and SAR exploration. European Journal of Medicinal Chemistry. 2018;**149**:257-268. DOI: 10.1016/j.ejmech.2018.02.020

[113] Ferreira LG, Andricopulo AD. Targeting cysteine proteases in trypanosomatid disease drug discovery. Pharmacology & Therapeutics. 2017;**180**:49-61. DOI: 10.1016/j.pharmthera.2017.06.004

[114] Siqueira-Neto JL, Debnath A, McCall L-I, Bernatchez JA, Ndao M, Reed SL, et al. Cysteine proteases in protozoan parasites. PLoS Neglected Tropical Diseases. 2018;**12**:e0006512. DOI: 10.1371/journal.pntd.0006512

[115] Martinez-Mayorga K, Byler KG, Ramirez-Hernandez AI, Terrazas-Alvares DE. Cruzain inhibitors: Efforts made, current leads and a structural outlook of new hits. Drug Discovery Today. 2015;**20**:890-898. DOI: 10.1016/j.drudis.2015.02.004

[116] Nicoll-Griffith DA. Use of cysteine-reactive small molecules in drug discovery for trypanosomal disease. Expert Opinion on Drug Discovery. 2012;**7**:353-366. DOI: 10.1517/17460441.2012.668520

[117] Engel JC, Doyle PS, Hsieh I, McKerrow JH. Cysteine protease inhibitors cure an experimental *Trypanosoma cruzi* infection. The Journal of Experimental Medicine. 1998;**188**:725-734

[118] Doyle PS, Zhou YM, Engel JC, McKerrow JH. A cysteine protease inhibitor cures Chagas' disease in an Immunodeficient-mouse model of infection. Antimicrobial Agents and Chemotherapy. 2007;**51**:3932-3939. DOI: 10.1128/AAC.00436-07

[119] Chen YT, Brinen LS, Kerr ID, Hansell E, Doyle PS, McKerrow JH,

et al. In vitro and in vivo studies of the trypanocidal properties of WRR-483 against *Trypanosoma cruzi*. PLoS Neglected Tropical Diseases. 2010;**4**:e825. DOI: 10.1371/journal.pntd.0000825

[120] Salomão K, De Castro SL, editors. Recent Advances in Drug Development for Chagas Disease: Two Magic Words, Combination and Repositioning. Differ. Asp. Chemother. Trypanos. New York: Leon L & Torres-Santos EC; 2017. pp. 181-226

[121] Mott BT, Ferreira RS, Simeonov A, Jadhav A, Ang KK-H, Leister W, et al. Identification and optimization of inhibitors of trypanosomal cysteine proteases: Cruzain, Rhodesain, and TbCatB. Journal of Medicinal Chemistry. 2010;**53**:52-60. DOI: 10.1021/jm901069a

[122] Beaulieu C, Isabel E, Fortier A, Massé F, Mellon C, Méthot N, et al. Identification of potent and reversible cruzipain inhibitors for the treatment of Chagas disease. Bioorganic & Medicinal Chemistry Letters. 2010;**20**:7444-7449. DOI: 10.1016/j.bmcl.2010.10.015

[123] Ndao M, Beaulieu C, Black WC, Isabel E, Vasquez-Camargo F, Nath-Chowdhury M, et al. Reversible cysteine protease inhibitors show promise for a Chagas disease cure. Antimicrobial Agents and Chemotherapy. 2014;**58**:1167-1178. DOI: 10.1128/AAC.01855-13

[124] Burtoloso ACB, de Albuquerque S, Furber M, Gomes JC, Gonçalez C, Kenny PW, et al. Anti-trypanosomal activity of non-peptidic nitrile-based cysteine protease inhibitors. PLoS Neglected Tropical Diseases. 2017;**11**:e0005343. DOI: 10.1371/journal.pntd.0005343

[125] Salas-Sarduy E, Landaburu LU, Karpiak J, Madauss KP, Cazzulo JJ, Agüero F, et al. Novel scaffolds for inhibition of Cruzipain identified from high-throughput screening of anti-kinetoplastid chemical boxes. Scientific Reports. 2017;**7**:12073. DOI: 10.1038/s41598-017-12170-4

[126] Pauli I, Ferreira LG, de Souza ML, Oliva G, Ferreira RS, Dessoy MA, et al. Molecular modeling and structure–activity relationships for a series of benzimidazole derivatives as cruzain inhibitors. Future Medicinal Chemistry. 2017;**9**:641-657. DOI: 10.4155/fmc-2016-0236

[127] de Souza AS, de Oliveira MT, Andricopulo AD. Development of a pharmacophore for cruzain using oxadiazoles as virtual molecular probes: Quantitative structure-activity relationship studies. Journal of Computer-Aided Molecular Design. 2017;**31**:801-816. DOI: 10.1007/s10822-017-0039-0

[128] Kaiser M, Mäser P, Tadoori LP, Ioset J-R, Brun R. Antiprotozoal activity profiling of approved drugs: A starting point toward drug repositioning. PLoS One. 2015;**10**:e0135556. DOI: 10.1371/journal.pone.0135556

[129] Bellera CL, Balcazar DE, Alberca L, Labriola CA, Talevi A, Carrillo C. Identification of levothyroxine antichagasic activity through computer-aided drug repurposing. Scientific World Journal. 2014;**2014**:1-9. DOI: 10.1155/2014/279618

[130] Bellera CL, Balcazar DE, Vanrell MC, Casassa AF, Palestro PH, Gavernet L, et al. Computer-guided drug repurposing: Identification of trypanocidal activity of clofazimine, benidipine and saquinavir. European Journal of Medicinal Chemistry. 2015;**93**:338-348. DOI: 10.1016/j.ejmech.2015.01.065

[131] Sbaraglini ML, Bellera CL, Fraccaroli L, Larocca L, Carrillo

C, Talevi A, et al. Novel cruzipain inhibitors for the chemotherapy of chronic Chagas disease. International Journal of Antimicrobial Agents. 2016;**48**:91-95. DOI: 10.1016/j.ijantimicag.2016.02.018

[132] Palos I, Lara-Ramirez EE, Lopez-Cedillo JC, Garcia-Perez C, Kashif M, Bocanegra-Garcia V, et al. Repositioning FDA drugs as potential Cruzain inhibitors from *Trypanosoma cruzi*: Virtual screening, in vitro and in vivo studies. Molecules (Basel, Switzerland). 2017;**22**:E1015. DOI: 10.3390/molecules22061015

[133] Branquinha MH, Marinho FA, Sangenito LS, Oliveira SSC, Goncalves KC, Ennes-Vidal V, et al. Calpains: Potential targets for alternative chemotherapeutic intervention against human pathogenic trypanosomatids. Current Medicinal Chemistry. 2013;**20**:3174-3185

[134] Saez ME, Ramirez-Lorca R, Moron FJ, Ruiz A. The therapeutic potential of the calpain family: New aspects. Drug Discovery Today. 2006;**11**:917-923. DOI: 10.1016/j.drudis.2006.08.009

[135] Donkor IO. An updated patent review of calpain inhibitors (2012-2014). Expert Opinion on Therapeutic Patents. 2015;**25**:17-31. DOI: 10.1517/13543776.2014.982534

[136] Ennes-Vidal V, Menna-Barreto RFS, Santos ALS, Branquinha MH, d'Avila-Levy CM. MDL28170, a calpain inhibitor, affects *Trypanosoma cruzi* metacyclogenesis, ultrastructure and attachment to Rhodnius prolixus midgut. PLoS One. 2011;**6**:e18371. DOI: 10.1371/journal.pone.0018371

[137] Ennes-Vidal V, Menna-Barreto RFS, Santos ALS, Branquinha MH, d'Avila-Levy CM. Effects of the calpain inhibitor MDL28170 on the clinically relevant forms of *Trypanosoma cruzi* in vitro. The Journal of Antimicrobial Chemotherapy. 2010;**65**:1395-1398. DOI: 10.1093/jac/dkq154

[138] Motta MCM. Kinetoplast as a potential chemotherapeutic target of trypanosomatids. Current Pharmaceutical Design. 2008;**14**:847-854

[139] Soeiro M de NC, de Castro SL. Screening of potential anti-*Trypanosoma cruzi* candidates: In vitro and in vivo studies. Open Medicinal Chemistry Journal. 2011;**5**:21-30. DOI: 10.2174/1874104501105010021

[140] Delain E, Brack C, Riou G, Festy B. Ultrastructural alterations of *Trypanosoma cruzi* kinetoplast induced by the interaction of a trypanocidal drug (hydroxystilbamidine) with the kinetoplast DNA. Journal of Ultrastructure Research. 1971;**37**:200-218

[141] Vannier-Santos MA, De Castro SL. Electron microscopy in antiparasitic chemotherapy: A (close) view to a kill. Current Drug Targets. 2009;**10**:246-260. DOI: 10.2174/138945009787581168

[142] Girard RMBM, Crispim M, Stolić I, Damasceno FS, Santos da Silva M, Pral EMF, et al. An aromatic diamidine that targets kinetoplast DNA, impairs the cell cycle in *Trypanosoma cruzi*, and diminishes trypomastigote release from infected mammalian host cells. Antimicrobial Agents and Chemotherapy. 2016;**60**:5867-5877. DOI: 10.1128/AAC.01595-15

[143] Balaña-Fouce R, Álvarez-Velilla R, Fernández-Prada C, García-Estrada C, Reguera RM. Trypanosomatids topoisomerase re-visited. New structural findings and role in drug discovery. International Journal for Parasitology: Drugs and Drug Resistance. 2014;**4**:326-337. DOI: 10.1016/j.ijpddr.2014.07.006

[144] Podestá D, Stoppani A, Villamil SF. Inactivation of *Trypanosoma cruzi* and *Crithidia fasciculata* topoisomerase I by Fenton systems. Redox Report. 2003;**8**:357-363. DOI: 10.1179/135100003225003366

[145] Wilson WD, Nguyen B, Tanious FA, Mathis A, Hall JE, Stephens CE, et al. Dications that target the DNA minor groove: Compound design and preparation, DNA interactions, cellular distribution and biological activity. Current Medicinal Chemistry—Anti-Cancer Agents. 2005;**5**:389-408

[146] Werbovetz K. Diamidines as antitrypanosomal, antileishmanial and antimalarial agents. Current Opinion in Investigational Drugs (London, England: 2000). 2006;7:147-157

[147] Silva CF, Meuser MB, De Souza EM, Meirelles MNL, Stephens CE, Som P, et al. Cellular effects of reversed amidines on *Trypanosoma cruzi*. Antimicrobial Agents and Chemotherapy. 2007;**51**:3803-3809. DOI: 10.1128/AAC.00047-07

[148] Silva CF, Batista MM, Mota RA, de Souza EM, Stephens CE, Som P, et al. Activity of "reversed" diamidines against *Trypanosoma cruzi* "in vitro". Biochemical Pharmacology. 2007;**73**:1939-1946. DOI: 10.1016/j.bcp.2007.03.020

[149] Santos CC, Lionel JR, Peres RB, Batista MM, da Silva PB, de Oliveira GM, et al. *In vitro*, *in silico*, and *in vivo* analyses of novel aromatic amidines against *Trypanosoma cruzi*. Antimicrobial Agents and Chemotherapy. 2017;**62**:e02205-17. DOI: 10.1128/AAC.02205-17

[150] Das A, Dasgupta A, Sengupta T, Majumder HK. Topoisomerases of kinetoplastid parasites as potential chemotherapeutic targets. Trends in Parasitology. 2004;**20**:381-387. DOI: 10.1016/j.pt.2004.06.005

[151] Chowdhury SR, Godinho JLP, Vinayagam J, Zuma AA, Silva STDM, Jaisankar P, et al. Isobenzofuranone derivative JVPH3, an inhibitor of *L. donovani* topoisomerase II, disrupts mitochondrial architecture in trypanosomatid parasites. Scientific Reports. 2018;**8**:11940. DOI: 10.1038/s41598-018-30405-w

[152] Chowdhury SR, Kumar A, Godinho JLP, De Macedo Silva ST, Zuma AA, Saha S, et al. Voacamine alters Leishmania ultrastructure and kills parasite by poisoning unusual bi-subunit topoisomerase IB. Biochemical Pharmacology. 2017;**138**:19-30. DOI: 10.1016/j.bcp.2017.05.002

[153] Brunoro GV-F, Caminha MA, Menna-Barreto RF. From proteins to molecular targets: *Trypanosoma cruzi* proteomic insights in drug development. In: Protozoan Parasitism: From Omics to Prevention and Control. Caister Academic Press; 2018;**1**:1-30. DOI: 10.21775/9781910190838.01

[154] Magalhães AD, Charneau S, Paba J, Guércio RAP, Teixeira ARL, Santana JM, et al. Trypanosoma cruzi alkaline 2-DE: Optimization and application to comparative proteome analysis of flagellate life stages. Proteome Science. 2008;**6**:24. DOI: 10.1186/1477-5956-6-24

[155] Queiroz RML, Charneau S, Bastos IMD, Santana JM, Sousa MV, Roepstorff P, et al. Cell surface proteome analysis of human-hosted *Trypanosoma cruzi* life stages. Journal of Proteome Research. 2014;**13**:3530-3541. DOI: 10.1021/pr401120y

[156] de Jesus LTC, Calderano SG, Vitorino FN d L, Llanos RP, Lopes M d C, de Araújo CB, et al. Quantitative proteomic analysis of replicative and nonreplicative forms reveals important insights into chromatin biology of *Trypanosoma cruzi*. Molecular & Cellular Proteomics. 2017;**16**:23-38. DOI: 10.1074/mcp.M116.061200

[157] Alves MJM, Kawahara R, Viner R, Colli W, Mattos EC, Thaysen-Andersen

M, et al. Comprehensive glycoprofiling of the epimastigote and trypomastigote stages of *Trypanosoma cruzi*. Journal of Proteomics. 2017;**151**:182-192. DOI: 10.1016/j.jprot.2016.05.034

[158] Atwood JA, Minning T, Ludolf F, Nuccio A, Weatherly DB, Alvarez-Manilla G, et al. Glycoproteomics of *Trypanosoma cruzi* trypomastigotes using subcellular fractionation, lectin affinity, and stable isotope labeling. Journal of Proteome Research. 2006;**5**:3376-3384. DOI: 10.1021/pr060364b

[159] de Godoy LMF, Marchini FK, Pavoni DP, Rampazzo R de CP, Probst CM, Goldenberg S, et al. Quantitative proteomics of *Trypanosoma cruzi* during metacyclogenesis. Proteomics. 2012;**12**:2694-2703. DOI: 10.1002/pmic.201200078

[160] Cordero EM, Nakayasu ES, Gentil LG, Yoshida N, Almeida IC, da Silveira JF. Proteomic analysis of detergent-solubilized membrane proteins from insect-developmental forms of *Trypanosoma cruzi*. Journal of Proteome Research. 2009;**8**:3642-3652. DOI: 10.1021/pr800887u

[161] Bayer-Santos E, Aguilar-Bonavides C, Rodrigues SP, Cordero EM, Marques AF, Varela-Ramirez A, et al. Proteomic analysis of Trypanosoma cruzi secretome: Characterization of two populations of extracellular vesicles and soluble proteins. Journal of Proteome Research. 2013;**12**:883-897. DOI: 10.1021/pr300947g

[162] de Pablos Torró LM, Retana Moreira L, Osuna A. Extracellular vesicles in Chagas disease: A new passenger for an old disease. Frontiers in Microbiology. 2018;**9**:E1190. DOI: 10.3389/fmicb.2018.01190

[163] Brunoro GVF, Caminha MA, Ferreira AT d S, Leprevost F d V, Carvalho PC, Perales J, et al. Reevaluating the *Trypanosoma cruzi* proteomic map: The shotgun description of bloodstream trypomastigotes. Journal of Proteomics. 2015;**115**:58-65. DOI: 10.1016/j.jprot.2014.12.003

[164] Brunoro GVF, Faça VM, Caminha MA, Ferreira AT d S, Trugilho M, de Moura KCG, et al. Differential gel electrophoresis (DIGE) evaluation of naphthoimidazoles mode of action: A study in *Trypanosoma cruzi* bloodstream trypomastigotes. PLoS Neglected Tropical Diseases. 2016;**10**:e0004951. DOI: 10.1371/journal.pntd.0004951

[165] Paba J, Santana JM, Teixeira ARL, Fontes W, Sousa MV, Ricart CAO. Proteomic analysis of the human pathogen *Trypanosoma cruzi*. Proteomics. 2004;**4**:1052-1059. DOI: 10.1002/pmic.200300637

[166] Ritagliati C, Alonso VL, Manarin R, Cribb P, Serra EC. Overexpression of cytoplasmic TcSIR2RP1 and mitochondrial TcSIR2RP3 impacts on *Trypanosoma cruzi* growth and cell invasion. PLoS Neglected Tropical Diseases. 2015;**9**:e0003725. DOI: 10.1371/journal.pntd.0003725

[167] Moretti NS, da Silva Augusto L, Clemente TM, Antunes RPP, Yoshida N, Torrecilhas AC, et al. Characterization of *Trypanosoma cruzi* Sirtuins as possible drug targets for Chagas disease. Antimicrobial Agents and Chemotherapy. 2015;**59**:4669-4679. DOI: 10.1128/AAC.04694-14

[168] Gaspar L, Coron RP, KongThoo Lin P, Costa DM, Perez-Cabezas B, Tavares J, et al. Inhibitors of *Trypanosoma cruzi* Sir2 related protein 1 as potential drugs against Chagas disease. PLoS Neglected Tropical Diseases. 2018;**12**:e0006180. DOI: 10.1371/journal.pntd.0006180

[169] Wan X, Wen J-J, Koo S-J, Liang LY, Garg NJ. SIRT1-PGC1α-NFκB pathway of oxidative and inflammatory stress during *Trypanosoma cruzi* infection: Benefits of SIRT1-targeted therapy in improving heart function

in Chagas disease. PLoS Pathogens. 2016;**12**:e1005954. DOI: 10.1371/journal.ppat.1005954

[170] Urbina JA. Specific chemotherapy of Chagas disease: Relevance, current limitations and new approaches. Acta Tropica. 2010;**115**:55-68. DOI: 10.1016/j.actatropica.2009.10.023

[171] Filardi LS, Brener Z. Susceptibility and natural resistance of *Trypanosoma cruzi* strains to drugs used clinically in Chagas disease. Transactions of the Royal Society of Tropical Medicine and Hygiene. 1987;**81**:755-759. DOI: 10.1016/0035-9203(87)90020-4

[172] Chatelain E, Ioset J-R. Phenotypic screening approaches for Chagas disease drug discovery. Expert Opinion on Drug Discovery. 2018;**13**:141-153. DOI: 10.1080/17460441.2018.1417380

[173] Moraes CB, Franco CH. Novel drug discovery for Chagas disease. Expert Opinion on Drug Discovery. 2016;**11**:447-455. DOI: 10.1517/17460441.2016.1160883

[174] Sueth-Santiago V, Decote-Ricardo D, Morrot A, Freire-de-Lima CG, Lima MEF. Challenges in the chemotherapy of Chagas disease: Looking for possibilities related to the differences and similarities between the parasite and host. World Journal of Biological Chemistry. 2017;**8**:57. DOI: 10.4331/wjbc.v8.i1.57

[175] Moraes CB, Giardini MA, Kim H, Franco CH, Araujo-Junior AM, Schenkman S, et al. Nitroheterocyclic compounds are more efficacious than CYP51 inhibitors against *Trypanosoma cruzi*: Implications for Chagas disease drug discovery and development. Scientific Reports. 2014;**4**:4703. DOI: 10.1038/srep04703

[176] Cal M, Ioset J-R, Fügi MA, Mäser P, Kaiser M. Assessing anti-*T. cruzi* candidates in vitro for sterile cidality. International Journal for Parasitology: Drugs and Drug Resistance. 2016;**6**:165-170. DOI: 10.1016/j.ijpddr.2016.08.003

[177] Ferraz ML, Gazzinelli RT, Alves RO, Urbina JA, Romanha AJ. Absence of CD4+ T lymphocytes, CD8+ T lymphocytes, or B lymphocytes has different effects on the efficacy of posaconazole and benznidazole in treatment of experimental acute *Trypanosoma cruzi* infection. Antimicrobial Agents and Chemotherapy. 2009;**53**:174-179. DOI: 10.1128/AAC.00779-08

[178] Chatelain E, Ioset J-R. Drug discovery and development for neglected diseases: The DNDi model. Drug Design, Development and Therapy. 2011;**5**:175-181. DOI: 10.2147/DDDT.S16381

[179] Friggeri L, Hargrove TY, Rachakonda G, Blobaum AL, Fisher P, de Oliveira GM, et al. Sterol 14α-demethylase structure-based optimization of drug candidates for human infections with the protozoan trypanosomatidae. Journal of Medicinal Chemistry. 2018;**61**:10910-10921. DOI: 10.1021/acs.jmedchem.8b01671

[180] Machado-Silva A, Cerqueira PG, Grazielle-Silva V, Gadelha FR, Peloso E de F, Teixeira SMR, et al. How *Trypanosoma cruzi* deals with oxidative stress: Antioxidant defence and DNA repair pathways. Mutation Research. Reviews in Mutation Research. 2016;**767**:8-22. DOI: 10.1016/j.mrrev.2015.12.003

[181] Tomás AM, Castro H. Redox metabolism in mitochondria of trypanosomatids. Antioxidants & Redox Signaling. 2013;**19**:696-707. DOI: 10.1089/ars.2012.4948

[182] Sykes ML, Avery VM. Approaches to protozoan drug discovery: Phenotypic screening. Journal of

Medicinal Chemistry. 2013;**56**:7727-7740. DOI: 10.1021/jm4004279

[183] Peña I, Pilar Manzano M, Cantizani J, Kessler A, Alonso-Padilla J, Bardera AI, et al. New compound sets identified from high throughput phenotypic screening against three kinetoplastid parasites: An open resource. Scientific Reports. 2015;**5**:8771. DOI: 10.1038/srep08771

[184] Chatelain E, Konar N. Translational challenges of animal models in Chagas disease drug development: A review. Drug Design, Development and Therapy. 2015;**9**:4807-4823. DOI: 10.2147/DDDT.S90208